MOLECULAR BASIS OF VIRUS DISEASE

SYMPOSIA OF THE
SOCIETY FOR GENERAL MICROBIOLOGY*

* Published by the Cambridge University Press, except for the first Symposium, which was published by Blackwell's Scientific Publications Limited.

MOLECULAR BASIS OF VIRUS DISEASE

EDITED BY

W. C. RUSSELL AND J. W. ALMOND

FORTIETH SYMPOSIUM OF
THE SOCIETY FOR
GENERAL MICROBIOLOGY
HELD AT
THE UNIVERSITY OF ST ANDREWS
APRIL 1987

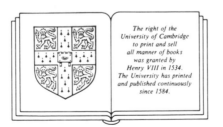

The right of the
University of Cambridge
to print and sell
all manner of books
was granted by
Henry VIII in 1534.
The University has printed
and published continuously
since 1584.

Published for the Society for General Microbiology

CAMBRIDGE UNIVERSITY PRESS
CAMBRIDGE
LONDON NEW YORK NEW ROCHELLE
MELBOURNE SYDNEY

Published by the Press Syndicate of the University of Cambridge
The Pitt Building, Trumpington Street, Cambridge CB2 1RP
32 East 57th Street, New York, NY 10022, USA
10 Stamford Road, Oakleigh, Melbourne 3166, Australia

© The Society for General Microbiology Limited 1987

First published 1987

Printed in Great Britain at The Bath Press, Avon

British Library cataloguing in publication data

Society for General Microbiology. *Symposium*
(40:1984: University of St. Andrews)
Molecular basis of virus disease: Fortieth
Symposium of the Society for General
Microbiology, held at the University of St.
Andrews, April 1987 – (Symposia of the
Society for General Microbiology; 40)
1. Virology
I. Title II. Russell, W. C. III. Almond, J. W.
IV. Series
576'.64   QR360

Library of Congress cataloguing in publication data

Society for General Microbiology. Sympsoium (40th:
1987: University of St. Andrews)
Molecular basis of virus disease.

(Symposia of the Society for General Microbiology; 40)
1. Virus diseases – Congresses. 2. Pathology,
Molecular – Congresses. 3. Virus diseases of plants –
Congresses. I. Russell, W. C. II. Almond, J. W.
III. Title. IV. Series: Society for General Micro-
biology. Symposium. Symposia of the Society for
General Microbiology; 40. [DNLM: 1. Virus diseases –
Microbiology-congresses. W3 SO59J 40th 1987m/WC 500
S678 1986m]
CR1.S6233 no. 40 [RC114.5] 576 s [616'.0194] 86–31000

ISBN 0 521 33105 6

CONTRIBUTORS

BEATY, B. J. Department of Microbiology, Colorado State University, Fort Collins, Colorado, USA

BISHOP, D. H. L., NERC, Institute of Virology, Mansfield Road, Oxford, OX1 3SR, UK

CAMPO, M. S. Beatson Institute for Cancer Research, Garscube Estate, Switchback Road, Bearsden, Glasgow G61 1BD, UK

COULON, P. Laboratoire de Génétique des Virus, Centre National de la Recherche Scientifique, 91190 Gif sur Yvette, France

DIALLO, A. Laboratoire de Génétique des Virus, Centre National de la Recherche Scientifique, 91190 Gif sur Yvette, France

DOLIVO, M. Institut de Physiologie de la Faculté de Médecine, Université de Lausanne, Ch-1011 Lausanne, Switzerland

FLAMAND, A. Laboratoire de Génétique des Virus, Centre National de la Recherche Scientifique, 91190 Gif sur Yvette, France

GINSBERG, H. Department of Microbiology, Institute of Cancer Research, Columbia University College of Physicians and Surgeons, New York, NY 10032, USA

HARRISON, B. D. Scottish Crop Research Institute, Invergowrie, Dundee DD2 5DA, UK

JARRETT, W. F. H. Department of Veterinary Pathology, Veterinary School, University of Glasgow, Garscube Estate, Bearsden, Glasgow, UK

KELLY, D. C. Centre for Applied Microbiology and Research, Porton, Wiltshire, UK

KUCERA, P. Institut de Physiologie de la Faculté de Médecine, Université de Lausanne, Ch-1011 Lausanne, Switzerland

LUNDHULM-BEAUCHAMP, U. Department of Microbiology, Institute of Cancer Research, Columbia University College of Physicians and Surgeons, New York, NY 10032, USA

MARSDEN, H. S., MRC Virology Unit, Institute of Virology, Church Street, Glasgow G11 5JR, UK

McCAULEY, J. W. Animal Virus Research Institute, Pirbright, Surrey GO24 0HF, UK

NOMOTO, Akio, Department of Microbiology, Faculty of Medicine, University of Tokyo, Tokyo, Japan

PREHAUD, L. Laboratoire de Génétique des Virus, Centre National de la Recherche Scientifique, 91190 Gif sur Yvette, France

PRINCE, G. Laboratory of Infectious Diseases, National Institute of Health, Bethesda, Maryland 20892, USA

PRINGLE, Craig, Department of Biological Sciences, University of Warwick, Coventry CV4 7AL, UK

RIGBY, Peter W. J. Cancer Research Campaign, Eukaryotic Molecular Genetics Research Group, Department of Biochemistry, Imperial College of Science and Technology, London SW7 2AZ, UK

SHOPE, R. E. Yale Arbovirus Research Unit, Yale University Department of Epidemiology & Public Health, New Haven, Connecticut, USA

VLOTEN-DOTING, Lois van, Department of Biochemistry, Rijksuniversiteit, Wassenaarseweg 64, 2333 AL Leiden, The Netherlands

WEISS, R. A. Institute of Cancer Research, Chester Beatty Laboratories, Fulham Road, London SW3 6JB, UK

WILDY, P. Department of Pathology, University of Cambridge, Tennis Court Road, Cambridge CB2 1QP, UK

WIMMER, Eckard, Department of Microbiology, School of Medicine, Health Sciences Center, State University of New York, Stony Brook, NY 11794, USA

CONTENTS

EDITORS' PREFACE

The first meeting of the Committee of the Virus Group of the Society for General Microbiology took place at Leeds on Wednesday, 11 July, 1962. Present were Sir Christopher Andrewes, Drs Bryan Harrison, Alick Isaacs and David Tyrrell and they decided to hold symposia in the following year – 'one on a basic subject and one on an applied subject'.

Twenty-five years later the Society has decided to mark that occasion by sponsoring a symposium which will attempt to bridge the gap that had been evident at that first meeting and had been so pertinently described by Sir Christopher Andrewes as a function of whether a virologist could be ascribed to the 'steam' or 'dream' variety.

In this symposium we have tried to highlight those advances in our understanding of disease processes which have accrued from the application of techniques in molecular biology. It is particularly appropriate that the first contribution describes the pioneering role of Edward Jenner in tackling the problem of virus disease since the great man obtained his medical degree from the University of St Andrews, the venue of this symposium, some 195 years ago. The historical survey provided by Peter Wildy serves also to remind us of the incredible acceleration in the advance of virology in the modern era and the spectacular progress that has been made since the Virus Group's inception. Molecular biology has played a dominant role in that progress and, in the past few years especially, molecular virology has become increasingly focussed on disease aspects. It will be evident from reading these chapters that, not only has our knowledge of the molecular architecture of viruses, of their genetic properties and of the regulatory mechanisms which determine their fate in infected cells advanced dramatically, but that interpretation of this information in terms of how diseases are caused is increasingly the order of the day. The symposium also gives coverage to aspects of plant and insect virus disease, the understanding of which is likely to have implications for control of disease in agriculture and horticulture.

Allied to parallel advances in other biological disciplines, knowledge of the molecular details of virus disease processes should allow a much more rational attack (but hopefully just as an effective one as Jenner's) on the virus diseases which are still prevalent in many communities throughout the world.

In compiling the ensuing chapters we are grateful to the authors for producing their manuscripts (mainly) on time, for the help and patience of staff at Cambridge University Press and for the forbearance shown by our secretaries.

JENNER, GENES, VACCINES AND BLACK BOXES

PETER WILDY

Department of Pathology, Tennis Court Road, Cambridge

INTRODUCTION

This chapter has been written in response to a request to mark the 25th year of the virus group of the Society for General Microbiology, to form the prelude to a virus symposium and to feature Edward Jenner who may be said to have fathered the sister disciplines of virology and immunology. This topic is especially appropriate since Jenner received his doctorate in medicine from the University of St Andrews.

PRE-JENNER

The notion of infectious disease and of immunity can be recognised in Thucydides's account of the plagues of Athens in the fifth century BC. From time to time there were other more or less fragmentary ideas, the best formed of which may have been those of Girolamo Fracastoro of Verona who, in the fifteenth century, clearly envisaged seeds of disease which could multiply. It is strange, therefore, that the first practical form of immunisation seems to have come from China around the tenth century; although their philosophy of health and disease did not admit either infectiousness or immunity (Needham, 1980). One of the more advanced means of protecting individuals against smallpox was to gather material from suitable mild cases; this was stored in a small container under the physician's clothing for a specified period, and, the season being propitious, applied on a piece of cotton to the inside of the nostril. The proportion of 'takes' was great, the degree of protection excellent and the adverse effects minimal.

From China the practice of variolation moved westward in the wake of smallpox, presumably along the trade routes, and was further disseminated by warfare, reaching the Middle East in the seventeenth century. According to Fisk (1959) it was brought to Britain by Dr Emanuel Timoni who read a paper in 1713 to the Royal Society. This is generally credited to Lady Mary Wortley Montagu, the intrepid wife of our Ambassador Extraordinary (with the mission of bringing about a reconciliation between Turkey and Austria, then on the very brink of war).

In her correspondence Lady Mary, herself a victim of the smallpox which had marked her face, wrote in 1717 of the 'engrafting procedure':

There is a set of old women who make it their business to perform the operation every autumn, in the month of September, when the great heat is abated. People send to one another to know if any of the family has a mind to have the smallpox: they make parties for this purpose; and when they are met (commonly fifteen or sixteen together) the old woman comes with a nutshell full of the matter *of the best sort of* smallpox and asks what veins you please to have opened. She immediately rips open what you offer with a large needle which gives you no more pain than a common scratch and puts into the vein as much venom as can lie upon the head of a needle, and after binds up the little wound with a hollow shell. . . .

The consequence of this procedure:

The children or young patients play together all the rest of the day, and are in perfect health until the eighth. Then fever begins to seize them, and they keep their beds two days, very seldom three. They have very rarely above twenty or thirty in their faces, which never mark, and in eight days they are as well as before their illness.

Lady Mary caused her son to be inoculated while in the Middle East and, upon return to Britain during the smallpox epidemic of 1721, also had her daughter attended to. Her energetic championship of inoculation extended into high society and for a time the practice flourished. Unfortunately, it soon became plain that variolation was somewhat hazardous; of 845 persons inoculated between 1722 and 1730, 17 (as much as 2%) died of the disease. Variolation became unpopular for a time but the increase in the incidence of the dreaded smallpox had so terrified people, that the practice was revived. There were some practitioners who refined the procedure in their own secret way, much as the Chinese had done; such were the Suttons, who practised at Ingatestone, Essex, in the mid-eighteenth century. The overall pattern was much more serious. It became plain a generation later that variolation was actually spreading smallpox; Benjamin Franklin attested that in 1753–4 more than 1:100 died from inoculation (Fisk, 1959).

EDWARD JENNER

Edward Jenner was born in Berkeley on 17 May 1749, the sixth surviving child of the Rev. Stephen Jenner, A.M. (Oxon). The Jenner family were country gentlemen, parsons, farmers and medical practioncrs. They lived the leisurely busy life of the upper class countrymen of the day. Both his elder brothers entered the church and so might Edward (had funds allowed). The other two openings were the Army (but the price of commissions was high) or a profession. For this reason Edward was plighted to the dishonorable role of a quack (Baron, 1838; Creighton, 1889; Fisk, 1959).

As a boy he was an apt pupil and mastered such formal learning as was presented to him. But his obsession was natural history. From the time of his parents' death (he was only 5 years old) he was cared for by his eldest brother Stephen and his sister Mrs Mary Black. He attended Mr Clissold's school at Wooten-under-Edge and later joined the Grammar School at Cirencester under Dr Washbourne, where he learned the elements of Latin; yet he was reasonably free to indulge his absorption with natural history.

At the age of 14 he left school becoming apprenticed to the surgeon Daniel Ludlow at Chipping Sodbury. Biographers have little to tell us about these years but, most probably, the seeds of his great discovery were sown at this time (see later). In 1770, when he was 21 years old, Jenner became the pupil of that intellectual powerhouse John Hunter (probably because Ludlow's son had been a pupil of his). Brother Stephen financed him.

The association with Hunter had the most profound effect on Jenner and a friendship developed which was to influence him for many years. At the outset Hunter was a surgeon at St George's Hospital; he also had a menagerie at Brompton and was continually enquiring into matters of natural history and medicine. The nature of these enquiries was certainly experimental and, within the limitations of available methods, was reasonably critical. The rigour that Jenner absorbed at this time was to stand him in very good stead in later years. While with Hunter, Jenner rubbed shoulders with many of London's intelligentsia, amongst whom was Sir Joseph Banks, recently returned from Captain Cook's first expedition. It fell to Jenner to help with the arrangement of Banks's biological collection and to make many detailed dissections of the specimens. He was so adept at this that it was suggested that he should join Cook's second expedition as the naturalist; in fact he did not.

Jenner returned to Berkeley in 1772 when he set up in practice from his brother Stephen's house. There, in addition to his practice, he continued pottering about the countryside watching animals and birds and cogitating all the while on natural phenomena. Until Hunter died, in 1793, there was continual correspondence between the two men. The evidence for this is the collection of Hunter's letters to Jenner which have survived in plenty. Most of Jenner's replies have been destroyed.

One may surmise from the biographies that Jenner was an easy-going, imaginative man but rather idle. He was fond of music, playing the fiddle and the flute. He loathed card games. He was clever with words and produced a stream of witty poems and epigrams exemplified by the following:

> On the death of a miser
> Tom at last has laid by his old niggardly forms,
> And now gives good dinners; to whom pray? – the worms.

He was sociable, enjoyed discussing ideas and organised social evenings that seem to have combined seminars, music and dinner.

As to his character, we find biographers quite divided. Baron (1838), who was a personal friend and his self-appointed Boswell had nothing bad to say. Jenner was a genius full of every kind of virtue. On the other hand Creighton (1889) set out to be a debunker; he insisted that what had attracted Jenner to the cow-pox : smallpox analogy was the fact that both names ended in 'pox'. 'Jenner was just the loose-thinking, imaginative sort of person to deal with the matter in a merely verbal way'. In other words, it was fortunate that he did not include chicken pox or perhaps plum pox in his thinking. Having adduced evidence that cowpox was quite like smallpox, he states: 'It is difficult to acquit Jenner of recklessness or even culpable laxity from the inception of his idea'. Nonetheless, nearly 100 years later we see that, even if lured by verbal imagery, Jenner did gather what evidence he could in support of his grand idea.

Undoubtedly, Jenner must have been a lazy man. He was continually being stimulated and almost hounded by Hunter, who was for ever demanding specimens and suggesting experiments. John Hunter was constantly complaining about lack of response from Jenner; perhaps we may be permitted a feeling of sympathy here, for, had the telephone been invented, Jenner's life would have become a misery.

Whoever was the mainspring of many diverse discourses, we can only be amazed that Jenner provided original observations on coronary artery disease, hypothermia, the preparation of tartar emetic and the hibernation of hedgehogs and bats. He experimented with a hydrogen balloon and, after chasing it to its landing point, found himself on the way to romance and eventually marriage to the desirable Miss Kingscott (Fisk, 1959).

Jenner's first coherent piece of research was on the natural history of the cuckoo; his observations of 1786 suggested that the foster parents of the young cockoo were responsible for the rejection of their own progeny. The observations, which included some experimental results, were sent to John Hunter for communication to the Royal Society. For some unknown reason Hunter delayed passing on Jenner's paper; it was as well because in the following year Jenner withdrew it. Further experiments, in which eggs and chicks were placed in the foster parents nest, showed that it was the infant cuckoo that, in the first 10 days of its life, ejected the eggs containing its step-siblings and not the foster parents. Indeed, at this stage of immaturity, the infant cuckoo possessed a depression in its back that was used for the purpose. Beyond 10 days, the young cuckoo became disinterested in eggs or other chicks that might inhabit the nest. The amended paper was resubmitted to the Royal Society.

Uncharitable critics have drawn attention to Jenner's intense craving for scientific and medical respectability. It is obvious that he very much wanted the FRS, for he wrote to Hunter about his observations on the cuckoo as follows:

I beg leave to lay before you the result of my observations with a hope that they may tend to illustrate a subject hitherto not sufficiently investigated; and should what is offered prove, in your opinion, deserving the attention of the Royal Society, you will do me the honour of presenting it to that learned body.

The revised paper did earn Jenner the FRS (Fisk, 1959).

We next find him ambitious for the degree of MD. Owing to his wife's ill-health, Jenner took her in 1792 (and for several years subsequently) to Cheltenham, where (if he were to practise gainfully) the degree of MD was important. This was the passport to the enlistment of the maximum number of wealthy and influential patients. Fortunately, the matter was little problem; degrees could be obtained by recommendation. All Jenner had to do was to apply with the backing of two respected medical practitioners. In those days certain universities granted degrees with minimum palaver and, at the University of St Andrews, the requirement for the degree of MD was that the candidate be recommended by 'physicians of repute'. Dr C. H. Parry and Dr J. H. Hickes, graduates of Edinburgh University, were apparently very well known at that time; they were also fortunately lifelong friends of Jenner. The citation for Edward Jennings (*sic*) is shown in facsimile (Fig. 1). The degree of MD was granted in 1792. There seems to have been no other connection between these eminent men and St Andrews except that they signed a second testimonial for another MD in the same year, 1792, and Parry later signed three more (two of them with Jenner as co-signatory; R. N. Smart, personal communication).

JENNER'S VACCINE

As a baseline we may assume that the phenomenon of acquired immunity was well perceived when Jenner became apprenticed to Ludlow. Besides the general experience of immunity to diseases such as plague and the naturally acquired immunity to smallpox, there was the artificially acquired immunity after variolation. Jenner refers in some of his writings to the Suttons. We also know that an 'old wives' tale was probably widespread that sufferers of the cowpox somehow became immune to smallpox. Jenner must have been made aware of this early on in his career since he had been told by a dairymaid, while still with Mr Ludlow, that she could not contract smallpox because she had already had the cowpox.

An extract from *The Continuing Story of the Sodburys* refers to the town accounts of 1782 (Couzens, 1972).

Feby. 7th, paid to Mr. Eb. Ludlow for inoculating the poor belonging to this borough by order of the Committee appointed for that purpose . . . £45.

This Ebenezer Ludlow was a surgeon with a considerable reputation and it was to him that a young a man by the name of Jenner had been apprenticed some twenty years before. Throughout this century the townspeople enjoyed the services of members of the Ludlow and Wallis families as apothecaries and doctors. Doctor Hardewicke of Bristol was also a local product and it seems not unlikely that between them they evolved an efficient antidote for smallpox. This statement is likely to raise the ire of that doughty champion of Dr. Jenner, Canon Gethin Jones. With respect therefore it can only be emphasised that when in 1795 the disease was rampant in Bristol, the pest house at Sodbury was not required for its original purpose and was serving as ordinary accommodation for the poor.

The likely interpretation of this extract is that Ludlow and his apprentice had become adept variolators since inoculation was in those times synonymous with grafting variolous matter. But one is left wondering whether the 'efficient antidote' may have meant a totally new procedure, perhaps vaccination (Couzens, 1973).

(*a*)

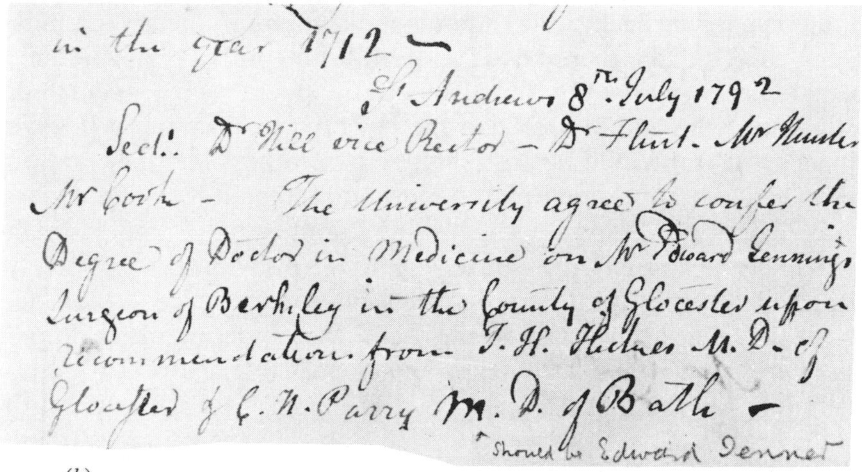

in the year 1712 —

St Andrews 8th July 1792

Sedt. Dr Hill vice Rector — Dr Flint. Mr Hunter
Mr Cook — The University agreed to confer the
Degree of Doctor in Medicine on Mr Edward Jennings*
Surgeon of Berkley in the County of Glocester upon
Recommendation from J. H. Hickes M.D. of
Glocester & C. H. Parry M.D. of Bath —

* Should be Edward Jenner

(*b*)

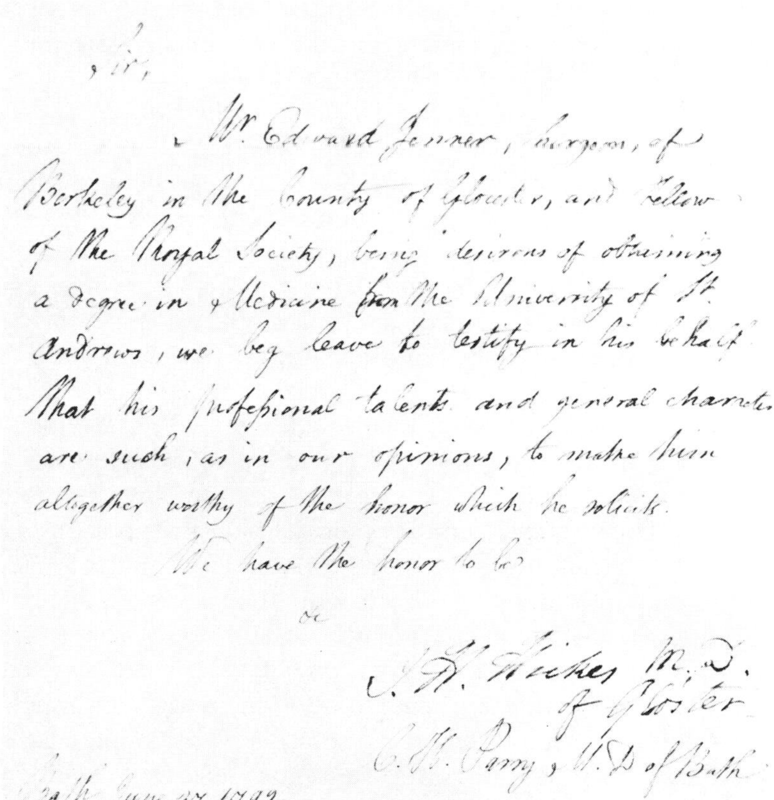

Sir,

Mr Edward Jenner, Surgeon, of Berkeley in the County of Glocester, and Fellow of the Royal Society, being desirous of obtaining a degree in Medicine from the University of St Andrews, we beg leave to testify in his behalf that his professional talents and general character are such, as in our opinions, to make him altogether worthy of the honor which he solicits.

We have the honor to be
&c

J. H. Hickes M.D. of Gloster
C. H. Parry, M.D. of Bath

Bath June 27, 1792

Fig. 1. Facsimiles of (*a*) the relevant part of the Minutes of the Senate of the University of St Andrews of 8 July 1792. (*b*) the letter of recommendation for the award of the degree of MD to Edward Jenner.

We are on certain ground in discussing Benjamin Jesty, a farmer in the Isle of Purbeck, who had married a young women when middle-aged and had fathered two children. When, in 1774, an epidemic of smallpox broke out in the village, many people were inoculated. But Jesty, who had himself had the cowpox, had reason to believe that he was protected against smallpox from personal observation of his acquaintances and their obvious immunity after similar experience. He wanted to protect his young family, and vaccinated them with material straight from the cow. Though his wife was excessively ill, an occurrence that made him the centre of public derision, all three survived both the procedure and the epidemic of smallpox. Jesty was undoubtedly 20 years ahead of Jenner in practice but, of course, nowhere in the scientific stakes. His gravestone at Worth Matravers in Dorsetshire records 'He was an upright, honest man, particularly noted for having been the first person (known) who introduced the cowpox inoculation, and who, for his great strength of mind, made the experiment from the cow on his wife and two sons in the year 1774'.

Jenner's *Inquiry into Cause and Effects of the Variolae Vaccinae* was dedicated to Dr Parry and was published in 1798. One cannot be anything but astonished by the length of time Jenner required to incubate his ideas and to formulate his thesis; but this leaves out of account the manifold uncertainties of the time. The diagnosis of smallpox and its differentiation from, for instance, measles was not so long past and the diagnosis of cowpox was more primitive still. Indeed Jenner was obliged to establish 'true' from 'false' cowpox (the latter must have embraced a number of conditions including mastitis). Gradually, as his critical faculty gained ascendancy, the results fell more perfectly into place and his confidence grew. The *Inquiry* provided positive evidence that (*a*) the cowpox could be transmitted from person to person and (*b*) that such persons would not take variolation, nor were they susceptible to smallpox. He and his entourage eventually vaccinated some thousands of persons with success and without untoward side-effects.

It was unfortunate that in an excess of zeal, when vaccination had become popular, certain practitioners began vaccinating actually within smallpox hospitals where variolation was also practised. Dr Woodville, who practised at the smallpox hospital at Paddington attending to a fair number of patients, began vaccinating people without taking the precaution of isolating them from smallpox patients. His account of the results differed from Jenner's; though some of the patients went through episodes, much as Jenner's had done, others developed generalised rashes indistinguishable from those of smallpox. The most probable inference is clear to us now. The confusion was greatly augmented by Dr Pearson who, after collecting pus from Woodville's patients, advertised that he would supply cowpox matter to medical practitioners requiring it (Fisk, 1959).

According to Creighton (1889), Woodville's stock was the origin of the world's vaccine so that his, and presumably other possible contaminations, have confused the provenance of vaccinia virus which has been in use for so long; but one thing is certain, vaccinia is different in many respects from the virus of cowpox (Downie, 1965).

Jenner's ambition was to wipe out smallpox by means of vaccination and, though the practice burgeoned and must have protected many people in Jenner's lifetime and reduced the smallpox rate in those countries where vaccination was seriously practised, the eradication of the disease had to await the multinational approach of a few years ago organised by the World Health Organization.

POST-JENNER

The next stage in the vaccine story came 84 years after Jenner's first publication on the subject. It is interesting to compare the background knowledge available to Jenner and Pasteur and it will later be salutory to consider how we fare in this respect today. Jenner was equipped with two pieces of general knowledge: (a) subjects who had suffered a particular infectious disease were refractory to subsequent disease; the mechanism for this was for him a black box; (b) the processes underlying the recognisable disease must be determinable, but they too represented a black box. For the rest he had his theories, his observations and (above all) the scientific rigour that Hunter had taught him. At the outset, when approaching the anthrax problem in 1881 Pasteur had very little more, although he understood the role of germs and he had the reassurance of Jenner's success; indeed, he made the term 'vaccine' a general one (e.g. *le premier vaccin* for anthrax). The notion of attenuation was developed further and is clear enough in Pasteur's successful attempts (1885) to make a vaccine against rabies. Again the mechanisms are enshrouded in blackness.

From that time onward, despite advancing knowledge of the properties of viruses and bacteria, the development of attenuated live vaccines remained essentially empirical. Examples are BCG (Calmette & Guérin, 1905), attenuated poliomyelitis vaccine (Sabin, 1955), yellow fever 17D strain (Theiler & Smith, 1937) and even recent vaccines against measles and rubella. The general approach has been to give the virus or bacterium a difficult time *in vitro* and then to select mutants that lack virulence and immunogenicity. This blunderbuss method no longer has scientific respectability, though clearly the above-mentioned vaccines have served humanity well.

The empirical approach outlined above has been valuable, introducing a library of experience which has contributed to the science of immunology, a particularly good example being Koch's phenomenon; over the years enlightenment has come to the black box of protective immunity. In some areas, such as antitoxic immunity, a working understanding came early (von Behring & Wernicke, 1892) and with it the realisation that it is important to immunise against a virulence or damage factor and unnecessary and undesirable to provoke immunity to irrelevant material. The many bacteriological and virological studies that now fill our library shelves have, over the years, illuminated an important branch of pathology, the black box of pathogenesis.

In the past 50 years or so we have seen a variety of microbial molecules investigated for their possible worth as protective immunogens – pneumococcal capsular substance, the adhesins of the enterobacteria, the glycoproteins of several enveloped viruses and so on. These are the putative 'subunit' vaccines. The approach requires the production of huge quantities of microbes, their purification and the chemical fractionation of the biologically important molecules. Not surprisingly, this approach is limited to those microbes that multiply abundantly; it is also constrained very much by the immunogenicity of the purified molecules. We thus return to the other black box (protective immunity). Much is now understood about humoral and cell-mediated immunity and about the mechanisms underlying them. We are coming to understand more and more about the functions and behaviour of T-lymphocytes and B-lymphocytes and our ability to distinguish subsets of T-cells by their surface antigens continually improves. These marked subsets are generally regarded as performing distinct, specific functions (such as cytotoxicity, delayed hypersensitivity, help for other lymphocytes and the suppression of their induction or performance). Now that clones of lymphocytes

are under study, it is becoming clear that the rather simplistic relationship between markers and functions may need revision. A wide
variety of protective immunological processes have been identified,
mainly from experimentation with the mouse (Nash, Leung & Wildy,
1985). The mouse represents mankind better than we deserve, but
this is fortuitous. We make use of the creature because it is inexpen-
sive, readily nurtured, can be maintained easily and has been exhaus-
tively researched both genetically and microbiologically so that, for
example, adoptive transfer experiments can be done. There is some
danger that we shall come to regard mice as miniature humans and
draw unwarranted conclusions. This becomes clearer from studies
on other vertibrate species which show considerable variation of
their immunological systems. One more caveat needs mentioning,
diseases differ in the dominant effector arms of immunity that are
important for protection. For example, antibodies are all-important
in diseases mediated by exotoxins and in some virus diseases (such
as poliomyelitis or foot-and-mouth disease) but for others (caused
by the herpes virus group of infections and mycobacterial diseases),
cell-mediated immunity is important. However, the particular arm
of cell-mediated immunity that is important is not always clear-cut
(Nash *et al.*, 1985).

THE IMPACT OF MOLECULAR BIOLOGY

Molecular biology began somewhere about the mid-1950s. The solution to the structure and function of DNA is the dominant discovery
heralding its inception; but, it has been a combination of multiple
advances by biochemists, geneticists, microbiologists and others that
has provided us with this immensely powerful set of tools. In consequence, intransigent problems of the past have been profitably
approached and the black boxes that we had just begun to peep
inside of are now unrecognisably translucent. We are now on the
crest of the wave; when the next big step has to be taken, a bewildering array of new black boxes will emerge.

In the context of vaccine design, most benefit has accrued from
the exploitation of DNA cloning and the use of monoclonal antibodies. These have necessarily been supplemented by other modern
techniques such as the sequencing and synthesis of DNA and polypeptides, and have drawn heavily on, for example, physicochemical
data allowing the prediction of probable chain folding and tertiary
configurations.

The advantages of DNA cloning are obvious; individual genes may be isolated and replicated in a defined manner. They may then be analysed further, modified if required, and sequenced; this enables the primary structure of the expressed gene to be deduced. The genes may then be expressed in prokaryote or in eukaryote systems providing antigen preparations of defined purity and, in particular, free from other toxic microbial products or (in the case of viruses, especially) of host-cell protein or nucleic acids. A practical advantage is that quite small amounts of DNA or RNA can serve as starting material; amplification is provided by the technology, so that the manipulation of genes from difficult systems (where it is otherwise impossible to collect sufficient material) can be engineered. The production of hepatitis B virus antigens (cf. Murray, 1983) furnishes a good example. Other organisms are being profitably exploited in this manner such as the rotaviruses, EB virus, the schistosomes and plasmodia.

The usefulness of the technology may be augmented by ingenuity and low cunning. For instance, stitching a desirable antigen gene to a marker gene (such as that specifying β-galactosidase) results in the expression of a fusion protein marked by its enzymic activity, bearing the desired antigenic determinants and of sufficient size that it is by itself immunogenic. This device has been used, for example, with hepatitis B virus core antigen (Stahl et al., 1982) and the glycoprotein D of herpes simplex virus (Weis et al., 1983). A final example of the production of a modified protein is provided by the expression in and secretion from eukaryotic cells of a truncated form of the same glycoprotein – the segment of the gene encoding the hydrophobic portion having been deleted (Lasky et al., 1984).

The concept that antigenic molecules possess a number of more or less discrete determinants or epitopes has been known for some time; indeed, Anderer (1963) showed that such a determinant on tobacco mosaic virus was generated by a very short run of amino acid residues. In recent years we have seen the introduction of oligopeptide vaccines containing a specific run of from 7 to 20 amino acids strung together synthetically and linked to a carrier protein such as haemocyanin. Such oligopeptides have been shown to elicit antibodies and to immunise against diphtheria toxin (Audibert et al., 1982). cholera toxin (Jacob, Sela & Arnon, 1983) and hepatitis B virus (Lerner et al., 1981; Gerin, Purcell & Lerner, 1984). In most of these studies the identification of the protective epitopes was painstakingly achieved; but there have been some ingenious approaches to the more certain delineation of epitopes. The startling success of Bittle *et al.* (1982) depended on the pre-knowledge of the nucleotide sequences specifying the VP-1 proteins of two sub-types of foot-and-mouth disease virus. It was argued that the two stretches unique to each subtype most likely encoded specific protective antigens. Oligopeptides were made according to the predicted sequences; these induced neutralising antibody and protected guinea pigs against the appropriate strain.

An elaboration has grown from the above. Beginning with neutralising monoclonal antibodies against herpes simplex virus, Holland *et al.* (1983) isolated mutant clones of virus resistant to neutralisation by them. Such mutants arise quite frequently. The sequences of the wild-type and mutant genes are next determined and comparison of the result allows localisation of the amino acid change responsible for the altered phenotype of the mutant. A reasonable guess can then be made as to which oligopeptides are worth investigating. This approach was used by Minson *et al.* (1986); unfortunately, the generation of epitopes depends less upon the primary sequence of amino acids in the polypeptide than on the specific configuration at the surface of the folded protein. Thus, Minson and colleagues found only one out of four epitopes studied was amenable to localisation by this means. A great deal is now being learnt about the structure of epitopes and how they arise; monoclonal antibodies are cardinal reagents for the purpose.

We come now to the use of genetic manipulation in the definition of attenuation and immunogenicity. The Sabin vaccine has for many years provided a satisfactory bulwark against poliomyelitis in a good many countries. But recently it has become plain that the very few paralytic cases in Britain seem to be vaccine-associated. The fruitful collaboration between workers at the National Institute of Biological Standards and Control and the University of Leicester provides a brilliant analysis of the evolutionary events of the type 3 strain (Leon). The original Leon strain, Sabin's attenuated derivative and revertant strains isolated from patients with vaccine-associated poliomyelitis were examined by making cDNA from the virus genomes and comparing their base sequences. The analysis showed that the original strain differed from the vaccine strain by ten base changes and the revertant strain differed from the vaccine strain by seven base changes. By examining a number of presumed revertants, it became clear that the biologically significant alterations reduced to three base changes, the most constant of which, occurred in a non-coding region of the genome. Virulence tests confirmed that this should account for much of the attenuation and it is suggested that the mutation might act by changing the secondary structure of the RNA of this virus. The rate of reversion to virulence in the human gut is rapid, showing that this mutation is strongly selected for (Almond *et al.*, 1985). These studies provide valuable information on the basis of neurovirulence, on the performance of a vaccine in practice and they inform us of what had happened in the original process of attenuation.

Other work exemplifies the prospective use of gene manipulation in the formulation of vaccines. There are two types of herpes simplex virus which share considerable sequence homology. Each has a number of glycoproteins in the envelope, important in protective immunity; some of these bear type-common epitopes predominantly, others are type-specific. Both types replicate using a strictly regulated cascade feedback programme such that three groups of transcriptional events are switched by successive products. It is also known that the enzyme thymidine kinase, a virus-specified enzyme, is important for performance of the viruses *in vivo*. Artificial strains have been constructed in which type 1 and 2 glycoproteins are combined in the same strain, the thymidine kinase gene is deleted or placed under different controls and other segments of the genome are deleted. These strains were found to be less virulent for mice, failed to induce latency and induced varying degrees of immunity against challenge with either type of virus. Though none of the constructs has proved a satisfactory vaccine, the prospective approach is worth remark. (Roizman *et al.*, 1982; cf. Meignier, 1985).

The final topic concerns the presentation of vaccine antigens. Oligopeptides must be linked to a carrier protein to be immunogenic – no more needs to be written here. Artificial live vaccines are being constructed in which protective immunogenic genes from a variety of pathogens are cobbled into plasmids or genomes of bacteria and viruses. The best example is the use of vaccinia virus as a vaccine vector of this kind. Vaccinia has a long history of success in the prevention of smallpox; though it is not a wholly blameless vaccine, it can be attenuated further by suitable manipulation. It has genetical accommodation for several genes and it is theoretically possible to produce a vaccine against a battery of pathogens. A possible disadvantage is lack of effectiveness in subjects who have been previously vaccinated. Already, genes encoding hepatitis B surface antigen, influenza haemagglutinin, herpes virus glycoprotein D, the glycoprotein of rabies and a sporozoite antigen of *Plasmodium knowlesii* have been expressed in cells infected with such recombinants and some of these have been used successfully to immunise suitable animal models (Moss, 1983).

We have discussed enough, albeit superficially, to note that the advent of genetic engineering has considerably enlivened the world of vaccines. At present all is in the future and there are no definite accomplishments. Unfortunately, the present-day lily-livered attitude is getting in the way of adventure. Who will dare to stick out his neck?

THEN AND NOW

Let us compare the predicaments of Jenner and ourselves. He possessed a minimum of background knowledge; he had only an appreciation of the overall behaviour of pathogenesis and immunity to infectious diseases. These two black boxes were supplemented by an old wives' tale, the rudiments of scientific method and his personal acumen. We, on the other hand, possess an unmanageable volume of background knowledge; there is a whole army of us concerned in some way with immunisation and, for each one of us, there are areas of competence and areas of ignorance. We comprise a workforce so heterogeneous that communication is seriously impaired and we lack the ability to focus effort, which (for Jenner) was no problem. For example, we sometimes find quite brilliant feats of molecular engineering spoilt by inadequate biological back-up. We try to meet this by improving the dissemination of information; the technology to do this is available but there remains the bottleneck of human capacity. It may be impossible for one human mind to encompass a full enough understanding of all that should be taken into account when thinking constructively. We therefore resort to the use of the ever more frequent multidisciplinary efforts. Plainly, it is good sound sense so to do; no longer can individuals expect to compete effectively with the concerted efforts of large teams of scientists. But, as Eliot says 'Flesh and blood are weak and frail, susceptible to nervous shock'; it must be recognised that multidisciplinary efforts are susceptible to all the human frailties such as personal ambitions, jealousies, differences in upbringing, temperament and so on. For this reason, successful teams, however loosely or closely knit require a personality, not necessarily the scientific leader nor the paymaster, whose overriding role is to hold the collective effort together and ensure a proper focus.

It is difficult to be sure quite how long Jenner took to evolve vaccination; it was probably of the order of 20–25 years. That says a good deal for his persistence and patience. In terms of man hours, this represents considerable efficiency – especially when one takes account of the almost total absence of fixed points from which he could take reliable bearings. He was obliged to prove true or false many contemporary tenets and many of his own wilder hypotheses; he had to establish criteria for the proper diagnosis of smallpox and the distinction of true from false cowpox. Much of the time he had to await natural events all the while looking out for experiments of nature; he had to wait for outbreaks of cowpox and though he could challenge vaccinated persons by variolating them, appreciation of their immunity to the natural disease required observation of their escape from severe epidemics. The problems we face are diametrically different. Experiments, simple or elaborate, can be done with comparative ease but they are becoming increasingly expensive. Jenner's immediate costs must have been minimal.

One final constraint on Jenner's activities (and on our own today) concerns the reactions of society to innovation, especially when it concerns medical matters. Jenner had the advantage that he was attempting to eliminate a terrifying menace. To be sure, variolation was already practised, but this was also killing an unacceptable proportion of people. Appreciation of the safety of vaccination, initially at least, earned acclaim. But when vaccine became contaminated by the foolishness of others, the fatalities were greeted with antagonistic derision by medical men and engendered a public outcry. The same sort of reaction had ostracised Jesty 30 years before. Jenner was obliged to seek the protection of various prominent people, including several members of the nobility and even King George himself. We are nowadays protected from such witch hunting but at what cost?

The modern social structure has so many interest groups such as consumer associations, professional bodies and trades unions that we are clogged with committees for our self-regulation. Committees must be serviced, so they become backed by secretariats. Secretariats tend to become self-perpetuating so that a permanent civil service of ever-increasing power and longevity develops. Nobody can deny the truth of all the trite old tags – 'prevention is better than cure', 'it is better to be safe than sorry' and so on – and it follows that regulation must not only be accepted but welcomed. In welcoming it, we must count the cost not so much in monetary terms but rather in terms of the detriment to science. In the present context we can point to the regulation of genetic engineering, the regulation of research on dangerous pathogens and the regulation of general laboratory safety. All these matters now consume an unacceptable proportion of the scientist's energies. Perhaps worse, one becomes increasingly aware of public suspicion as to what we are about; perhaps people believe that the vaccines we create will somehow get out of hand (like Frankenstein or the stuff of Quatermass); some may even think we are deliberately contriving evil. If so, this is nothing new, as Jenner found in his more difficult times: one of the great fears was that vaccine would transform people into cattle as depicted by Gilray in 1802; 'The Cow-Pock – or the Wonderful Effects of the New Inoculation' (Fig. 2).

Fig. 2. 'The Cow-Pock . . . or . . . the Wonderful Effects of the New Inoculation.'

ACKNOWLEDGEMENTS

It is a pleasure to thank Professor W. C. Russell and Mr R. N. Smart for archival research. I am grateful to Messrs Zenith and Juki for editing and typing the manuscript.

REFERENCES

ALMOND, J. W., WESTROP, G. D., CANN, A. J., STANWAY, G., EVANS, D. M. A., MINOR, P. D. & SCHILD, G. C. (1985). Attenuation and reversion to neurovirulence of the Sabin poliovirus type-3 vaccine. In *Vaccines 85*, ed. R. A. Lerner, R. M. Chanock & F. Brown, pp. 271–7. New York: Cold Spring Harbor Laboratory.

ANDERER, F. A. (1963). Versuche zur Bestimmung der serologisch determinten Gruppen des Tobakmosaikvirus. *Zeitschrift für Naturforschung, Teil B*, **18**, 1010–14.

AUDIBERT, F., JOLIVET, M., CHEDID, L., ARNON, R. & SELA, M. (1982). Successful immunization with a totally synthetic diphtheria vaccine. *Proceedings of the National Academy of Sciences, USA*, **79**, 5042–6.

BARON, J. (1838). *The life of Edward Jenner, MD, LLD, FRS*. London: Henry Colburn.

BEHRING, E. VON & WERNICKE. (1892). Uber Immunisierung und Heilung von Veirsuchsthieren bei der Diphtherie. *Zeitschrift für Hygien und Infektionskrankheiten*, **12**, 10–44.

BITTLE, J. L., HOUGHTEN, R. A., ALEXANDER, H., SHINNICH, T. M., SUTCLIFFE, J. G., LERNER, R. A., ROWLANDS, D. J. & BROWN, F. (1982). Protection against foot and mouth disease by immunization with a chemically synthesized peptide predicted from the viral nucleotide sequence. *Nature, London*, **298**, 30–3.

CALMETTE, A. & GUÉRIN, C. (1905). Origine intestinale de la tuberculose pulmonare. *Annals de l'Institut Pasteur*, **19**, 601–18.

COUZENS, P. A. (1972). *The Continuing Story of the Sodburys*, pp. 145–6 Gloucester Couzens.

COUZENS, P. A. (1973). Letter. *The Sodbury Gazette*, Dursley: Bailey & Sons.

CREIGHTON, C. (1889). *Jenner and vaccination*. London: Swan Sonnenschein.

DOWNIE, A. W. (1965). Poxvirus group. In *Viral and Rickettsial Disease of Man*, ed. F. L. Horsfall & I. Tamm, pp. 932–64. Philadelphia: J. B. Lippincatt Company.

FISK, D. (1959). *Doctor Jenner of Berkeley*. London: Heineman.

GERIN, J. L., PURCELL, R. H. & LERNER, R. A. (1984). Recombinant DNA and synthetic peptide approaches to HBV vaccine development: immunogenicity and protective efficacy in chimpanzees. In *Modern Approaches to Vaccines: Molecular and Chemical Basis of Virus Virulence and Immunogenicity*, ed. R. M. Chanock & R. A. Lerner, pp. 212–17. New York: Cold Spring Harbor Laboratory.

HOLLAND, T. C., MARLIN, S. D., LEVINE, M. & GLORIOSO, J. (1983). Antigenic variants of herpes simplex virus selected with glycoprotein-specific monoclonal antibodies. *Journal of Virology*, **45**, 672–82.

JACOB, C. O., SELA, M. & ARNON, R. (1983). Antibodies against synthetic peptides of the beta subunit of cholera toxin: crossreaction and neutralization of the toxin. *Proceedings of the National Academy of Sciences, USA*, **80**, 7611–15.

LASKY, L. A., DOWBENKO, C. C., SIMONSEN, C. C. & BERMAN, P. W. (1984). Protection of mice from lethal herpes simplex virus infection by vaccination with a secreted form of cloned glycoprotein D. *Biotechnology*, **2**, 527–32.

LERNER, R. A., GREEN, N., ALEXANDER, H., LIU, F. T., SUTCLIFFE, J. G. & SHINNICK, T. M. (1981). Chemically synthesized peptides predicted from the nucleotide sequence of the hepatitis B virus genome elicit antibodies reactive with the native envelope protein of Dane particles. *Proceedings of the National Academy of Sciences, USA*, **78**, 3403–7.

MEIGNIER, B. (1985). Vaccination against herpes simplex virus infections. In *The*

Herpesviruses, ed. B. Roizman & C. Lopez, vol. 4, pp. 265–96. New York: Plenum Press.

MINSON, A. C., HODGMAN, T. C., DIGARD, P., HANCOCK, D. G., BELL, S. C. & BUCKMASTER, E A. (1986). An analysis of the biological properties of monoclonal antibodies against glycoprotein D of herpes simplex virus and identification of amino acid substitutions that confer resistance to neutralisation. *Journal for General Virology*, (in press).

MOSS, B. (1983). Infectious vectors for vaccines: The use of virus vectors for vaccines. In *New Approaches to Vaccine Development*, ed. R. Bell & G. Torrigiani, pp. 167–77. Basel: Schwabe & Co.

MURRAY, K. (1983). New and improved techniques for polypeptide synthesis. In *New Approaches to Vaccine Development*, ed. R. Bell & G. Torrigiani, pp. 5–27. Basel: Schwabe & Co.

NASH, A. A., LEUNG, K. N. & WILDY, P. (1985). The T-cell-mediated immune response of mice to herpes simplex virus. In *The Herpesviruses*, vol 4, ed. B. Roizman & C. Lopez, pp. 87–102. New York: Plenum Press.

NEEDHAM, J. (1980). *China and the Origins of Immunology*. Hong Kong: Centre of Asian Studies, University of Hong Kong.

PASTEUR, L. (1885). Méthode pour prévenir la Rage après morsure. *Comptes rendues de l'Acadamie Digan*, **101**, 765–72.

ROIZMAN, B., WARREN, J., THUNING, C. A., FANSHAW, M. S., NORRILD, B. & MEIGNIER, B. (1982). Application of molecular genetics to the design of live herpes simplex virus vaccines. *Developments in Biological Standardization (Basel)*, **53**, 287–304.

SABIN, A. (1955). Characteristics and genetic potentialities of experimentally produced and naturally occurring varients of poliomyelitis virus, *Annals of the New York Academy of Science*. **61**, 924–38.

STAHL, S., MACKAY, P., MAGAZIN, M., BRUCE, S. A. & MURRAY, K. (1982). Hepatitis B virus core antigen: Synthesis in *Escherichia coli* and application in diagnosis. *Proceedings of the National Academy of Sciences, USA*, **79**, 1606–10.

THEILER, M. & SMITH, H. H. (1937). The effect of prolonged cultivation *in vitro* upon the pathogenicity of yellow fever virus. *Journal of Experimental Medicine*, **65**, 767–86.

WEIS, J. H., ENDQUIST, L. W., SALSTROM, J. S. & WATSON, R. J. (1983). An immunologically active chimaeric protein containing herpes simplex type 1 glycoprotein D. *Nature*, **302**, 72–4.

INFLUENZA VIRUS GLYCOPROTEIN VARIATION AND ITS ROLE IN DISEASE

JOHN W. McCAULEY

Animal Virus Research Institute, Pirbright, Surrey GU24 0NF

The impact any virus has on a host population reflects a combination of the prevalence of infection and the severity of disease inflicted. The prevalence of infection is due, in part, to the ease of spread of virus and also to the proportion of susceptible individuals in the population. The severity of disease is determined by the ability of the virus directly to cause tissue damage, as well as the response of the host to infection. Influenza viruses cause epidemic and sporadic outbreaks in man and in many species of mammals and birds. They show variation in properties that affect virus spread within a population and the degree of pathology seen in an infected individual. This chapter details the properties of the influenza virus glycoproteins which may influence the frequency and severity of influenza virus infection.

VIRUS STRUCTURE

The influenza virus genome consists of distinct single-stranded RNA pieces which are associated with virus polypeptides in the virion and the infected cell and form a ribonucleoprotein complex. Three serotypes of influenza virus, A, B and C, can be defined on the basis of the antigenicity of their ribonucleoprotein. Complement-fixing antibodies raised against ribonucleoprotein of all type A viruses cross-react, whereas no cross-reaction is observed between virus types.

Most work on the structure and replication of influenza virus has been carried out on serotypes A and B, which appear very similar in both structure and replication (Lamb & Choppin, 1983; McCauley & Mahy, 1983), but viruses of the C serotype show substantial differences. The genome of influenza virus types A and B consists of eight segments of RNA, in contrast to type C viruses, which contain seven. The RNA in the ribonucleoprotein complex is associated with four polypeptides: the nucleoprotein polypeptide (NP) and three larger molecular weight polypeptides, PA, PB_1 and PB_2, which together form an RNA polymerase complex. The virion ribonucleoprotein is surrounded by a shell of matrix protein which underlies a host-derived lipid bilayer into which the spike-like glycoproteins are inserted.

The virions of influenza types A and B contain two sorts of glycoprotein, the haemagglutinin and the neuraminidase. Influenza type C viruses contain only a single glycoprotein, which has both haemagglutinating and receptor-destroying activity, and the following description will refer only to types A and B virus. The haemagglutinin glycoprotein consists of two polypeptide chains, HA_1 and HA_2, which are linked by a single disulphide bond (Waterfield, Scrace & Skehel, 1981) and are assembled into a trimer of approximately 225 000 molecular weight (Wilson, Skehel & Wiley, 1981). The neuraminidase is approximately 200 000 molecular weight, and consists of four monomers of a single polypeptide chain (Varghese, Laver & Colman, 1983).

Both glycoproteins contain a soluble domain which can be removed intact from the virus envelope by proteolysis and the crystal structures of the released portions of the HA and NA have been solved to 0.3 and 0.29 nm resolution, respectively (Wilson et al., 1981; Varghese et al., 1983). The electron density maps show that the HA contains a globular head region, which contains a sialic acid binding site held away from the virus envelope by a long alpha-helical domain consisting predominantly of the HA_2 polypeptide chain (Fig. 1). The neuraminidase tetramer is held away from the membrane by a stalk, but the stalk is lost upon proteolytic release from the virus. The head consists of four monomers, each made up of a number of beta-sheets arranged in a propeller formation and surrounds the catalytic (sialic acid binding) site (Fig. 2).

Two non-structural polypeptides are encoded by the smallest RNA segment, segment 8 (Lamb & Choppin, 1979), and, in addition to the matrix protein of the virion of type A viruses, RNA segment 7 encodes a second polypeptide, M2, which is expressed on the surface of infected cells (Lamb, Zebedee & Richardson, 1985) but is not glycosylated. No counterpart has been found in influenza B virus-infected cells, although the coding capacity of RNA segment 7 to produce a second polypeptide product from a spliced mRNA is maintained (Breidis, Lamb & Choppin, 1982). Two gene products can

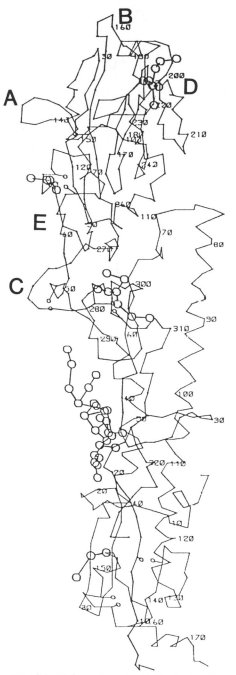

Fig. 1. A schematic tracing of the alpha-carbon atoms of a single HA monomer. The locations of antigenic sites A to E of the H3 sub-type haemagglutinin are shown. (Courtesy of J. J. Skehel and D. C. Wiley.) Details of the structure are provided by Wiley *et al.* (1981) and Wilson *et al.* (1981).

Fig. 2. Diagram of the polypeptide folding pattern in a neuraminidase tetramer viewed from above looking down the four-fold symmetry axis. Indicated on the diagram in the top left are disulphide bonds (—); bottom left, carbohydrate attachment sites (●), bottom right, two metal binding ligands (↑), and top right, highly conserved amino acids (● and △) in different neuraminidase sub-types which surround the sialic acid binding site (∗). Taken from Varghese *et al.* (1983) and Colman *et al.* (1983). Reprinted by permission of *Nature*, Vol. 303, pp. 35–44. © 1983 Macmillan Journals Ltd.

be detected which are synthesised from a single bi-cistronic mRNA complementary to RNA segment 6 of influenza B viruses. Segment 6 codes for the neuraminidase but a second glycoprotein (NB), which does not become incorporated into virus, can be found in infected cells (Shaw & Choppin, 1984; Shaw, Choppin & Lamb, 1983). As far as is known, all the other RNA segments are complementary to only one mono-cistronic mRNA which is translated to produce a single primary translation product.

The segmented nature of the influenza virus genome affords the virus a high degree of genetic flexibility through recombination (re-assortment of genes), which can be readily demonstrated between viruses of the same type both *in vivo* (Webster, Campbell & Granoff, 1971) and *in vitro* (for example, Hirst, 1973). While a high frequency of recombination is found between viruses of the same serotype, no recombination between viruses of different serotypes has been recorded. The high frequency of recombination and the segmented nature of the genome can be considered consequences of the virus replication strategy in the cell (reviewed by McCauley & Mahy, 1983).

THE ORIGIN OF PANDEMIC INFLUENZA

Variation in antigenicity of the haemagglutinin and neuraminidase is the basis for classification of influenza A virus into subtypes. In all, including influenza viruses of horses, pigs and birds, there are thirteen H subtypes and nine N subtypes (*WHO memorandum*, 1980; Hinshaw *et al.*, 1982). Introduction of new subtypes into the human population results in pandemics (extensive world-wide outbreaks) of human influenza. Since the influenza virus pandemic of 1918, which resulted in 20 million deaths in the space of thirteen months (Hoyle, 1968), influenza pandemics have occurred in 1957 (Asian influenza) and 1968 (Hong Kong influenza) but between these pandemics epidemics occur almost annually. Human viruses isolated between 1933, when laboratory infection of ferrets with nasal washings resulted in the first isolation of human influenza virus (Smith, Andrewes & Laidlaw, 1933), and 1956 are designated H1N1; those of the 'Asian' type isolated from 1957 to 1967 are designated H2N2; and those isolated since 1968 of the 'Hong Kong' type, H3N2. Since 1977, when re-introduction of H1N1 viruses occurred, H3N2 and H1N1 have been co-circulating in the human population.

The means of introduction of new subtypes into the population and their spread has been studied extensively since the emergence of H3N2 in man (reviewed by Webster *et al.*, 1982*b*). Molecular analysis of the HA and NA genes has revealed a high degree of sequence diversity between the glycoproteins of different subtypes: amino acid sequence homology between typical H1, H2 and H3 haemagglutinins is 35% for H1 and H3 viruses and 40% for H2 and H3 viruses (Gething *et al.*, 1980; Min-Jou *et al.*, 1980; Verhoyen *et al.*, 1980). H1 and H2 haemagglutinins are more highly conserved; 70% homology is found (Winter, Fields & Brownlee, 1981), which presumably reflects a more recent common ancestor.

A similar degree of variation is shown between neuraminidase molecules: no more than 41% homology is found between N1 and N2 neuraminidase subtypes (Bentley & Brownlee, 1982; Markoff & Lai, 1982; van Rompuy *et al.*, 1982). This observed degree of variation in the glycoproteins can result in viruses that are unaffected by antibody directed against the glycoproteins of other virus subtypes.

Prior to the introduction of H3N2 virus into humans, examples of H3 viruses had been isolated from birds (for example, A/Duck/Ukraine/1/63 H3N8) and horses (A/equine/Miami/1/63 H3N8). It is considered highly likely that H3 subtype viruses arose by recombination between a human virus and an animal influenza virus to produce a virus with growth (and pathogenic) potential for man. The HA gene is thought to have arisen from an avian host, since a higher degree of homology is seen between avian and human H3 haemagglutinins (Fang *et al.*, 1981) than between equine and human haemagglutinins (Daniels *et al.*, 1985a). Avian and human haemagglutinins show 90% homology, whereas the equine and human counterparts show only 80% homology. The other genes in the earliest human isolates were shown by RNA hybridisation to be more closely related to those of the H2N2 viruses of the preceding subtype than to those from an avian background (Scholtissek *et al.*, 1978b). The introduction of a novel HA gene into a non-immune population would explain the pandemic of 1968 and 1969; this process has been termed antigenic *shift*. It is not clear, however, why the introduction of H3N2 viruses in man should result in the subsequent failure to isolate H2N2 viruses.

The re-emergence of H1N1 viruses in 1977 resulted in the re-introduction of a subtype into a partially immune population. Those born before 1957 had experienced H1N1 viruses previously and epidemics were confined to the susceptible, post-1957, age-group. H1N1 viruses, in contrast to the H2N2 and H3N2 viruses, did not replace the previous subtype and, as referred to earlier, both H1N1 and H3N2 viruses are currently (1986) co-circulating. The origin of the recent H1N1 strains is unknown, although all the genes are closely related to those of viruses isolated from man in 1950 (Nakajima, Desselberger & Palese, 1978; Scholtissek, von Hoyningen & Rott, 1978a; Kendal *et al.*, 1978). The re-emergence of H1N1 viruses demonstrates that new virus can be introduced into and circulate in a partially immune population, but a substantial part of the population may need to consist of individuals who are totally non-immune and who have contact with a large number of other non-immune individuals (such as in school or military camp).

EPIDEMIC DISEASE FOLLOWING PANDEMIC INFLUENZA

The ability of influenza subtypes to persist and cause epidemic disease in the population is considered to be due to frequent mutation, which results in changes in antigenicity of the glycoproteins, a phenomenon referred to as antigenic *drift*. Diversity in the amino acid sequence and changes in antigenicity of the haemagglutinin and the neuraminidase are associated with epidemics of influenza. The production and selection of novel virus must occur in infected individuals so that virus can infect people who were previously immune.

Virus strains can be differentiated by a number of criteria, but the most frequent means to demonstrate differences is by the reaction of virus with antibodies raised against reference virus strains. Convalescent ferret antiserum or monoclonal antibodies are commonly used to detect variation of virus strains. Antibodies directed against the haemagglutinin are usually used in haemagglutination inhibition tests and those raised against the neuraminidase are used to block its enzymatic activity. From year to year, strains are found to circulate and then be replaced by new strains, although co-circulation of distinct strains may occur. Current information indicates that the strains circulating during the winters of 1984–85 and 1985–86 are similar to A/Christchurch/4/85 (H3N2), A/Mississippi/1/85 (H3N2) and A/Chile/1/83 (H1N1) (*Weekly Epidemiological Record*, 1986).

Nucleotide sequence analysis of the genes that code for the HA and NA glycoproteins from a large number of antigenically distinct strains has been used to identify regions of the molecules that undergo changes associated with epidemics of the disease. Selection in the laboratory of virus mutants which can overcome neutralisation by monoclonal antibodies identify antigenic sites on the molecule. Single nucleotide changes in the glycoprotein genes are found, which define neutralising epitopes on the virus glycoproteins. Location of these regions of amino acid sequence variation within the three-dimensional structure of the glycoproteins defines five antigenic sites on the H3 haemagglutinin (sites A to E, Fig. 1) (Wiley *et al.*, 1981), four sites on the H1 molecule (Caton *et al.*, 1982; Raymond *et al.*, 1983, 1986) and at four sites on the influenza B haemagglutinin (Krystal *et al.*, 1983). From these studies it has been concluded that change in at least three different sites is necessary in order that an epidemiologically significant strain of H3 virus can emerge (Wiley, Wilson & Skehel, 1981). Similar changes are necessary for the emergence of new strains of H1 virus. An insufficient number of B virus strains has been examined to allow firm conclusions to be drawn about the number of changes needed for a new epidemic strain of influenza B to emerge.

Experimental documentation that viruses produced during infection of one individual may contain antigenic variants comes from analysis of viruses isolated during epidemics in semi-closed communities or in defined geographical areas. In an outbreak of influenza A (H3N2) in Christ's Hospital School, a boys' boarding school in the UK, thirty groups of antigenically distinct viruses could be detected, although two groups predominated (Oxford *et al.*, 1986). As the epidemic progressed, increased numbers of antigenically distinguishable viruses were isolated, but the changes seen were considered minor in comparison to those of new epidemic strains. Viruses had been grown in eggs infected with nasal washings of boys and some or all of the changes may be due to the selection of mutants during isolation rather than reflecting heterogeneity in the infected boy (see, for example, Schild *et al.*, 1983; Robertson *et al.*, 1985, 1986). Nucleotide sequence analysis of the HA and NA genes should reveal the exact nature of the heterogeneity of the virus population during the course of the epidemic. Similar results have been found for an influenza B virus epidemic from the same school (Oxford *et al.*, 1983). Extensive antigenic heterogeneity has also been seen in H1N1 viruses isolated in a single urban community from 1977–81 in Houston (Six *et al.*, 1983), although in this case the introduction of new viruses from outside the area is considered likely. The epidemiological significance of the virus variants isolated in Houston or in the boys' school is unknown but, even if they do not represent production of novel strains, they may represent early stages in the evolution of such viruses.

It is known that influenza variants also arise easily on laboratory passage (Brand & Palese, 1980) and this is thought to be due to error-prone replication by the RNA polymerase complex (Holland *et al.*, 1982). No *in vitro* measurements have been reported which would indicate the fidelity of replication of influenza viruses in the absence of possible selective pressure. Such experiments have to be carried out to compare error rates with those of other RNA replication complexes as well as for DNA replication. The absence of 'proof-reading' in RNA transcription may cause low-fidelity replication, but the contribution of proof-reading to the precision of replication of DNA *in vitro* varies considerably and depends on the enzyme used (Shi & Fersht, 1984).

The immunological response of the host animal is also an important component in the generation of virus variants for, while error-prone replication may create the variants, they must overcome the individual's immunity in order to be selected. The specificity of the immune response is therefore an important feature in the selection of viruses which evolve to cause epidemic disease. Individuals show a large degree of variation in susceptibility and resistance to infectious agents, in which the immune response plays a major role (Bodmer, 1980). In a study to estimate the variety of anti-influenza virus antibodies in post-infection human sera, it was found that sera differed considerably in their ability to compete with monoclonal antibodies directed against defined antigenic sites on the HA (Wang, Skehel & Wiley, 1986) and that individuals respond to infection by producing antibodies with different ranges of specificity. The results also show that antibody in serum can recognise more than one antigenic site and that no site is immunodominant: serum competes with monoclonal antibodies directed against differing epitopes. In reciprocal experiments, some post-infection human sera fail to recognise antigenically variant viruses selected with a single monoclonal antibody directed against the HA. This result would suggest that some sera show a highly restricted spectrum of anti-HA response and that antibody to a single immunodominant site may be present in certain individuals (Natali, Oxford & Schild, 1981).

As described above, the nucleotide sequence data indicate that at least one change at each of three antigenic sites is necessary to produce a virus capable of causing widespread disease and infecting people who were previously immune to the circulating strains. The sequence data, antibody specificity data and evidence of production of virus variants are consistent with the idea that novel viruses arise by sequential selection in partially immune individuals. In time, this results in the production of novel strains which are sufficiently different from previously widespread viruses to overcome immunity in the population and give rise to epidemics. It is not known whether selection of variants occurs solely during the initiation of infection of a partially immune individual or whether the generation of new immunity influences and selects for the production of novel viruses.

The role of the neuraminidase in antigenic shift or drift is less clear-cut. The introduction of H3N2 viruses did not result in the introduction of a new N subtype, and yet pandemic influenza occurred. Antibody against the NA is only neutralising at high concentration *in vitro* (Webster, Hinshaw & Laver, 1982*a*), yet antigenic variation is seen in nature (summarised in Colman, Varghese & Laver, 1983) and is, to a large extent, concentrated at antigenic sites defined by neutralising monoclonal antibodies (Colman *et al.*, 1983; Colman & Ward, 1985). Although the lack of experimental data precludes firm conclusions from being drawn, sequential selection of variants probably occurs by a similar mechanism to that which results in variation in the HA.

ANTIGENIC VARIATION IN ANIMAL INFLUENZA

Besides man, influenza virus commonly infects horses, pigs and birds, and serious disease has been reported in seals (Lang, Gagnon & Gerachi, 1981), mink (Klingeborn *et al.*, 1985) and whales (Hinshaw *et al.*, 1986). Viruses from all 13H and 9N subtypes can be isolated from avian species alone (Alexander, 1982).

H1N1 viruses have been isolated from swine suffering from epizootics of respiratory disease. H3N2 and influenza C viruses also have been isolated from pigs but they have not been associated with disease. Antigenic variation of recent swine H1 viruses can be detected and at least two distinct antigenic variants currently co-circulate in pigs (Hinshaw *et al.*, 1984). It is not known, however, whether antigenic drift in these pig viruses is occurring and whether the observed variations in the amino acid sequence of the glycoproteins are associated with outbreaks of disease. Co-circulating antigenic variants do not necessarily imply antigenic drift in the pig population, but novel H1 glycoproteins may have been introduced into the pig population by a process more akin to antigenic shift in man, from another animal reservoir. There is evidence, indeed, that viruses may be spread from turkeys to pigs or vice versa (Hinshaw *et al.*, 1981; Aymard *et al.*, 1985).

H3N8 and H7N7 viruses both cause epidemic outbreaks of coughing disease in stabled horses. No epidemics of equine influenza have been reported in the British Isles since 1979, but previously both H3 and H7 equine viruses regularly caused outbreaks of disease (Burrows *et al.*, 1981). As with swine viruses, it is clear that equine viruses also vary (Hinshaw *et al.*, 1983; Daniels *et al.*, 1985*a*), and antigenic variants of the same subtype co-circulate in the population (Hinshaw *et al.*, 1983). Nucleotide sequence analysis has revealed that variation between A/equine/Miami/63 and A/equine/Fontainebleau 79 H3 haemagglutinins occurs in antigenic sites A, B, C and D (Daniels, Skehel & Wiley, 1985*a*). However, it is not known whether this represents antigenic drift or mutation in the absence of immune selection. Such variation may be insufficient to cause widespread epizootic disease (Burrows *et al.*, 1981; Hinshaw *et al.*, 1983) and outbreaks could arise merely by infection of non-immune or poorly immune animals.

Influenza viruses are widespread in birds but serious disease appears to occur most frequently in flocks of domestic fowl. Variation in the glycoproteins resulting in the introduction of novel viruses which can overcome immunity is certainly feasible, either by antigenic drift or transmission across a species barrier, if one exists. Besides the finding that transmission of virus may be possible between turkeys and pigs (Aymard *et al.*, 1985; Mohan *et al.*, 1981), transmission of viruses between avian species is considered likely (Alexander, 1982). Viruses of low pathogenicity for domestic fowl can be considered as enzootic on some commercial duck farms in UK and repeated introduction into turkey flocks are thought to occur (Alexander, 1982). Antigenic drift is not thought to play a major role in disease of turkeys in such circumstances. Influenza viruses that are highly pathogenic for chickens and turkeys have only been associated with H5 and H7 virus subtypes, although by no means all H5 and H7 viruses are pathogenic. Again, antigenic variation is not considered to play a major role in the outbreak of diseases caused by highly pathogenic avian influenza virus strains, but cross-species transmission from ducks, which are refractory for the disease (Alexander *et al.*, 1978; R. G. Webster, personal communication), is considered much more likely.

Influenza viruses also kill mink, seal and whale. These viruses, which have been occasionally isolated from dying or dead animals, are closely related antigenically and genetically to viruses from avian sources (Webster *et al.*, 1981; Klingeborn *et al.*, 1985; Hinshaw *et al.*, 1986). Subtypes isolated from infected seals were H7N7; mink, H10N4; and whales, H13N9 and H13N2. Epidemic infection of these

animals is thought to have occurred by virus crossing a species barrier. Again, like the introduction of highly pathogenic avian strains
into flocks of chickens and turkeys, infection of these mammals may
be more analogous to human pandemic influenza (antigenic shift)
rather than human epidemic disease and antigenic drift.

VIRUS REPLICATION IN EXPERIMENTAL MODELS

In the preceding discussion concerning the role of glycoprotein variation in epidemics of influenza, little attention was paid to the mechanism of virus neutralisation or to variation in properties, other than
antigenicity, that may be important in disease. For example, the
ability to cross species barriers and then produce infectious virus
in sufficient quantity to form the nucleus of an epidemic is of clear
importance during the introduction of pandemic influenza. Successful establishment of viruses in a new host may call for considerable
adaptation. Moreover, virus variation may occur within a single host
and result in the emergence of viruses with enhanced virulence and
pathogenicity. A number of animal model systems, as well as virus
growth in organ culture and tissue culture, have been used to study
influenza virus virulence and pathogenicity. In terms of the molecular
biology of the virus, to understand the significance of results from
work carried out using organ culture and experimental animals
detailed knowledge is required of the synthesis and the role of virus
polypeptides in replication (reviewed by McCauley & Mahy, 1983),
their interaction with the host cell (exemplified by host range
mutants, cited in Mahy, 1984) and their stimulation of host immunity
(e.g. Townsend *et al.*, 1984; Ada, Leung & Ertl, 1981).

The synthesis and properties of the glycoproteins

Influenza virus types A and B bind to sialoligosaccharides on the
surface of cells. The HA is responsible for this binding and the sialic
acid binding site has been identified as a cleft or pocket close to
the top of the HA trimer distal to the viral envelope (Wilson *et
al.*, 1981). The exact specificity of the HA for a particular receptor
can vary (Rogers & Paulson, 1983), and amino acid changes associated with receptor variants of natural or laboratory-selected H3
viruses are located in the pocket (Rogers *et al.*, 1983). The neuraminidase removes sialic acid from suitable substrates and as such contains
potential receptor destroying activity. However, it has been speculated that the neuraminidase may play a role in reducing abortive

binding of virus (Burnet, 1948) – for example, to sialic acid-rich glycoconjugates in nasal mucins – but there is no firm evidence to support this notion. Neuraminidase may also prevent progeny virus from sticking to each other (Palese *et al.*, 1974) or be involved in release of virus from the cell (Sato & Rott, 1966). The catalytic site of the neuraminidase has been located in the three-dimensional structure by diffusion of sialic acid into neuraminidase crystals and observing change in electron density (Varghese *et al.*, 1983).

Once bound to the cell surface, virus-receptor complexes undergo receptor-mediated endocytosis and fusion of the viral and cellular membranes takes place in the acidic vesicles of the endosome (reviewed by Marsh, 1984). A large body of evidence indicates that the HA is the mediator of fusion. In summary, viruses with a changed pH optimum for fusion have mutations that map to the HA gene (Daniels *et al.*, 1985*b*); cells that express only the HA gene in isolation from other viral genes can cause syncytia formation upon incubation at pH 5.0 (White, Helenius & Gething, 1982); HA released from virus by the protease, bromelain, acquires the ability to bind lipid or detergent and undergoes a conformational change when exposed to low pH (Skehel *et al.*, 1982); and specifically solubilised HA_2 can be shown to partition as an amphipathic polypeptide (Doms, Helenius & White, 1985). The neuraminidase has been reported to be necessary for fusion to cell membranes when the HA and NA are incorporated into artificial membranes (Huang *et al.*, 1980*a*, *b*; Huang, Diftsch & Rott, 1985) but a role in viral-cell fusion remains doubtful.

During biosynthesis and assembly into trimer and tetramer, both the HA and NA undergo extensive post-translational modification. Both are intrinsic membrane proteins co-translationally inserted into the host cell membranes; in the case of the HA, a cleavable signal sequence is present at the amino-terminus of HA_1 (McCauley *et al.*, 1979) and a hydrophobic polypeptide domain at the carboxy-terminus of HA_2 serves as a membrane anchor (Skehel & Waterfield, 1975). The neuraminidase is anchored in the bilayer by an amino-terminal hydrophobic domain (Blok *et al.*, 1982), which presumably also serves as a signal sequence. Both the HA and NA undergo glycosylation of asparagine residues during intra-cellular transport to the cell surface (e.g. Allen, Skehel & Yuferof, 1977; Keil *et al.*, 1985). HA is also modified by acylation to form an adduct with palmitic acid during transport to the cell membrane (Schmidt, 1982). Assembly into trimers and tetramers occurs at some stage prior to their appearance at the plasma membrane but the exact location for this assembly has yet to be elucidated. Both the haemagglutinin and the neuraminidase reach the apical surface of polarised epithelial-like cells (Rodrigues-Boulin & Prendergast, 1980), whether synthesised during virus infection or from recombinant DNA introduced into the cell (Roth et al., 1983; Jones et al., 1985). The resultant cellular localisation is, therefore, an inherent property of each glycoprotein.

During production of infectious virus, the haemagglutinin undergoes further proteolytic cleavage of the polypeptide chain, which results in conversion of the primary translation product by a cellular protease of the HA gene to HA_1 and HA_2 linked by a disulphide bond (Lazarowitz & Choppin, 1975; Klenk et al., 1975) and the removal of a connecting peptide or single arginine residue (Bosch et al., 1981). Only certain virus and cell combinations will result in the production of infectious virus in vitro. Although virus particles can be isolated from cells which fail to cleave the HA, they are not infectious but treatment of such virus with trypsin results in cleavage of the HA and the virus becomes infectious. Conversion of HA to HA_1 and HA_2 leads to a conformational change in the HA (Flanagan & Skehel, 1977) and is necessary for the low-pH-induced fusion activity of the HA (White, Matlin & Helenius, 1981). It is not known whether uncleaved HA can undergo the low-pH-induced conformational change without the concomitant exhibition of fusion properties.

Variation of virus growth in vitro and in vivo

To understand how variation in the structure of the glycoproteins may affect virulence in nature, correlation needs to be found between in vitro phenomena (growth in tissue culture, organ culture, differential effects of non-specific inhibitors) and in vivo animal models (ferret or monkey virulence, neurovirulence in mice, growth in hens' eggs) with the ability of a virus to cause disease in the natural host. Even in the absence of experiments in the natural host, comparison of disease in animal models and virus behaviour in vitro provides some scope for understanding the disease processes that may be involved.

(i) Variation in virus interaction with cell receptors
The earliest event of virus replication in vitro is attachment to cellular receptors and, as mentioned above, variation in receptor specificity

is known. H3 viruses preferentially bind either sialic acid alpha 2,6 galactose (SA 2,6 Gal) or sialic acid alpha 2,3 galactose (SA 2,3 Gal) linkages (Rogers & Paulson, 1983; Rogers *et al.*, 1983). Selected receptor variants of human viruses which could bind to SA 2,3 Gal differed from those that bind SA 2,6 Gal by a single amino acid at residue 226 of HA_1, located in the sialic acid binding site (Rogers *et al.*, 1983); leucine is found in SA 2,6 Gal binding strains and glutamine in SA 2,3 Gal binding. A methionine at residue 226 confers equal binding to sialic acid 2,3 and sialic acid 2,6 linkages. Avian and equine H3 viruses all show SA 2,3 Gal specific binding but, by selection, an avian virus capable of binding SA 2,6 Gal has been isolated. The haemagglutinins of these avian variants differed from the parent at residue 226 and, like the human strains, glutamine showed SA 2,3 Gal specificity and leucine SA 2,6 Gal specificity. The avian SA 2,6 Gal viruses were stable on tissue culture passage but reverted in hens' eggs to the parental sequence (Rogers *et al.*, 1985). These results suggest that the membrane receptors that interact with the HA are involved in host-mediated selection of viruses.

Further evidence for host cell selection of viruses is provided by laboratory growth of virus from human clinical specimens. Antigenically distinct viruses can be differentially selected when samples are grown exclusively in mammalian tissue culture, on the one hand, or in eggs, on the other (Schild *et al.*, 1983; J. C. De Jong, personal communication). Analysis of the variation of the HA in both H1N1 and B type viruses shows amino acid substitutions which are located on the periphery of the receptor binding site (Robertson *et al.*, 1985, 1986). The significance of the variation seen in these human strains needs to be established and it is not possible to say whether selection of these viruses is caused primarily by cellular receptors either *in vivo* or *in vitro* or by antibody in the infected individuals.

Mutants of swine influenza viruses have also been isolated which show differences in their ability to grow in MDCK (Madin–Darby canine kidney) cells and chick embryos (Kilbourne, 1978). Recombinants were produced that contained seven genes from human virus (A/PR8/8/34 H1N1) and only the HA gene of the swine virus. Those that grew poorly in chick embryos (L) were infectious for swine but the corresponding virus which yielded high titre in eggs (H) was not (Kilbourne, McGregor & Easterday, 1979). Sequence analysis of the HA genes cloned in *E. coli* showed that a single amino acid residue close to the receptor binding site at the tip of the HA (residue 158) predicts either H or L phenotype and infectivity for swine (Both, Shi & Kilbourne, 1983). The additional effects of other mutations in the HA, however, cannot be ruled out.

Amino acid variation in the sialic acid binding site also has been shown to influence virus replication in ducks. A series of recombinant viruses were made by crossing A/Mallard/NY/6750/78 and A/Udorn/307/72 and tested for their ability to grow in the duck intestine. Viruses that had exchanged only the HA gene of Udorn into Mallard/NY were selected and a recombinant was isolated which could replicate in ducks. The HA gene had two mutations: one at residue 226 of HA_1 and the other at 228 of HA_1. Residue 226 changed from leucine (characteristic for sensitivity to inhibitors of SA 2,6 Gal binding) to glutamine, as normally found in avian viruses; and residue 228 changed from serine to glycine (Naeve, Hinshaw & Webster, 1984). Both changes are thought to be associated with receptor binding.

Thus, examples of variation within the HA in the vicinity of the receptor binding site can show biological differences and the potential for changes in pathogenicity or virulence. The location of amino acid residues on the HA molecule, which are associated with these changes, is shown in Fig. 3.

(ii) Viruses with defective intracellular processing of the glycoproteins
Variation in processing of the influenza virus glycoproteins has been studied extensively in the laboratory, particularly in viruses which show temperature-sensitive phenotypes (e.g. see Mahy, 1983; Naruse, Scholtissek & Klenk, 1986). Temperature-sensitive virus mutants of fowl plague virus, in general, are apathogenic in chickens (Rott, Orlich & Scholtissek, 1982), and ts-lesions in the HA are no exception. The result is not surprising since, in the temperature sensitivity assay, the restrictive temperature was the body temperature of the hen. Much more surprising though, is the observation made using a cold-sensitive virus mutant which exhibits a defect in neuraminidase synthesis, even at the permissive temperature (Breuning & Scholtissek, 1986). In this virus, synthesis of the neuraminidase is inhibited at 40.5 °C and, although virus is produced, the NA is not incorporated into infectious particles. This virus can replicate in chickens and cause fowl plague, which suggests that the neuraminidase is dispensable for biological activity of avian influenza viruses. These results must be considered with some caution since no evidence has been presented to rule out reversion to the normal

phenotype when the virus was grown *in vivo*; nor is it clear that the defective synthesis of the NA was not overcome *in vivo* in the absence of reversion – the defect may be tissue-specific.

Adaptation of virus to growth in different cell types *in vitro* can be determined by the ability to produce infectious virus with cleaved HA in the virion. A single base mutation has been associated with the ability of X-31 virus, a human H3 recombinant virus, to grow productively in MDCK cells in the absence of trypsin. This amino acid substitution, which changed amino acid 17 from histidine to arginine (see Fig. 3), changed the extent of glycosylation of the HA and susceptibility to cleavage (Rott *et al.*, 1984). No *in vivo* biological consequences of this mutation have been defined for this virus, but far-reaching consequences of a similar *in vitro* phenotype have been seen in avian influenza viruses.

The major determinant of pathogenicity for chickens infected with fowl plague virus is the HA (Rott *et al.*, 1982). Both pathogenic and non-pathogenic strains replicate in fowl, but replication of non-pathogenic strains is restricted to the respiratory and intestinal tracts, whereas pathogenic virus can be recovered from many other organs of the body (Bosch *et al.*, 1979; D. J. Alexander, personal communication). The restricted growth of non-pathogenic avian viruses correlates with an inability to produce infectious virus in tissue culture. Virus produced by non-pathogenic strains contain uncleaved HA, whereas only pathogenic virus produced fully infectious virus with cleaved HA (Bosch *et al.*, 1979). Examination of H5 viruses isolated from an outbreak of disease in Pennsylvania during 1983 confirmed this conclusion (Kawaoka, Naeve & Webster, 1984; Webster, Kawaoka & Bean, 1986) and by use of recombinant virus and drug sensitivity, it was possible to relate the ability to cause disease and produce infectious virus *in vitro* to a single amino acid change in HA_1. This change was at position 23 of HA_1 and is thought to affect glycosylation of the HA at residue 20 of HA_1 (see Fig. 3). A carbohydrate moiety at position 20 is located sufficiently close to the HA_1/HA_2 cleavage site for it to influence cleavage. In the H7 strains of avian virus, changes in protein sequence at the cleavage site probably account for changes in susceptibility to cleavage (Table 1). Pathogenic H7 viruses have a series of basic amino acids at the HA cleavage site, whereas non-pathogenic strains contain a single arginine residue at this site, in common with human and seal virus isolates. It should be noted that non-pathogenic H5 viruses from Pennsylvania contained a series of basic amino acids at the cleavage site, but in these

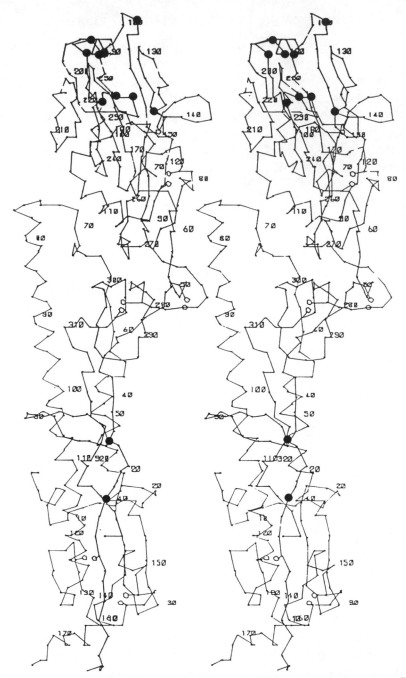

Fig. 3. A stereo-pair of the alpha-carbon backbone of an H3 haemagglutinin monomer. Residues indicated are those which vary in mutants which are thought to exhibit changes in receptor binding (indicated at the top of the molecule) or those which differ in their ability to produce infectious virus in tissue culture (indicated closer to the bottom). (Data from Both *et al.*, 1982; Rogers *et al.*, 1983; Naeve *et al.*, 1984; Rott *et al.*, 1984; Kawaoka *et al.*, 1984; Robertson *et al.*, 1985, 1986.)

Table 1. *Nucleotide sequences of the haemagglutinin gene of avian influenza virus strains of the H7 sub-type. The nucleotide sequence and deduced amino acid sequence in the region of the HA cleavage site is shown. Pathogenic strains have a series of basic amino-acids at the HA_1/HA_2 cleavage site, whereas non-pathogenic strains contain a single arginine residue. The amino terminal residues of HA_2 are underlined.*

Pathogenic viruses

A/FPV/ROSTOCK/34
MetLysAsnValProGluProSerLysLysArgLysLysArgGlyLeuPhe
ATGAAGAACGTTCCCGAACCTTCCAAAAAAAGGAAAAAAAGAGGCCTNTTT

A/FPV/DOBSON/27
MetLysAsnValProGluLeuProLysLysArgArgLysArgGlyLeuPhe
ATGAAGAATGTTCCCGAACTTCCCAAAAAAAGAAGAAAAAGAGGCCTNTTT

A/FPV/EGYPT/45
MetLysAsnValProGlyPheSerLysLysArgArgLysArgGlyLeuPhe
ATGAAAAATGTTCCCGAATTTTCCAAAAAAAGAAGAAAGAGGGGACTATTT

Non-pathogenic viruses

A/TURKEY/ENGLAND/192-329/79
MetLysAsnValProGlyIleProLysGlyArgGlyLeu
ATGAAGAACGTTCCTGAAATTCCAAAAGGGAGAGGACTAT

A/TURKEY/ENGLAND/192-328/79
MetLysAsnValProGlyIleProLysGlyArgGlyLeuPhe
ATGAAGAACGTTCCTGAAATTCCAAAAGGGAGAGGACTATTT

A/CHICKEN/ENGLAND/71/82
MetLysAsnValProGlyIleProLysGlyArgGlyLeuPhe
ATGAAGAACGTTCCTGAAATTCCAAAAGGGAGAGGCCTATTT

A/PARROT/NORTHERN IRELAND/VG7367/73
MetLysAsnValProGlyIleProLysGlyArgGlyLeuPheGly
ATGAAGAACGTTCCTGAAATTCCAAAAGCAAGAGGCCTATTTGGTG

J. W. McCauley and D. J. Alexander, unpublished results.

cases attenuation of pathogenicity was presumably mediated by glycosylation. No evidence has been found for such a mechanism of attenuation in H7 viruses, but the possible occurrence of H5 viruses with a single arginine residue at the HA cleavage site cannot be ruled out. The ability of virus to establish productive infections in organs other than the intestinal tract therefore leads to serious disease in fowl (death) and restriction of replication is, to a large extent, due to the failure to produce infectious virus outside the gastrointestinal tract.

(iii) Recombinant virus

The high frequency of recombination of influenza viruses has been used extensively to dissect the genetic basis of distinct virus phenotypes, though caution needs to be exercised when recombinants are used. Virus polypeptides interact with one another as well as polypeptides of the host cell and restricted replication may be seen when polypeptides are in combination with a particular polypeptide of another virus or of a specific cell type. Furthermore, virus mutants may be selected during laboratory passage and bear no relationship to the properties of the parental virus. Recombinants made between viruses whose phenotypes are multi-factorial in nature may also lead to results in which the conclusions are valid only in the context of the viruses used. For example, consider the ability of virus to grow or not in a particular cell: several genes may determine this property, and an inability to grow may be located in one or more of these genes. Thus, when viruses of two phenotypes and recombinants made from them are compared, an important determinant will only be detected if the parent viruses differed with respect to it.

Neurovirulence in mice is a property of some strains of influenza virus. The genetic basis of neurovirulence has been examined using recombinant viruses and, taken together, the results from a number of studies implicate all gene segments in neurotropism (Scholtissek *et al.*, 1979; Suguira & Ueda, 1980; Bonin & Scholtissek, 1983). Of particular interest here is the observation that the HA of pathogenic avian virus strains, when present in recombinant viruses of non-pathogenic avian or human strains, is a major determinant of neurovirulence (Bonin & Scholtissek, 1983). In addition, it is notable that the neuraminidase of A/WSN/33 can confer neurovirulence on A/Hong Kong/1/68 (Suguira & Ueda, 1980).

In addition to neurotropism, some influenza viruses can replicate in the mouse liver. Schulman (1983) has used recombinant viruses to examine hepatotropism in mice. Here an optimal genotype for hepatotropism required genes that encoded PA, HA, NP and RNA segment 7, which encodes both matrix protein and M_2, from the hepatotrophic parent.

Recombinant viruses have also been used to examine virulence of human strains of virus. Ferrets are susceptible to infection by human influenza virus strains and the disease syndrome in ferrets is similar to that in man. Virulence in ferrets may be assessed by four criteria: minimal infectious dose, the degree of virus replication in the upper respiratory tract, lung involvement in infection and pyrexia. Using these criteria it can be seen that virulence in ferrets parallels differences in virulence seen on infection of human volunteers (Toms *et al.*, 1977). In the ferret, marked differences are seen in the ability of virus to infect the lower respiratory tract, where avirulent virus strains show low degrees of lung infection (for example, see Sweet *et al.*, 1981, 1985). The glycoprotein genes of two human viruses (A/Okuda/57, H2N2 and A/Finland/4/74, H3N2) have been shown to confer differences in ferret virulence (pyrexia and levels of virus growth in the respiratory tract) in recombinants made between the two parents (Campbell *et al.*, 1982). Furthermore, the haemagglutinin gene of A/England/939/69 (H3N2), when present in a recombinant virus which contained all the remaining genes of A/PR/8/34 (H1N1) dramatically increased the ferrets' inflammatory response to virus infection and caused a corresponding high fever (Coates *et al.*, 1985). Still assessment of virulence is difficult: virus growth in the lower respiratory tract and fever production do not always covary; indeed recent H1N1 viruses elicit fever in ferrets (Coates *et al.*, 1985) and yet show poor infection in the lower respiratory tract (Sweet *et al.*, 1985). Clearly, although the role of other genes cannot be ruled out, glycoproteins do play a role in the response of the ferret to infection and may be of particular importance in the generation of fever.

VARIATION IN RECOVERY

People's recovery from influenza has not been examined in detail, but it is likely that some individuals may recover from infection more easily than others, and feasible that infection with certain strains of virus is more easily overcome. It has been proposed that, in man, a correlation exists between the level of the response to infection by the production of cytotoxic T cells (Tc) and the recovery from infection (McMichael *et al.*, 1983). In mice, polyclonal and monoclonal Tc cells have been found to reduce virus titres in infected recipient animals (Yap, Ada & McKenzie, 1978; Lin & Askonas, 1981). Although Tc cells recognise other virus determinants on the cell surface in addition to the glycoproteins (Bennink, Yewdell & Gerhard, 1982; Townsend & Skehel, 1982, 1984; Yewdell *et al.*, 1985), the extent of the contribution of glycoprotein variation to this process is not clear at present. The role of delayed-type hypersensitivity (T_{DH}) cells in murine infection has also been examined. The level of T_{DH} cells found in the lungs of infected mice correlated with death from viral pneumonia (Leung & Ada, 1980). Here again, a potential for virus variation exists: the host response can influence the outcome of infection and virus variation can thus potentially stimulate, to a greater or lesser degree, the host response.

CHEMOTHERAPY

Chemotherapy may be thought of as acting at the level of the individual or at the level of the population. Reduction in virus production from an individual may be enough to halt an epidemic, especially if susceptible individuals in close contact are protected by prophylaxis. Virus virulence may therefore be modulated by its susceptibility to anti-virus drugs.

Chemotherapy of influenza virus infection is currently restricted to amantadine (and its structural analogue, rimantadine) and two drugs that inhibit RNA synthesis, ribavirin and phosphonoformic acid (Galbraith, 1985; Streissle, Paessens & Oediger, 1986). The mode of action *in vitro* of amantidine has recently been established (Hay *et al.*, 1985) and it now appears that the glycoproteins influence susceptibility only to a minor degree. At high concentrations (0.1 mM), amantadine inhibits the membrane fusion activity of the virus HA (Daniels *et al.*, 1985*b*), but its action is non-specific; other weak bases also show this property, which is mediated by an elevation of the pH in the endosome. At low concentrations (1–5 μM), however, drug and strain-specific inhibition of replication can be seen. Amantadine results in the inhibition of infection or assembly of virus, and resistance can be mapped to the transmembrane region of the M_2 polypeptide (Hay *et al.*, 1985). The HA may also play a role since recombinants made between two avian virus strains and an amantadine-resistant variant of A/Bel/42 (H1N1) show that the HA gene of the H1N1 virus can confer some resistance, but maximal resistance requires the M_2 gene from a resistant donor virus.

It is not known whether virus variation would be sufficient to overcome any containment of outbreaks in which anti-viral chemotherapy was practised, but clearly strain variation of susceptibility to drugs suggests that the potential to do so exists.

CONCLUSIONS

It is clear that variation in the glycoproteins of human influenza viruses can result in epidemic disease, and that the introduction of glycoprotein genes from non-human strains can lead to pandemic influenza. Although these two events are examples of virus overcoming the host's immunity, changes in virulence *per se* have been found in avian viruses, with dramatic consequences, and infection of seals and mink with avian strains has also led to lethal infection in the wild. Is such an event capable of occurring in man? During adaptation of virus for growth in different cell types *in vitro*, changes in the HA are seen. Whether *in vivo* changes in tissue tropism will be seen in viruses that infect man is not known, but the possible consequences of such an event are disturbing.

ACKNOWLEDGEMENTS

Amongst others, I should like to thank Drs T. Barrett, C. Bostock, P. Mahony and C. Penn for criticism of the manuscript and Drs J. C. de Jong, J. S. Robertson, J. J. Skehel and D. J. Alexander for discussion.

REFERENCES

ADA, G. L., LEUNG, K. N. & ERTL, H. (1981). An analysis of effector T cell generation and function in mice exposed to influenza A or Sendai viruses. *Immunological Reviews*, **58**, 5–24.

ALEXANDER, D. J. (1982). Avian influenza – recent advances. *Veterinary Bulletin*, **52**, 341–59.

ALEXANDER, D. J., ALLEN, W. H., PARSONS, D. & PARSONS, G. (1978). The pathogenicity of four avian influenza viruses for fowls, turkeys and ducks. *Research in Veterinary Science*, **24**, 242–7.

ALLEN, A. K., SKEHEL, J. J. & YUFEROF, V. J. (1977). The amino acid and carbohydrate composition of the neuraminidase of B/Lee/40 influenza virus. *Journal of General Virology*, **37**, 625–8.

AYMARD, M., DOUGLAS, A. R., FONTAINE, M., GOURREAU, J. M., KAISER, C., MILLION, J. & SKEHEL, J. J. (1985). Antigenic characterization of influenza A(H1N1) viruses recently isolated from pigs and turkeys in France. *Bulletin of the World Health Organization*, **63**, 537–42.

BENNINK, J. R., YEWDELL, J. W. & GERHARD, W. (1982). A viral polymerase involved in recognition of influenza virus-infected cells by a cytotoxic T-cell clone. *Nature, London*, **296**, 75–8.

BENTLEY, D. R. & BROWNLEE, G. G. (1982). Sequence of the neuraminidase from influenza virus A/NT/60/68. *Nucleic Acid Research*, **10**, 5033–42.

BLOK, J., AIR, G. M., LAVER, W. G., WARD, C. W., LILLEY, G. G., WOODS, E. F., ROXBURGH, C. M. & INGLIS, A. S. (1982). Studies on the size, chemical composition and partial sequence of the neuraminidase (NA) from type A influenza viruses show that the N-terminal region of the NA is not processed and serves to anchor the NA in the viral membrane. *Virology*, **119**, 109–21.

BODMER, W. F. (1980). The HLA system and disease. *Journal of the Royal College of Physicians (London)*, **14**, 43–50.

BONIN, J. & SCHOLTISSEK, C. (1983). Mouse neurotropic recombinants of influenza A viruses. *Archives of Virology*, **75**, 255–68.

BOSCH, F. X., GARTEN, W., KLENK, H-D. & ROTT, R. (1981). Proteolytic cleavage of influenza virus hemagglutinins: primary structure of the connecting peptide between HA1 and HA2 determines proteolytic cleavability and pathogenicity of avian influenza virus. *Virology*, **113**, 725–35.

BOSCH, F. X., ORLICH, M., KLENK, H-D. & ROTT, R. (1979). The structure of the haemagglutinin, a determinant for the pathogenicity of influenza viruses. *Virology*, **95**, 197–297.

BOTH, G. W., SHI, C. H. & KILBOURNE, E. D. (1983). Hemagglutinin of swine influenza virus: a single amino acid change pleitropically affects viral antigenicity and replication. *Proceedings of the National Academy of Sciences, USA*, **80**, 6996–7000.

BRAND, C. & PALESE, P. (1980). Sequential passage of influenza virus in embryonated eggs or tissue culture: selection of mutants. *Virology*, **107**, 424–33.

BREIDIS, D. J., LAMB, R. A. & CHOPPIN, P. W. (1982). Sequence of RNA segment 7 of the influenza B virus genome: partial amino acid homology between membrane proteins (M_1) of influenza A and B viruses and conservation of a second open reading frame. *Virology*, **116**, 581–8.

BREUNING, A. & SCHOLTISSEK, C. (1986). A reassortment between influenza A viruses (H7N2) synthesising an enzymatically inactive neuraminidase at 41° which is not incorporated into infectious particles. *Virology*, **150**, 65–74.

BURNET, F. M. (1948). Mucins and mucoids in relation to influenza virus action. IV. Inhibition by purified mucoid of infection and haemagglutinin with the virus strain WSE. *Australian Journal of Experimental Biology and Medical Science*, **26**, 381–7.

BURROWS, R., DENYER, M., GOODRIDGE, D. & HAMILTON, F. (1981). Field and laboratory studies of equine influenza viruses isolated in 1979. *The Veterinary Record*, **109**, 353–6.

CAMPBELL, D., SWEET, C., HAY, A. J., DOUGLAS, A., SKEHEL, J. J., MASON, T. J. & SMITH, H. (1982). Genetic composition and virulence of influenza virus: differences in facets of virulence in ferrets between two pairs of recombinants with RNA segments of the same parental origin. *Journal of General Virology*, **58**, 387–98.

CATON, A. J., BROWNLEE, G. G., YEWDELL, J. W. & GERHARD, W. (1982). The antigenic structure of the influenza virus A/PR/8/34 hemagglutinin (H1 subtype). *Cell*, **31**, 297–301.

COATES, D. M., SWEET, C., QUARLES, J. M., OVERTON, H. A. & SMITH, H. (1985). Antigens, pyrexia, level of nasal virus and inflammatory response in the ferret. *Journal of General Virology*, **66**, 1627–31.

COLMAN, P. M., VARGHESE, J. N. & LAVER, W. G. (1983). Structure of the catalytic and antigenic sites in influenza virus neuraminidase. *Nature, London*, **303**, 41–4.

COLMAN, P. M. & WARD, C. W. (1985). Structure and diversity of influenza virus neuraminidase. *Current Topics in Microbiology and Immunology*, **114**, 177–255.

DANIELS, R. S., SKEHEL, J. J. & WILEY, D. C. (1985a). Amino acid sequences of haemagglutinins of influenza viruses of the H3 subtype isolated from horses. *Journal of General Virology*, **66**, 457–64.

DANIELS, R. S., DOWNIE, J. C., HAY, A. J., SKEHEL, J. J., WANG, H. L. & WILEY, D. C. (1985b). Fusion mutants of the influenza virus haemagglutinin glycoprotein. *Cell*, **40**, 431–9.

DOMS, R. W., HELENIUS, A. & WHITE, J. (1985). Membrane fusion activity of the influenza virus haemagglutinin. The low pH-induced conformational change. *Journal of Biological Chemistry*, **260**, 2973–81.

FANG, R., MIN-JOU, W., HUYLEBROECK, D., DEVOS, R. & FIERS, W. (1981). Complete structure of A/duck/Ukraine/63 influenza haemagglutinin gene: animal virus as a progenitor of human H3 Hong Kong 1968 influenza haemagglutinin. *Cell*, **25**, 315–23.

FLANAGAN, M. T. & SKEHEL, J. J. (1977). The conformation of the influenza virus haemagglutinin. *FEBS Letters*, **80**, 57–60.

GALBRAITH, A. W. (1985). Influenza – recent development in prophylaxis and treatment. *British Medical Bulletin*, **41**, 381–5.

GETHING, M-J., BYE, J., SKEHEL, J. & WATERFIELD, M. D. (1980). Cloning and DNA sequence of double-stranded copies of haemagglutinin genes from H2 and H3 strains elucidates antigenic shift and drift in human influenza virus. *Nature, London*, **287**, 301–6.

HAY, A. J., WOLSTENHOLME, A. J., SKEHEL, J. J. & SMITH, M. H. (1985). The molecular basis of the specific anti-influenza action of amantadine. *EMBO Journal*, **4**, 3021–4.

HINSHAW, V. S., AIR, G. M., GIBBS, A. J., GRAVES, L., PRESCOTT, B. & KARUNAKARAN, D. (1982). Antigenic and genetic characterisation of a novel haemagglutinin subtype of influenza A viruses from gulls. *Journal of Virology*, **42**, 865–72.

HINSHAW, V. S., ALEXANDER, D. J., AYMARD, M., BACHMANN, P. A., EASTERDAY, B. C., HANNOUN, C., KIDA, H., LIPKIND, M., MACKENZIE, J. S., NEROME, K., SCHILD, G. C., SCHOLTISSEK, C., SEENE, D. A., SHORTRIDGE, K. F., SKEHEL, J. J. & WEBSTER, R. G. (1984). Antigenic comparisons of swine influenza-like H1N1 isolates from pigs, birds and humans: an international collaborative study. *Bulletin of the World Health Organization*, **62**, 871–8.

HINSHAW, V. S., BEAN, W. J., GERACI, J., FIORELLI, P., EARLY, G. & WEBSTER, R. G. (1986). Characterisation of two influenza A viruses from a pilot whale. *Journal of Virology*, **58**, 655–6.

HINSHAW, V. S., NAEVE, C. W., WEBSTER, R. G., DOUGLAS, A., SKEHEL, J. J. & BRYANS, J. (1983). Analysis of antigenic variation in equine 2 influenza A viruses. *Bulletin of the World Health Organization*, **61**, 153–8.

HINSHAW, V. S., WEBSTER, R. G., EASTERDAY, B. C. & BEAN, W. J. (1981). Replication of avian influenza A viruses in mammals. *Infection and Immunity*, **34**, 354–61.

HIRST, G. K. (1973). Mechanism of influenza recombination. 1. Factors influencing recombination rates between temperature-sensitive mutants into complementation–recombination groups. *Virology*, **55**, 81–93.

HOLLAND, J., SPINDLER, K., HORODYSKI, F., GRABAU, E., NICHOL, S. & VANDEPOL, S. (1982). Rapid evaluation of RNA genomes. *Science*, **215**, 1577–85.

HOYLE, L. (1968). In *The Influenza Viruses*. Virology Monographs, Vol. 4. New York: Springer-Verlag.

HUANG, R. T. C., DIFTSCH, E. & ROTT, R. (1985). Further studies on the role of neuraminidase and the mechanism of low pH dependence in influenza virus-induced membrane fusion. *Journal of General Virology*, **66**, 295–301.

HUANG, R. T. C., ROTT, R., WAHN, K., KLENK, H-D. & KOHAMA, T. (1980a). Function of neuraminidase in membrane fusion induced by myxoviruses. *Virology*, **107**, 313–19.

HUANG, R. T. C., WAHN, K., KLENK, H-D. & ROTT, R. (1980b). Fusion of liposomes containing the glycoproteins of influenza virus with tissue culture cells. *Virology*, **104**, 294–302.

JONES, L. V., COMPANS, R. W., DAVIS, A. R., BOS, T. J. & NAYAK, D. P. (1985). Surface expression of influenza virus neuraminidase, an amino-terminally anchored viral membrane glycoprotein, in polarised epithelial cells. *Molecular and Cellular Biology*, **5**, 2181–9.

KAWAOKA, Y., NAEVE, C. W. & WEBSTER, R. G. (1984). Is virulence of H5N2 influenza viruses in chickens associated with loss of carbohydrate from the haemagglutinin? *Virology*, **139**, 303–16.

KEIL, W., GEYER, R., DABROWSKI, J., DABROWSKI, U., NIEMANN, H., STIRM, S. & KLENK, H-D. (1985). Carbohydrates of influenza virus. Structural elucidation of the individual glycans of the FPV haemagglutinin by two-dimensional ^{3}H n.m.r. and methylation analysis. *EMBO Journal*, **4**, 2711–20.

KENDAL, A. P., NOBLE, G. R., SKEHEL, J. J. & DOWDLE, W. R. (1978). Antigenic similarity of influenza A (H1N1) viruses from epidemics in 1977–1978 to 'Scandinavian' strains isolated in epidemics of 1950–1951. *Virology*, **89**, 632–6.

KILBOURNE, E. D. (1978). Genetic dimorphism in influenza viruses: characterisation of stably associated haemagglutinin mutants differing in antigenicity and biological properties. *Proceedings of the National Academy of Sciences, USA*, **75**, 6258–62.

KILBOURNE, E. D., McGREGOR, S. & EASTERDAY, B. C. (1979). Haemagglutinin mutants of swine influenza virus differing in replication characteristics in their natural host. *Infection and Immunity*, **26**, 197–201.

KLENK, H-D., ROTT, R., ORLICH, M. & BLODORN, J. (1975). Activation of influenza A viruses by trypsin treatment. *Virology*, **68**, 426–39.

KLINGEBORN, B., EGLUND, L., ROTT, R., JUNTTI, N. & ROCKBORN, G. (1985). An avian influenza A virus killing a mammalian species – the mink. *Archives of Virology*, **86**, 347–51.

KRYSTAL, M., YOUNG, J. F., PALESE, P. & WILEY, D. C. (1983). Sequential mutations in haemagglutinins of influenza B virus isolates: definition of antigenic domains. *Proceedings of the National Academy of Sciences, USA*, **80**, 4527–31.

LAMB, R. A. & CHOPPIN, P. W. (1979). Segment eight of influenza virus is unique in coding for two polypeptides. *Proceedings of the National Academy of Sciences, USA*, **76**, 4908–12.

LAMB, R. A. & CHOPPIN, P. W. (1983). The gene structure and replication of influenza virus. *Annual Reviews of Biochemistry*, **52**, 467–506.

LAMB, R. A., ZEBEDEE, S. L. & RICHARDSON, C. D. (1985). Influenza virus M_2 protein is an integral membrane protein expressed on the infected cell surface. *Cell*, **40**, 627–33.

LANG, G., GAGNON, A. & GERACHI, J. R. (1981). Isolation of an influenza A virus from seals. *Archives of Virology*, **68**, 189–95.

LAZAROWITZ, S. G. & CHOPPIN, P. W. (1975). Enhancement of the infectivity of influenza A and B viruses by proteolytic cleavage of the haemagglutinin in polypeptide. *Virology*, **68**, 440–54.

LEUNG, K-N. & ADA, G. L. (1980). Cells mediating delayed type hypersensitivity in the lungs of mice infected with influenza A virus. *Scandinavian Journal of Immunology*, **12**, 393–400.

LIN, Y-L. & ASKONAS, B. A. (1981). Biological properties of an influenza A virus-specific killer T cell clone. *Journal of Experimental Medicine*, **154**, 225–34.

McCAULEY, J. W. & MAHY, B. W. J. (1983). Structure and function of the influenza virus genome. *Biochemical Journal*, **211**, 281–94.

McCAULEY, J. W., BYE, J., ELDER, K., GETHING, M-J., SKEHEL, J. J., SMITH, A. & WATERFIELD, M. D. (1979). Influenza virus haemagglutinin signal sequence. *FEBS Letters*, **108**, 422–8.

McMICHAEL, A. J., GOTCH, F. H., NOBLE, G. R. & BEARE, P. A. S. (1983). Cytotoxic T-cell immunity to influenza. *New England Journal of Medicine*, **309**, 13–17.

MAHY, B. W. J. (1983). Mutants of influenza virus. In *Genetics of Influenza Viruses*, ed. P. Palese & D. W. Kingsbury, pp. 192–254. New York: Springer-Verlag.

MARKOFF, L. & LAI, C-J. (1982). Sequence of the influenza A/Udorn/72 (H3N2) virus neuraminidase gene as determined from cloned full-length DNA. *Virology*, **119**, 288–97.

MARSH, M. (1984). The entry of enveloped viruses into cells by endocytosis. *Biochemical Journal*, **218**, 1–10.

MIN-JOU, W., VERHOEYEN, M., DEVOS, R., SAMAN, E., FANG, R., HUYLEBROECK, D., FIERS, W., THRELFALL, G., BARBER, C., CAREY, N. & EMTAGE, S. (1980). Complete structure of the haemagglutinin gene from the human A/Victoria/3/75 (H3N2) strain as determined from cloned DNA. *Cell*, **19**, 683–96.

MOHAN, R., SAIF, Y. M., ERICKSON, G. A., GUSTAFSON, G. A. & EASTERDAY, B. C. (1981). Serological and epidemiologic evidence of infection of turkeys with an agent related to swine influenza virus. *Avian Diseases*, **25**, 11–16.

NAEVE, C. W., HINSHAW, V. S. & WEBSTER, R. G. (1984). Mutations in the haemagglutinin receptor-binding site can change the biological properties of an influenza virus. *Journal of Virology*, **51**, 567–9.

NAKAJIMA, K., DESSELBERGER, U. & PALESE, P. (1978). Recent human influenza A (H1N1) viruses are closely related genetically to strains isolated in 1950. *Nature, London*, **274**, 334–9.

NARUSE, H., SCHOLTISSEK, C. & KLENK, H-D. (1986). Temperature-sensitive mutants of fowl plague virus defective in the intracellular transport of the haemagglutinin. *Virus Research* **5**, 293–305.

NATALI, A., OXFORD, J. S. & SCHILD, G. C. (1981). Frequency of naturally occurring antibody to influenza virus antigenic variants selected with monoclonal antibody. *Journal of Hygiene, Cambridge*, **87**, 185–90.

OXFORD, J. S., ABBO, H., CORCORAN, T., WEBSTER, R. G., SMITH, A. J., GRILLI, E. A. & SCHILD, G. C. (1983). Antigenic and biochemical analysis of field isolates of influenza B virus: evidence for intra- and inter-epidemic variation. *Journal of General Virology*, **64**, 2376–7.

OXFORD, J. S., SALUM, S., CORCORAN, T., SMITH, A. J., GRILLI, E. A. & SCHILD, G. C. (1986). An antigenic analysis using monoclonal antibodies of influenza A (H3N2) viruses isolated from an epidemic in a semi-closed community. *Journal of General Virology*, **67**, 265–74.

PALESE, P., TOBITA, K., UEDA, M. & COMPANS, R. W. (1974). Characterisation of temperature-sensitive influenza virus mutants defective in neuraminidase. *Virology*, **61**, 397–410.

RAYMOND, F. L., CATON, A. J., COX, N. J., KENDAL, A. P. & BROWNLEE, G. G. (1983). Antigenicity and evolution amongst recent influenza viruses of H1N1 subtype. *Nucleic Acids Research*, **11**, 7191–203.

RAYMOND, F. L., CATON, A. J., COX, N. J., KENDAL, A. P. & BROWNLEE, G. G. (1986). The antigenicity and evolution of influenza H1 haemagglutinin, from 1950–1957 and 1977–1983. *Virology*, **148**, 275–87.

ROBERTSON, J. S., BOOTMAN, J. S., OXFORD, J. S., NEWMAN, R., NAEVE, C. W., WEBSTER, R. G. & SCHILD, G. C. (1986). Characterisation of egg-adapted variants of human influenza viruses. In *Replication of Negative Strand Viruses*, ed. B. W. J. Mahy & D. Kolakofsky. Elsevier Biomedical Press (in press).

ROBERTSON, J. S., NAEVE, C. W., WEBSTER, R. G., BOOTMAN, J. S., NEWMAN, R. & SCHILD, G. C. (1985). Alterations in the haemagglutinin associated with adaptation of influenza B virus to growth in eggs. *Virology*, **143**, 166–74.

RODRIGUEZ-BOULIN, E. & PRENDERGAST, M. (1980). Polarised distribution of viral envelope glycoproteins in the plasma membrane of infected epithelial cells. *Cell*, **20**, 45–54.

ROGERS, G. N., DANIELS, R. S., SKEHEL, J. J., WILEY, D. C., WANG, X-F., HIGA, H. H. & PAULSON, J. C. (1985). Host-mediated selection of influenza virus receptor variants. Sialic acid-alpha 2,6 Gal-specific clones of A/Duck/Uk-raine/1/63 revert to sialic acid-alpha 2,3 Gal-specific wild type *in ovo*. *Journal of Biological Chemistry*, **260**, 7362–7.

ROGERS, G. N. & PAULSON, J. C. (1983). Receptor determinants of human and animal influenza virus isolates: differences in receptor specificity of the H3 hae-magglutinin based on species of origin. *Virology*, **127**, 361–73.

ROGERS, G. N., PAULSON, J. C., DANIELS, R. S., SKEHEL, J. J., WILSON, I. A. & WILEY, D. C. (1983). Single amino acid substitutions in influenza haemaggluti-nin change receptor binding specificity. *Nature, London*, **304**, 76–8.

ROTH, M. G., COMPANS, R. W., GIUSTI, L., DAVIS, A. R., NAYAK, D. P., GETHING, M-J. & SAMBROOK, J. (1983). Influenza virus haemagglutinin expression is polar-ised in cells infected with recombinant SV40 viruses carrying cloned haemaggluti-nin DNA. *Cell*, **33**, 435–43.

ROTT, R. (1982). Determinants of influenza virus pathogenicity. *Hoppe-Seyler's Zeitschrift fur Physiologische Chemie*, **363**, 1273–82.

ROTT, R., ORLICH, M., KLENK, H-D., WANG, M.-L., SKEHEL, J. J. & WILEY, D. C. (1984). Studies on the adaptation of influenza virus to MDCK cells. *EMBO Journal*, **3**, 3329–32.

ROTT, R., ORLICH, M. & SCHOLTISSEK, C. (1982). Differences in the multiplication at elevated temperature of influenza virus recombinants pathogenic and non-pathogenic for chicken. *Virology*, **120**, 215–24.

SATO, J. T. & ROTT, R. (1966). Functional significance of sialidase during influenza virus multiplication. *Virology*, **30**, 731–7.

SCHILD, G. C., OXFORD, J. S., DE JONG, J. C. & WEBSTER, R. G. (1983). Evidence for host-cell selection of influenza virus antigenic variants. *Nature, London*, **303**, 706–9.

SCHOLTISSEK, C., ROHDE, W., VON HOYNINGEN, V. & ROTT, R. (1978b). On the origin of human influenza subtypes H2N2 and H3N2. *Virology*, **87**, 13–20.

SCHOLTISSEK, C., VALLBRACHT, A., FLEHMIG, B. & ROTT, R. (1979). Correlation of pathogenicity and gene constellation of influenza A viruses. II. Highly neuro-virulent recombinants derived from non-neurovirulent or weakly neurovirulent parent virus strains. *Virology*, **95**, 492–500.

SCHOLTISSEK, C., VON HOYNINGEN, V. & ROTT, R. (1978a). Genetic relatedness between the new 1977 epidemic strains (H1N1) of influenza and human influenza strains isolated between 1947 and 1957 (H1N1). *Virology*, **89**, 613–17.

SCHMIDT, M. F. G. (1982). Acylation of viral spike glycoproteins: a feature of enveloped RNA viruses. *Virology*, **116**, 327–38.

SCHULMAN, J. L. (1983). Virus-determined differences in the pathogenesis of influenza virus infections. In *Genetics of Influenza Viruses*, ed. P. Palese & D. W. Kingsbury. New York: Springer-Verlag.

SHAW, M. W. & CHOPPIN, P. W. (1984). Studies on the synthesis of the influenza B virus N_B glycoprotein. *Virology*, **139**, 178–84.

SHAW, M. W., CHOPPIN, P. W. & LAMB, R. A. (1983). A previously unrecognised influenza B virus glycoprotein from a bicistronic mRNA that also encodes the neuraminidase. *Proceedings of the National Academy of Sciences, USA*, **80**, 4879–83.

SHI, J-P. & FERSHT, A. R. (1984). Fidelity of DNA replication under conditions for oligodeoxynucleotide directed mutagenesis. *Journal of Molecular Biology*, **177**, 269–78.

SIX, H. R., WEBSTER, R. G., KENDAL, A. P., GLEZEN, W. P., GRIFFIS, C. & COUCH, R. B. (1983). Antigenic analysis of H1N1 viruses isolated in the Houston metropolitan area during four successive seasons. *Infection and Immunity*, **42**, 453–8.

SKEHEL, J. J., BAYLEY, P. M., BROWN, E. B., MARTIN, S. R., WATERFIELD, M. D., WHITE, J. M., WILSON, I. A. & WILEY, D. C. (1982). Changes in the conformation of influenza virus haemagglutinin at the pH optimum of virus-mediated membrane fusion. *Proceedings of the National Academy of Sciences, USA*, **79**, 968–72.

SKEHEL, J. J. & WATERFIELD, M. D. (1975). Studies on the primary structure of the influenza virus haemagglutinin. *Proceedings of the National Academy of Sciences, USA*, **72**, 93–7.

SMITH, W. C., ANDREWES, C. H. & LAIDLAW, P. P. (1933). A virus obtained from influenza patients. *Lancet*, **ii**, 66–8.

STREISSLE, G., PAESSENS, A. & OEDIGER, H. (1986). New antiviral compounds. *Advances in Virus Research*, **30**, 83–138.

SUGUIRA, A. & UEDA, M. (1980). Neurovirulence of influenza in mice. I. Neurovirulence of recombinants between virulent and avirulent virus strains. *Virology*, **101**, 440–49.

SWEET, C., BIRD, R. A., COATES, D. M., OVERTON, H. A. & SMITH, H. (1985). Recent H1N1 viruses (A/USSR/90/77, A/Fiji/15899/83, A/Firenze/13/83) replicate poorly in ferret bronchial epithelium. *Archives of Virology*, **85**, 305–11.

SWEET, C., MACARTNEY, J. C., BIRD, R. A., CAVANAGH, D., COLLIE, M. H., HUSSEINI, R. H. & SMITH, H. (1981). Differential diagnosis of virus and histological damage in the lower respiratory tract of ferrets infected with influenza viruses of differing virulence. *Journal of General Virology*, **54**, 103–14.

TOMS, G. L., DAVIES, J. A., WOODWARD, C. G., SWEET, C. & SMITH, H. (1977). The relation of pyrexia and nasal inflammatory response to virus levels in the nasal washings of ferrets infected with influenza viruses of differing virulence. *British Journal of Experimental Pathology*, **58**, 444–58.

TOWNSEND, A. R. M., MCMICHAEL, A. J., CARTER, N. P., HUDDLESTON, J. A. & BROWNLEE, G. G. (1984). Cytotoxic T cell recognition of the influenza nucleoprotein and haemagglutinin expressed in transfected mouse cells. *Cell*, **39**, 13–25.

TOWNSEND, A. R. M. & SKEHEL, J. J. (1982). Influenza A specific cytotoxic T cell clones that do not recognise viral glycoproteins. *Nature, London*, **300**, 655–7.

TOWNSEND, A. R. M. & SKEHEL, J. J. (1984). The influenza A virus nucleoprotein gene controls the induction of both sub-type specific and cross-reactive cytotoxic T cells. *Journal of Experimental Medicine*, **160**, 552–63.

VAN ROMPUY, L., JOU, W. M., HUYLEBROECK, D. & FIERS, W. (1982). Complete nucleotide sequence of a human influenza neuraminidase gene of subtype N2 (A/Victoria/3/75). *Journal of Molecular Biology*, **161**, 1–11.

VARGHESE, J. N., LAVER, W. G. & COLMAN, P. M. (1983). Structure of the influenza virus glycoprotein antigen neuraminidase at 2.9 angstrom resolution. *Nature, London*, **303**, 35–40.

VERHOYEN, M., FANG, R., MIN-JOU, W., DEVOS, R., HUYLEBROECK, D., SAMAN, E. & FIERS, W. (1980). Antigenic drift between the haemagglutinin of the Hong Kong influenza strains A/Aichi/2/68 and A/Victoria/3/75. *Nature, London*, **286**, 771–6.

WANG, M-L., SKEHEL, J. J. & WILEY, D. C. (1986). Comparative analyses of the specificities of anti-influenza haemagglutinin antibodies in human sera. *Journal of Virology*, **57**, 124–8.

WATERFIELD, M., SCRACE, G. & SKEHEL, J. (1981). Disulphide bonds of haemagglutinin of Asian influenza virus. *Nature, London*, **289**, 422–4.

WEBSTER, R. G., CAMPBELL, C. H. & GRANOFF, A. (1971). The 'in vivo' production of 'new' influenza A viruses. 1. Genetic recombination between avian and mammalian influenza viruses. *Virology*, **44**, 317–28.

WEBSTER, R. G., HINSHAW, V. S., BEAN, W. J., VAN WYKE, K. L., GERACI, J. R., ST AUBIN, D. J. & PETURSSON, G. (1981). Characterisation of an influenza A virus from seals. *Virology*, **113**, 712–24.

WEBSTER, R. G., HINSHAW, V. S. & LAVER, W. G. (1982*a*). Selection and analysis of antigenic variants of the neuraminidase of N2 influenza viruses with monoclonal antibodies. *Virology*, **117**, 93–104.

WEBSTER, R. G., KAWAOKA, Y. & BEAN, W. J. (1986). Molecular changes in A/chicken/Pennsylvania/83 (H5N2) influenza virus associated with acquisition of virulence. *Virology*, **149**, 165–73.

WEBSTER, R. G., LAVER, W. G., AIR, G. M. & SCHILD, G. C. (1982*b*). Molecular mechanisms of variation in influenza viruses. *Nature, London*, **296**, 115–21.

WEEKLY EPIDEMIOLOGICAL RECORD (1986). Recommended composition of influenza virus vaccines for use in the 1986–1967 season. **61**, 61–4.

WHITE, J., HELENIUS, A. & GETHING, M-J. (1982). Haemagglutinin of influenza virus exposed from a cloned gene promotes membrane fusion. *Nature, London*, **300**, 658–9.

WHITE, J., MATLIN, K. & HELENIUS, A. (1981). Cell fusion by Semliki forest, influenza and vesicular stomatitis virus. *Journal of Cell Biology*, **89**, 674–9.

WILEY, D. C., WILSON, I. A. & SKEHEL, J. J. (1981). Structural identification of the antibody-binding sites of Hong Kong influenza haemagglutinin and their involvement in antigenic variation. *Nature, London*, **289**, 373–8.

WILSON, I. A., SKEHEL, J. J. & WILEY, D. C. (1981). Structure of the haemagglutinin membrane glycoprotein of influenza virus at 3 Å resolution. *Nature, London*, **289**, 366–73.

WINTER, G., FIELDS, S. & BROWNLEE, G. G. (1981). Nucleotide sequence of the haemagglutinin gene of a human influenza virus H1 subtype. *Nature, London*, **292**, 72–5.

WHO MEMORANDUM (1980). A revised system of nomenclature for influenza viruses. *Bulletin of the World Health Organization*, **58**, 585–91.

YAP, K. L., ADA, G. L. & McKENZIE, I. F. C. (1978). Transfer of specific cytotoxic T lymphocytes protects mice inoculated with influenza virus. *Nature, London*, **273**, 238–9.

YEWDELL, J. W., BENNINK, J. R., SMITH, G. L. & MOSS, B. (1985). Influenza A virus nucleoprotein is a major target antigen for cross-reactive anti-influenza A virus cytotoxic T lymphocytes. *Proceedings of the National Academy of Sciences, USA*, **82**, 1785–9.

PARAMYXOVIRUSES AND DISEASE

CRAIG R. PRINGLE

Department of Biological Sciences, University of Warwick, Coventry CV4 7AL, UK

THE PARAMYXOVIRUSES

The family *Paramyxoviridae* comprises three distinct groups of viruses which have been gathered into a single taxonomic unit by virtue of a common morphology and the fact that they are normally transmitted by the respiratory route and often associated with respiratory disease. The paramyxoviruses are enveloped viruses of pleomorphic form and variable dimensions possessing an elongate helical nucleocapsid with a single-stranded monomolecular negative sense RNA genome. They are the largest in size of the RNA viruses. The precise chain length of the genome of any paramyxovirus has not yet been determined but, if estimated from the relative molecular weight of the virion RNA to be in the range 15 000–16 000 nucleotides, the genome of these viruses may be close to the upper limit of 17 600 set by Reanney (1984) for a genetically stable RNA genome.

The three groups are ranked in current taxonomy as the genera *Morbillivirus*, *Paramyxovirus* and *Pneumovirus* (Kingsbury *et al.*, 1978). The second of the genera would be better designated as the genus *Parainfluenzavirus* and this terminology will be used here to avoid confusion. The individual members of three genera are listed in Table 1 together with their principal characteristics and disease potential.

The viruses included in the genus *Morbillivirus* are the most distinctive in disease-producing potential, being associated with both exanthematous and, less frequently, neurological disease in addition to respiratory disease. Measles virus and canine distemper virus are closely related, and studies of the evolutionary relationships of morbilliviruses using collections of monoclonal antibodies suggested that rinderpest virus is the common ancestor of these two morbilliviruses (Norrby *et al.*, 1985).

The genus *Parainfluenzavirus* is an assemblage of antigenically heterogenous viruses predominantly associated with respiratory disease in birds and mammals, together with mumps virus which is typically associated with parotitis, sometimes accompanied by widespread invasion of visceral organs and the central nervous system.

Table 1. *The disease-producting potential of paramyxoviruses*

Family Paramyxoviridae: Genus *Parainfluenzavirus*		
Serotype	Primary host (virus)	Associated disease
Parainfluenzavirus type 1	Man	URTI, rarely pneumonia in adults
	Mouse (Sendai)	Inapparent, latent in mice
Parainfluenzavirus type 2	Man	URTI, croup mainly in children
	Monkey (SV5)	URTI
	Dog	URTI
Parainfluenzavirus type 3	Man	URTI, laryngitis, bronchiolitis and pneumonia in children
	Cattle	Shipping fever
	Sheep	URTI
	Monkey	URTI, pneumonia
Parainfluenzavirus type 4	Man	URTI
Mumps virus	Man	Parotitis, gastroenteritis, oophoritis, orchitis, pancreatitis and other conditions
Avian parainfluenzavirus type 1	Chicken (Newcastle disease)	Inapparent to lethal viscerotropic and neurotropic (Meningoencephalitis in mink; URTI, pneumonitis and encephalitis in man)
Avian parainfluenzavirus type 2	Several avian spp (Yucaipa)	Inapparent
Avian parainfluenzavirus type 3	Several avian spp	Inapparent
Avian parainfluenzavirus type 4	Duck, chicken, geese	Inapparent
Avian parainfluenzavirus type 5	Budgerigar (Kunitachi)	Inapparent
Avian parainfluenzavirus type 6	Duck and chicken	Inapparent
Ungrouped viruses	Several avian species	Inapparent

Orvell, Rydbeck & Love (1986) concluded from studies of immunological relationships with panels of monoclonal antibodies that Sendai virus (murine parainfluenzavirus type 1) is a descendent of both parainfluenza virus type 3 and mumps virus, which have evolved separately and subsequently diverged. No cross reactivity in Elisa tests or by immunofluorescence was observed between any of these

Table 1. (*contd.*)

Genus *Morbillivirus*		
Serotype	Principal host (virus)	Associated disease
Measles virus	Man	URTI, rash, otitis media, bronchopneumonia, rarely (1/2000) encephalitis, exceptionally (1/1 000 000) SSPE
Canine distemper virus	Dog and mustelids	URTI, skin eruptions, bronchopneumonitis, keratitis of the feet (hard pad), old dog encephalitis, demyelination of neural tissue
Rinderpest virus	Cattle, sheep, goats, pigs, buffalo	Mucosal lesions, diarrhoea, bronchopneumonia
Peste des petits ruminants	Sheep, goats	Mucosal disease, diarrhoea, bronchopneumonia, abortion

Genus *Pneumovirus*		
Serotype	Principal host (virus)	Associated disease
Respiratory syncytial virus	Man	URTI, bronchiolitis and pneumonia in infants, otitis media
	Cattle, sheep, goats	URTI, pneumonia, bronchiolitis and emphysema in calves
Murine pneumonia virus	Mouse, Syrian hamster, cotton rats	Inapparent, latent in mice, rarely disease-producing in guineapigs

viruses and Newcastle disease virus (NDV) or parainfluenzavirus type 2. Human parainfluenza virus type 2 and the simian and canine virus SV5 do cross-react and are related, although perhaps no more so than types 1 and 3 (Goswami & Russell, 1982). Since SV5 is better characterised than the human virus it will be considered as representative of type 2 in this review. Goswami *et al*. (1984*b*) have presented evidence which suggests that SV5 can also be regarded as a human virus. Bovine parainfluenza virus type 3, associated with shipping fever in cattle, and human parainfluenza virus type 3 are not closely related. Only 2 of a panel of 20 anti-human haemagglutin

(HN) monoclonal antibodies cross-reacted with the bovine virus (Ray & Compans, 1986). NDV and the other five avian serotypes show little or no relationship to the mammalian parainfluenza viruses and only low-grade cross-reactions with one another (Alexander, 1980). However, a comparison of partial amino acid sequences of the HN proteins of NDV, Sendai virus and SV5 has shown that there is some homology at the genotypic level (Millar, Chambers & Emmerson, 1986).

Pneumonia virus of mice (PVM), one of the first paramyxoviruses to be isolated (Horsfall & Hahn, 1939), is now included together with respiratory syncytial (RS) virus in the genus *Pneumovirus*. PVM is prevalent as a latent infection in rodents (Horsfall & Hahn, 1940). RS virus is the principal cause of acute respiratory illness in infancy and it has the most clearly defined disease pattern of any paramyxovirus, being associated predominantly with pneumonia and bronchiolitis (reviewed by Chanock *et al.*, 1976; Hall, 1980; Stott & Taylor, 1985). It is also an important cause of acute respiratory illness in the elderly, the immunocompromised and in military personnel (see Belshe, Bernstein & Dansby, 1984). The closely related bovine RS virus is a major respiratory pathogen in intensively reared calves (Pirie *et al.*, 1981). PVM and RS virus resemble one another in terms of their complement of polypeptides and cytopathology (see below), and the nucleoproteins of these two viruses have cross-reactive antigens and were immunoprecipitated by the heterologous antisera (Gimenez, Cash & Melvin, 1984; R. Ling & C. R. Pringle, unpublished). Recent seroepidemiological studies suggest that PVM or an antigenically similar virus may also be responsible for some respiratory disease in man (Pringle & Eglin, 1986).

Of the human viruses, parainfluenzavirus type 3 and RS virus have their major impact as disease producing agents in infancy, whereas measles virus and mumps virus as inducers of disease are encountered later in childhood. For this reason and because satisfactory methods of control of measles virus and mumps virus are available, this chapter will concentrate on parainfluenzavirus type 3 and RS virus.

GENOME STRUCTURE AND GENE FUNCTION

In terms of genome structure, the morbilliviruses and the paramyxoviruses exhibit the same basic pattern, whereas the pneumoviruses are distinct from either of the other two. The genome of the pneumo-virus RS virus is a linear array of ten genes without overlap *(Collins, Huang & Wertz, 1984a), compared with the six–seven genes of the parainfluenza viruses and morbilliviruses where the phosphoprotein (P) gene includes an internally coded non-structural protein (C) gene. In the genomes of measles virus (Bellini et al., 1985) and Sendai virus (murine parainfluenzavirus type 1) (Giorgi, Blumberg & Kolakofsky, 1983; Shioda et al., 1983) this internal gene is encoded in an overlapping reading frame, whereas in mumps virus (Herrler & Compans, 1982), Newcastle disease virus (Collins et al., 1982) and SV5 (parainfluenzavirus type 2) (Peluso, Lamb & Choppin, 1977; Paterson, Harris & Lamb, 1984a) the internal gene is read from separate initiation sites in the same reading frame. Now that a second translated open reading frame has been identified in the NS (phosphoprotein) gene of vesicular stomatitis virus (VSV) (Herman, 1986; Hudson, Condra & Lazzarini, 1986), there are more similarities between the morbilliviruses and parainfluenzaviruses and the rhabdoviruses, than between these two paramyxoviruses groups and the third member of the family *Paramyxoviridae*, the pneumoviruses. There are, in addition, significant homologies between certain regions of the genomes of the parainfluenzaviruses and the rhabdoviruses. For example the leader RNA of NDV and the rhabdovirus VSV show extensive homology (Kurilla, Stone & Keene, 1985), and the sequences at the gene boundaries of Sendai virus and VSV show similarities. Sendai virus and VSV have specific tetranucleotides, AUUC and AUAC respectively, before the oligo polyU sequences initiating 3' terminal polyA synthesis (Gupta & Kingsbury, 1982, 1984). The untranscribed trinucleotide intergenic junction sequence GAA in Sendai virus resembles the dinucleotide GA of three of the four intergenic junctions of VSV. The intergenic junctions between the genes of other parainfluenzaviruses may differ from this pattern, however, those of SV5 extending to at least 22 nucleotides (Paterson et al., 1984a). The intergenic sequences of the RS virus genome show little conservation of length or sequence ranging from 1 to 46 nucleotides (Collins & Wertz, 1985b).

Paradoxically knowledge of the molecular biology of the pneumovirus RS virus, one of the least easily propagated, has progressed more rapidly than that of any other paramyxovirus, although some of the others are now fast catching up. Much of the nucleotide

---

* Sequencing of the A2 strain has revealed a short overlap of the end of the 22 K (M2) gene and the start of the L gene (P. L. Collins, personal communication).

sequence of the A2 strain of human RS virus has now been determined: nine of the ten messenger RNAs have been sequenced, accounting for 56% of the transcribed sequences leaving only the L protein messenger RNA and some of the untranscribed regions to be sequenced (Collins *et al.*, 1985). It is apparent that there is no significant sequence homology between RS virus and other paramyxoviruses, even for regions such as at the N-terminus of the F1 protein, which exhibit a high degree of conservation in different parainfluenza viruses and morbilliviruses (Varsanyi, Jornvall & Norrby, 1985). The 5'-terminal nine nucleotides of all nine RS virus messages examined so far show strict conservation, and the 3'-termini conform to a nine nucleotide consensus sequence, neither of these terminal sequences resembling those of other negative strand RNA viruses. However, faint traces of homology do exist. Chambers *et al.* (1986) have observed that the cleaved F1 fragment of the F0 protein of all paramyxoviruses examined so far were of identical amino acid chain length with strict conservation of the position of the cysteine residues.*

All viruses in the family possess a single promoter site adjacent to the 3'-terminus and a linear array of genes transcribed at decreasing frequency with increasing distance from the promoter site. The most obvious difference between RS virus and the other paramyxoviruses is the greater number of genes but, nonetheless, genes of homologous function are located in the same order relative to the termini (Fig. 1).

Unlike all other unsegmented genome, negative strand RNA viruses where the nucleoprotein gene is the promoter-proximal gene, two non-structural protein genes (1C and 1B) are located between the nucleoprotein gene of RS virus and the promoter site. In accordance with the expected progressive attenuation of transcription observed at increasing distance from the promoter, transcripts of the two promoter-proximal non-structural protein genes (1C and 1B) are the most abundant messenger RNAs in RS virus-infected cells (Collins & Wertz, 1983). In the RS virus genome a third presumptive non-structural gene is interposed between the matrix and glycoprotein genes, and there is an additional structural protein gene, possibly encoding a second unglycosylated membrane protein, located between the F and L protein genes. The functions of the three supernumary non-structural protein-coding genes are unknown. There are short overlapping open reading frames coding for proteins of 75

---

* Although the RS virus F protein showed few amino acid identities, three of the cysteines in the F1 region corresponded directly to Newcastle disease virus cysteines and five of the remaining eight cysteines were close.

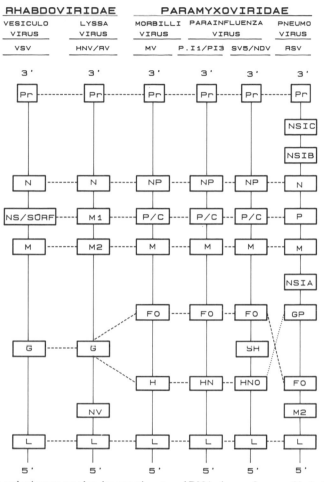

Fig. 1. Gene order in monomolecular negative strand RNA viruses. Sources: Vesicular stomatitis virus (VSV), Hudson *et al.*, 1986; rabies virus (RV), Tordo *et al.*, 1985; haematopoietic necrosis virus (HNV), Kurath *et al.*, 1985; measles virus (MV), Rima *et al.*, 1986, Dowling *et al.*, 1986; parainfluenzavirus type 1 (PI-1), Blumberg *et al.*, 1985*a, b*, Shioda, Iwasaki & Shibuta, 1986; SV5/parainfluenzavirus type 2 (PI-2), Paterson *et al.*, 1984*a*; Newcastle disease virus (NDV), Chambers *et al.*, 1986; respiratory syncytial virus (RSV), Collins *et al.*, 1984*a, b*. Pr = promoter site; HN0 and F0 code for precursor proteins.

and 90 amino acids respectively in the M and M2 genes, but no gene products have been identified (Satake & Venkatesan, 1984; Collins & Wertz, 1985*b*). Lambden (1985) has reported that a second open reading frame encoding 65 amino acids occurs in a different strain of RS virus (the Edinburgh strain). However this is not encoded internally, as in the parainfluenza viruses and morbilliviruses, but overlaps the 3' end and its significance is doubtful. Table 2 lists the known functions of the other RS virus genes.

Table 2. *Respiratory syncytial virus – gene size and gene products*

Gene		Gene product			
Order (5'–3')	ORF[a]	Molecular weight[b] (k)	Location	Modification	Function
L	6500	200	Core	None	Polymerase?
M2 (22 k)	957	22	Membrane	None	?
F	1899	68	Membrane	N-glycosylated cleaved to disulphide-linked 48 k F1 + 22 k F2 proteins. F2 is N-terminal to F1	Fusion (F1 carries major neutralisation epitopes)
G	918	33	Membrane	N- and O-glycosylated with apparent molecular weight of 84 k	Attachment? (carries minor neutralisation epitope)
NS1A	405	9.5	Non-structural	None	?
M(26 k)	952	26	Membrane	None	Matrix?
P	907	32	Core	Phosphorylated	Polymerase modifier or transport?
N	1197	42	Core	None	Nucleoprotein
NS1B	499	11	Non-structural	None	?
NS1C	528	14	Non-structural	None	?

[a] Number of nucleotides excluding polyA; ORF – open reading frame.
[b] From polyacrylamide gel electrophoretic analysis.

Which type of genome represents the higher form cannot be deduced from these data alone. Loss of genes by the parainfluenza viruses with progression towards the simplicity of the VSV pattern (Schubert, Harmison & Meier, 1984), or alternatively acquisition of genes with progression towards the greater complexity exhibited by the pneumoviruses, may equally be the direction of evolution in the family *Paramyxoviridae*. Nucleotide sequencing of the genome of the Pasteur strain of rabies virus has revealed a residual or evolving gene between the L and G protein (Tordo *et al.*, 1985) and the fish rhabdovirus haematopoietic necrosis virus encodes six unique proteins as opposed to the five of VSV and other vesiculoviruses (Kurath & Leong, 1985; Kurath *et al.*, 1985), which indicates that there is variation in gene number also within the rhabdovirus group

(Fig. 1). Similarly Hiebert, Paterson & Lamb (1985) have identified a previously unrecognised gene between the F and HN genes of SV5 virus coding for a small hydrophobic (SH) protein of 44 amino acids, and Millar *et al.* (1986) have identified an analogous open reading frame in a corresponding region of the genome of NDV coding in this case for a hydrophilic protein of 41 amino acids. For these reasons, conclusions about evolutionary relationships among negative strand RNA viruses are premature.

The essential conclusions from these features of genome structure are that the pneumoviruses are only distantly related to the other paramyxoviruses, and it is likely that sooner or later they will be given separate taxonomic status. It is one of the purposes of this chapter to consider how far this divergence in genetic organisation and information content is reflected in disease patterns and pathogenesis.

GROWTH CHARACTERISTICS AND CYTOPATHOLOGY *IN VITRO*

All the paramyxoviruses multiply in the cytoplasm, but with varying degrees of nuclear involvement, which again emphasises the distinctiveness of the pneumoviruses. The pneumoviruses can complete their growth cycle unhindered in enucleate cells, whereas the paramyxoviruses (Sendai virus) are dependent on the continued presence of the cell nucleus (Pringle, 1977). Whereas the actinomycin-D-sensitive influenza virus was released from nuclear dependence after 2 h, the actinomycin-D-resistant Sendai virus was still dependent on the presence of the nucleus up to at least 6 h post-infection (Pennington & Pringle, 1978). Intracellular polypeptide synthesis and haemadsorption occurred normally in enucleate cells, but no infectious particles were released. This suggests that the nuclear function is involved in the maturation of paramyxoviruses, rather than in messenger RNA synthesis as in the orthomyxoviruses. The presence of viral antigen in the nuclei of paramyxovirus-infected cells late in the infectious cycle, in the nuclei of persistently infected cells, or in the nuclei of cells infected with late temperature-sensitive (ts) mutants under restrictive conditions is possibly another aspect of this phenomenon. Measles virus appears similarly to be blocked in maturation, but the nuclear function is only essential early during the multiplication cycle, because, as with influenza virus, infectious virus was released from cells enucleated at 4–6 h after infection.

A further unique feature of pneumovirus-infected cells is the profuse extrusion of filament which occurs early after infection of some types of cells (Fig. 2).

The development of these processes is seen to best effect in Potoroo kidney cells which are uniquely devoid of surface feature (Fig. 2*a*). Cells infected with RS virus (Fig. 2*b–e*) or PVM (Fig. 2*f*) produced an abundance of slender filaments prior to fusion, whereas infection of these or other cells with syncytium-forming viruses such as parainfluenzavirus type 3 (Fig. 2*g*), a feline foamy retrovirus and a syncytial mutant of herpes simplex virus (Fig. 2*h*), or with a variety of non-syncytial viruses of most of the major virus groups did not induce filament production (Faulkner *et al.*, 1976; Parry, Shirodaria & Pringle, 1979). Staining with specific fluorescent antibody indicated that virus-specific surface antigen was located predominantly in these filaments. If such filaments are extruded from the cells lining the bronchioles of infants as a result of RS virus infection, it is possible that they play a role in pathogenesis by obstruction of the airways. This would provide a non-immunological explanation of the vulnerability of infants to RS virus infection. The human lung is not fully developed at birth and the diameter of the airways in infancy is half that in adulthood, although airway conductance relative to lung volume is higher in infancy (Motoyama, 1977). The declining severity of disease with age could be a consequence of the rapid increase in diameter of the airways which occurs during early childhood, thereby reducing the likelihood of obstruction and its consequences.

THE ROLE OF THE GLYCOPROTEINS IN PATHOGENESIS

The biology of the paramyxoviruses is dominated by the properties of the surface glycoproteins. The surface glycoproteins of paramyxoviruses are involved in both attachment and penetration (Choppin & Scheid, 1980). The HN protein of the parainfluenzaviruses has haemagglutinating (HA) and neuraminidase (NA) activities and is present on the surface of the virus as a disulphide-bonded dimer. In Sendai virus, SV5 and NDV the HN protein (like the neuraminidase protein of influenza A virus) is inserted into the lipid bilayer with the N-terminus located internally (Schuy *et al.*, 1984; Blumberg *et al.*, 1985*a, b*; Hiebert *et al.*, 1985; Millar *et al.*, 1986). The HA and NA functions are considered to mediate attachment and penetration (Choppin & Scheid, 1980). In the parainfluenzaviruses both functions reside in the same molecule, in contrast to the orthomyxoviruses where separate molecules perform these functions (see McCauley, this volume). The morbilliviruses, on the other hand, have no neuraminidase, and some pneumoviruses (human and bovine RS viruses) exhibit neither HA nor NA activities. PVM, however, can haemagglutinate mouse red blood cells. The second envelope glycoprotein of all the paramyxoviruses behaves as a fusion (F) protein and is inserted into the viral membrane in the more usual orientation with the COOH-terminus located internally (Paterson *et al.*, 1984*b*; Blumberg *et al.*, 1985*a*). The F proteins of the parainfluenzaviruses and the morbilliviruses are structurally analogous, but the F protein of pneumoviruses is distinct and lacks any close sequence homology (Collins *et al.*, 1984*b*). Figure 3 summarises the location of the HA, NA and F functions in the different groups of paramyxoviruses.

It is generally considered that the HA function represents a receptor-binding activity, and the F function is involved in penetration (see reviews by Choppin & Compans, 1975; Choppin & Scheid, 1980). The role of the neuraminidase is less clear. Absorption of parainfluenzaviruses involves attachment to neuraminic acid containing receptors. Since glycoproteins and glycolipids containing neuraminic acid are abundant in the membrane of vertebrate cells, attachment lacks the specificity to determine host range or tissue tropism. The lack of specificity of receptor binding contrasts with the restriction in host range of the different paramyxoviruses (Table 1). Perhaps, as in the orthomyxoviruses and rhabdoviruses, interaction of the virus polymerase with host cell factors is a major determinant of host range. The glycoproteins of neuraminidase-defective ts mutants of influenza A virus produced under restrictive conditions, unlike those of wild-type virus, contained neuraminic acid (Palese *et al.*, 1974). These virus particles were prone to form large aggregates, since they carry both the receptor and its attachment protein, The role of the neuraminidase, therefore, may be to eliminate these receptors and to promote dissemination of the virus by preventing aggregation.

Attachment of the morbilliviruses and pneumoviruses, which lack a neuraminidase, does not involve neuraminic-acid-containing receptors. The H protein of morbilliviruses is structurally (and probably functionally) analogous to the HN protein of the parainfluenzaviruses, although the precise chemical nature of the receptor is unknown. The G protein of the pneumoviruses is structurally quite

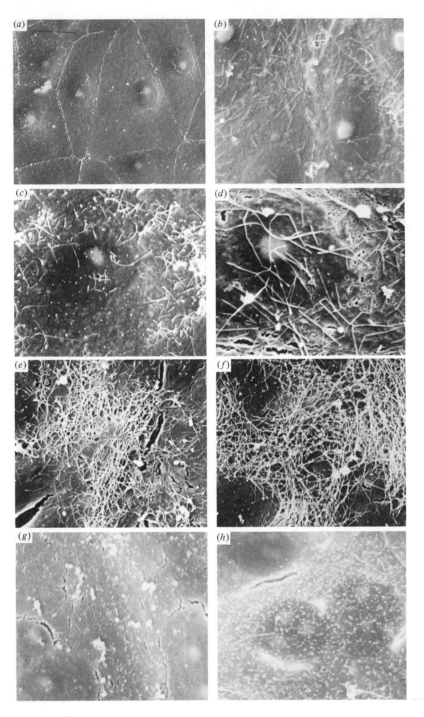

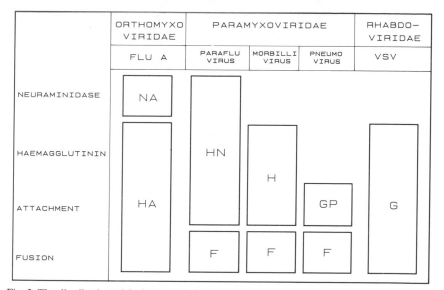

Fig. 3. The distribution of the haemagglutinin, neuraminidase, attachment and fusion functions among the membrane proteins of negative strand RNA viruses. The HA and NA proteins of influenza A viruses, the HN and F proteins of parainfluenzaviruses, the H and F proteins of morbilliviruses, the GP and F proteins of pneumoviruses and the G protein of rhabdoviruses are indicated by boxes.

different (see later) and unique when compared with other viral membrane proteins (Satake et al., 1985; Wertz et al., 1985b). The nature of the pneumovirus receptor is unknown but, as with other paramyxoviruses, it is unlikely to be instrumental in determining tropisms, since saturation was not achieved in receptor-binding experiments using purified ^{125}I-labelled G protein (Walsh, Schlesinger & Brandriss, 1984b).

Activation of biological function by proteolytic cleavage is an important property of the surface glycoproteins of paramyxoviruses. Cleavage is not necessary for particle formation and depends on the presence of an appropriate protease in the host cell. In the avirulent Ulster and Queensland strains of NDV an HN precursor (HN$_0$) is synthesised which is converted to the active HN by removal of a C-terminal glycopeptide. The HN gene of NDV has an unusually

Fig. 2. Scanning electron micrographs of the surface of cells infected with respiratory syncytial virus, murine pneumonia virus, parainfluenzavirus type 3 and herpes simplex virus type 1. (a) uninfected Potoroo kidney cells, (b) Potoroo kidney cells 33 h after infection with respiratory syncytial virus, (c) the same after 40 h, (d) the same after 48 h, (e) the same after 72 h, (f) Potoroo kidney cells 144 h after infection with murine pneumonia virus, (g) BSC-1 cells 72 h after infection with parainfluenzavirus type 3, (h) BSC-1 cells 72 h after infection with a syncytium-forming mutant of herpes simplex virus type 1. (Panels a–f from Pringle & Parry, 1980; reproduced with permission from the Journal of Virological Methods.)

long 3'-end non-coding region and it is possible that mutation could generate a longer open reading frame (Millar *et al.*, 1986), or more likely that mutation reduced the reading frame in virulent strains where the HN appears to be synthesised without precursor-cleavage. Cleavage of the F protein, however, plays a more direct role in determining the pathogenicity of NDV. For example, in avirulent strains of NDV such as La Sota and B1, F0 is not cleaved in certain cell lines resulting in decreased infectivity (Nagai, Klent & Rott, 1976; Nagai & Klenk, 1977).

The tropism of paramyxoviruses appears to be determined subsequent to adsorption during penetration and to be dependent on the properties of the F protein. The role of the F protein has been defined in classical experiments by Scheid and Choppin (reviewed in Choppin & Scheid, 1980) using protease activation (pa) mutants of Sendai virus. Cleavage of the F protein of Sendai virus is necessary for infection to proceed beyond the adsorption stage. This cleavage is mediated by a host cell enzyme with trypsin-like specificity. Proteases with different specificities cannot activate wild-type Sendai virus even if successfully adsorbed to susceptible cells. The pa mutants however were activated by different proteases, including chymotrypsin, elastase, plasmin or thermolysin, and may retain or have lost their sensitivity to trypsin. The existence of these mutants suggested that the host range and tissue tropism of Sendai virus was determined by the availability of the appropriate protease and this was confirmed directly by experiments in whole chick embryos. For example a mutant designated pa-c1 which was chymotrypsin-sensitive but trypsin-resistant, unlike the wild-type virus, fails to multiply in the allantoic sac. It can be activated, however, by injection of chymotrypsin into the allantoic sac. Other pa mutants respond to injection of the appropriate protease in like manner. The nucleotide sequences around the cleavage site of some of the pa mutants of Sendai virus have been determined (Hsu *et al.*, 1985). The wild type sequence is Val-Pro-Gln-Ser-Arg/Phe-Phe-Gly-Ala. Mutant pa-c1 has a single nucleotide change Arg→ Ile at the cleavage site, which renders the F protein resistant to cleavage by trypsin and sensitive to cleavage by elastase or chymotrypsin. Mutant pa-C2 has an additional change Gly→ Asp in another nucleotide adjacent to the cleavage site which renders the mutant resistant to chymotrypsin in addition to trypsin, but still sensitive to elastase.

The ease of isolation of the pa mutants of parainfluenzaviruses and the lack of specificity in adsorption together suggest that variants can rise under natural conditions which are able to multiply in a normally inviolate cell type (e.g. nervous or endocrine tissue) thereby altering the disease potential of the virus. A different consequence of the mutability of protease activation is that the properties of a virus propagated *in vitro* may be a reflection of the protease specificity of the cell type used for isolation of the virus. Although the chick embryo is not the normal host of Sendai virus, Tashiro & Homma (1983) showed that a trypsin-like activity in the bronchial epithelium of the mouse was responsible for activation and multiplication of Sendai virus in the mouse lung. It is likely but remains to be demonstrated that the host range and tropism of other paramyxoviruses are determined by protease activation of infectivity by cleavage of the F protein.

In the case of RS virus, since the fusion epitope has been located to the 22 k cleavage fragment and syncytium formation is not always observed, it is possible that cleavage of the F protein is not essential for infectivity. Spring & Tolpin (1983) concluded that although the F protein has at least two different protease-sensitive sites they were not normally cleaved. Failure of cleavage did not appear to be responsible for the low titre of RS virus preparations, since treatment of released virus with the appropriate proteases or incubation in their presence did not enhance infectivity.

GLYCOPROTEINS AND THE IMMUNE RESPONSE

Description of pathogenesis in molecular terms is remote for most paramyxoviruses. Of more immediate concern is analysis of the host immune response to virus infection and determination of what constitutes protection. Antibodies to the two surface proteins of parainfluenzaviruses and morbilliviruses are important in conferring immunity to infection. Antibody to the HN protein of parainfluenzaviruses and the H protein of morbilliviruses prevent adsorption to host cells, inhibit HA and NA activities (or HA only in the case of morbilliviruses), whereas antibody to the F protein inhibits haemolysis and membrane fusion. Antibodies to both proteins neutralise the infectivity of released virus, but only anti-F antibody inhibits the lateral spread of virus in a monolayer *in vitro* (Merz, Scheid & Choppin, 1980). The first formalin-inactivated vaccines used against measles, mumps and parainfluenzaviruses appeared to induce a defective immune response generating antibodies to only one, the attachment (HN or H) protein, of the two surface glycoproteins of these viruses. The transient immunity associated with such vaccines may have been responsible for the atypical disease response observed on subsequent exposure to these viruses (Norrby & Gollmar, 1975). The absence of antibody to F protein in the sera of recipients of inactivated vaccines and the requirement for anti-F antibody to eliminate lateral spread of virus in monolayers have together provided a convenient explanation of the atypical measles syndrome observed in a proportion of vaccinees on subsequent natural infection. On re-infection, the virus would be able to multiply initially because of absence of local immunity and of anti-F antibody to prevent spread by cell fusion. The resulting release of viral antigen could induce a secondary immune response to the complement of virus proteins other than F to which the host had been primed by the inactivated vaccine. Consequences of this might be formation of immune complexes in the lungs, inflammation and antibody-dependent cytotoxic cell proliferation.

Early attempts to vaccinate with a formalin-inactivated RS virus encountered similar difficulties with exacerbation of disease rather than protection in vaccinees exposed to subsequent natural RS virus infection. This occurred within a few months of vaccination, in contrast to the atypical measles response which was only apparent several years later when immunity had waned. These facts and the poor correlation between serum neutralising antibody titres and resistance to re-infection with RS virus in the paediatric and general populations suggested that there might be an immunopathological component in the aberrant response of vaccinees.

The functions of the surface glycoproteins of the pneumoviruses are less well defined than those of the parainfluenzaviruses or the morbilliviruses. The G protein of RS virus is considered to be the attachment protein (Fernie & Gerin, 1982; Walsh *et al.*, 1984*b*). Molecular cloning of the G protein gene and limited amino acid sequencing have revealed the uniqueness of this protein (Satake *et al.*, 1985; Wertz *et al.*, 1985*b*). A polypeptide of 33 000 molecular weight is transcribed from the G mRNA and glycosylated to become a glycoprotein of 80 000–90 000 molecular weight. There are seven putative N-linked glycosylation sites which contribute 10 000–15 000 to the molecular weight and a stable N-glycosylated 46 000 molecular weight intermediate (Gruber & Levine, 1985), is synthesised in a step-wise manner (Fernie *et al.*, 1985*a*). The remaining increment in molecular weight is tunicamycin-resistant and presumed to be due to 0-linked glycosylation since the molecule is rich in hydroxy-amino acids (91 of the 298 amino acids). The G protein may be released from the cell surface faster than the F protein, since in cultures of RS virus infected BSC-1 cells the G protein is found predominantly in the culture fluid (Fernie et al., 1985b). A hyperimmune polyclonal antibody to the purified G protein neutralised infectivity in the absence of complement, but did not inhibit cell fusion (Walsh et al., 1984b). Only one monoclonal antibody to the G protein neutralised infectivity, and then only in the presence of complement.

The other surface glycoprotein of RS virus is functionally, though not structurally, analogous to the F protein of other paramyxoviruses. One anti-F monoclonal antibody inhibits fusion of previously infected cells confirming the assignment of the fusion function to this protein, although other neutralising anti-F monoclonals do not inhibit fusion (Walsh & Hruska, 1983). The F protein is synthesised as a 70 000 molecular weight polypeptide which, unlike the G protein or the F protein of other paramyxoviruses, exists as a non-covalently-bonded 145 000 molecular weight dimer. In this respect it resembles the HN proteins of other paramyxoviruses. The 70 000 molecular weight F polypeptide is cleaved to disulphide-bond-linked 48 000 and 23 000 molecular weight products. Since cleavage still occurs in infected HeLa cells in the presence of monensin, which prevents migration of most glycoproteins from the Golgi apparatus to the plasma membrane, cleavage can occur intracellularly. Pulse-chase experiments, however, indicate that cleavage is a late event and that it can take place in the absence of glycosylation (Fernie et al., 1985a, b; Gruber & Levine, 1985). Polyclonal rabbit antiserum to the dimerised (145 000 molecular weight) F protein neutralises infectivity, but does not inhibit fusion activity. Antiserum to the monomer (70 000 molecular weight) form of the F protein both neutralised infectivity and inhibited fusion (Walsh, Brandriss & Schlesinger, 1985a). Only the anti-F monomer serum appeared to bind to the fusion epitope in vitro, and only this antiserum reacted with the 23 000 molecular weight cleavage product in Western blotting experiments. The neutralisation epitope(s) has been located to the 48 000 molecular weight region both by monoclonal antibody studies and analysis of immunogenicity in mice (Walsh & Hruska, 1983; Trudel et al., 1986).

Prince and colleagues (1986) working with cotton rats have concluded that the original vaccine preparation utilised in the trials in 1965/6 was responsible for the unfortunate outcome and that the exacerbation of disease observed following the use of inactivated RS virus in children was vaccine-batch-related, rather than an inevitable response to RS virus antigens. Studies with monoclonal antibodies showed that epitopes on both the F and G proteins were affected by formalin treatment. It is possible that the adverse response to RS virus inactivated vaccine was also a consequence of destruction of the antigenicity of the F protein by formalin treatment, but the experiments of Prince and colleagues (while clearly implicating formalin as a determining factor) did not identify the G or the F protein as the defective component. Furthermore, in view of the differing reactivities of antibodies to the dimeric and monomeric forms of the F protein of RS virus, use of purified F protein in any vaccine formulation will require careful biological assessment and monitoring.

Radioimmunoprecipitation has been used to analyse the antibody response to individual RS virus proteins. Ward *et al.* (1983) detected antibodies to the G, F and N proteins in the sera of adults and older children, but infants between 6 and 12 months appeared unable to mount a response to the G protein. This could be a consequence of the highly glycosylated state of the G protein, since, in infants, there is a deficiency of the subclass IgG-2 and IgG-4 antibodies which react with sugar polymers. In a survey of 40 sera from adults Pringle Wilkie & Elliott (1985) also found that antibodies to the F and N proteins predominated, but only one serum reacted strongly with the G protein. Vainionpaa, Meurmen & Sarkinen (1985), on the other hand – in an examination of sera from children with primary RS virus infections – reported that antibodies to the G, F and M proteins predominated. The discrepancies between these studies may reside in differences in technique, or they may be a consequence of virus strain variation. It is also apparent that individuals vary in their responses.

Monoclonal antibodies and an indirect enzyme-linked immunosorbent assay have been used to measure the relative amounts of antibodies to the G, F and N proteins in nasal secretions (Hendry *et al.*, 1985). Antibodies to the F protein predominated. Reconstruction experiments indicated that the antigenicity of the G protein was affected adversely by nasal secretions, which might explain the poor antigenic response to this protein. The absence of antibodies to the N protein which is the major intracellular protein in cultured cells is more puzzling and perhaps indicates tissue-specific modulation of viral gene expression or viral protein synthesis.

THE NATURE OF PROTECTION

The failure of the inactivated RS virus vaccine in children, despite the induction of high titres of neutralising antibodies in response to the vaccine, suggested that anti-RS virus antibodies might be harmful *per se*. There is no evidence from studies *in vitro*, however, that antibodies enhance virus infectivity. Furthermore, in the initial epidemiological studies, it was established that the most severe RS-virus-associated disease occurred in infants under 6 months of age (when maternally derived antibody was still present in the infants serum). This suggested that there might be an immunopathological component in RS virus disease related to the presence of maternally derived antibody. However more extensive epidemiological studies did not support this hypothesis. It was observed that infants up to the age of 2 months were relatively spared, and that disease symptoms were less frequent and less severe in infants with high levels of maternal antibodies.

Passively transferred antibody can protect experimental animals against RS virus infection. Passively administered monoclonal antibodies to certain epitopes on both the G and F proteins protected mice (Taylor *et al.*, 1983, 1984) and cotton rats (Walsh, Schlesinger & Brandriss, 1984c) against intranasal infection with RS virus, whereas monoclonal antibodies to the N and P proteins of RS virus did not provide protection. Therefore antibodies to either of the surface glycoproteins of RS virus can provide protection. Prince, Horswood & Chanock (1985b) subsequently estimated the level of passively acquired serum neutralising antibodies required to protect the respiratory tract of cotton rats from infection with RS virus. They found that virus replication in the respiratory tract was suppressed when serum neutralisation titres of 1:100 or greater were attained, but titres of 1:380 or greater were required to suppress virus replication in the lungs. These values corresponded well with the geometric mean titre of serum neutralising antibodies of 1:400 in the sera of healthy infants in the first month of life, compared with 1:200 in age-matched infants with bronchiolitis and 1:100 in infants with pneumonia (Parrott *et al.*, 1973). At the age of 2 months, when RS virus disease peaks, maternal antibody titres have declined by half. The conclusion from these studies is that the levels of serum neutralising antibodies required to provide protection are high, and the quantitative similarity between passive protection in the cotton rat and the human infant provides an experimental system for measuring the potential of new vaccines and passive immunising techniques.

Prince *et al.* (1985*a*) have shown also that pooled human convalescent sera or Sandoglobulin (a human IgG preparation suitable for intravenous administration with titres of RS virus neutralising titres of >1 : 10 000) could be used therapeutically to control RS virus infection in cotton rats. Animals treated with Sandoglobulin were resistant to re-infection with RS virus despite a depressed primary immune response. Intravenous administration of Sandoglobulin to RS virus infected owl monkeys significantly reduced the amount of virus shed from the nasal and tracheal epithelia and induced no clinical or histopathological evidence of exacerbation of disease (Hemming *et al.*, 1985). These experiments emphasise again the absence of any adverse immunopathological response to antibody. It was considered that the rapidity of the response to Sandoglobulin, where an appreciable reduction of RS virus in the lung was observed by 3 h, was the combined result of enhanced clearance of aggregated virus, antibody-dependent cellular cytotoxicity and complement-dependent cytolytic antibodies. Indeed, the absence of viraemia in RS virus infection and the predominantly cell-associated nature of RS virus infectivity *in vitro* suggest that complement-dependent cytolytic antibodies and antibody-dependent cellular cytotoxicity may play major roles in virus clearance.

It has been shown that fostering of newborn ferrets or cotton rats on immune mothers provided protection against RS virus infection. However, this protection was not mediated by specific antibody, but by some other immune mechanism (Suffin *et al.*, 1979; Prince *et al.*, 1983). Epidemiological studies of human infants have also suggested that breast feeding reduced the incidence of severe RS virus infection. As in experimental animals, protection of the human infant was not related to the titre of specific IgA antibody in the colostrum or breast milk and was presumed to be due to cellular components. It was suggested that this cellular reactivity was transferable to the immunologically immature mucosal cells of the infant possibly by pharmacological means (Scott, Scott & Toms, 1981). 39% of mothers breast feeding their offspring gave positive responses during the first 5 days post partum in a lymphocyte-transformation assay confirming the involvement of colostral cells (Scott, Scott & Toms, 1985). It remains to be demonstrated whether this reactivity can be transferred to the newborn infant and whether it is protective.

RS-virus-infected HeLa cells were able to activate both the classical and alternate complement pathways and increased complement activation occurred in the presence of antibody (Edwards, Snyder & Wright, 1986). In the presence of antibody, both the classical and alternate pathways were capable of lysis of RS-virus-infected cells. Local complement activation in the respiratory tract may contribute to RS-virus-induced bronchiolitis by release of chemotactic and anaphylactic factors.

Complement-dependent lysis occurred in the absence of effector cells. Cytolysis also occurred with the participation of lymphocytes by antibody-dependent cell-mediated cytotoxicity (Kaul, Welliver & Ogra, 1983), or in the absence of antibody by human neutrophils in the presence of complement (Kaul *et al.*, 1984). Cytolysis of RS virus-infected cells could be both beneficial by destroying infected cells prematurely before release of large amounts of infectious virus, and detrimental by causing necrosis of respiratory tissue.

PREVALENCE AND DIVERSITY

A striking feature of some paramyxoviruses is their ubiquity, while others are restricted in occurrence and in host range. An example of the latter is measles virus which requires a minimum community size of around 250 000 for its continued survival (Black, 1966). Consequently measles may be a human disease of relatively recent origin since communities of a size necessary to sustain it have only existed for some 200–300 generations. Similarly other paramyxoviruses may be human pathogens of recent origin, since their respiratory mode of transmission necessitates a high host density.

Parainfluenzavirus type 3 and RS virus on the other hand are ubiquitous viruses. Antibodies against RS virus in particular are universal throughout the human population and no human community, however small or isolated, appears to be free of infection (Chanock *et al.*, 1976). Antibodies also occur widely in domestic animals: in limited surveys the only animal sera regularly without RS virus neutralising activity were sera from foetal animals and Australian marsupials (Lundgren, Magnuson & Clapper, 1969; Pringle & Cross, 1978, Richardson-Wyatt *et al.*, 1981; C. R. Pringle, unpublished data). It is not clear whether human RS virus is spilling over into animal populations or whether there are related viruses circulating in animals and these strains are contributing to the maintenance of the virus in the human population. If alternate animal hosts exist, the control of RS virus infection by vaccination will be more difficult

(a)   (b)

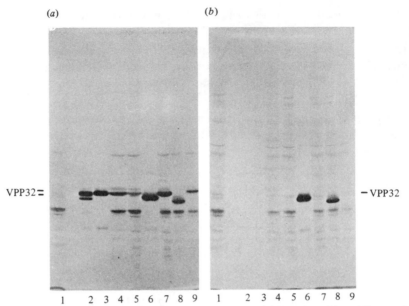

VPP32=   −VPP32

    1   2 3 4 5 6 7 8 9   1   2 3 4 5 6 7 8 9

Fig. 4. The discrimination of sub-types A and B of RS virus. (a) Reaction with anti-P monoclonal antibody 3-5; (b) reaction with anti-P monoclonal antibody 4-14. 1, uninfected cells;
2, RS A2 virus (subtype A); 3, RS Long virus (subtype A); 4, RS RSS-2 virus (subtype
A); 5, RSN 7335 (subtype A); 6, RS RSN-2 virus (subtype B); 7, RS RSG4988 virus (subtype
A); 8, RS RSN1599 virus (subtype B). The virus antigen used in lanes 2, 3 and 6 was prepared
from pelleted virus and the remainder from cell lysates. The reactions of the two monoclonal
antibodies with these antigens were investigated by SDS-polyacrylamide gel electrophoresis
and transfer to nitrocellulose membranes. These immunoblots show that only the P proteins
of the two subtype B strains react with monoclonal antibody 4-14, and both have faster
mobilities in SDS-PAGE than the P proteins of the subtype A strain. From Gimenez *et
al.*, 1986, reproduced with permission of the *Journal of General Virology*.

to achieve than in the case of measles virus, where the chain of
transmission can be more easily interrupted. Bovine RS virus however, despite its antigenic similarity to human RS virus, is probably
a distinct biological entity.

It is now recognised as a result of monoclonal antibody studies
that these apparently antigenically monotypic viruses do exhibit considerable antigenic diversity. Intratypic antigenic variation has been
reported for mumps virus, parainfluenza virus type 3, RS virus and
NDV. Two (or possibly three) subtypes of RS virus occur worldwide
and co-circulate in the same epidemics. The two subtypes were
defined initially by their reaction with an anti-G protein monoclonal
antibody (Anderson *et al.*, 1985; Mufson *et al.*, 1985). The
A and B subtypes can also be distinguished by a difference in P
protein electrophoretic mobility (Fig. 4) and reaction with a particular anti-P protein monoclonal antibody (Gimenez *et al.*, 1986). This

covariation suggests that the A and B subtypes may have diverged considerably. The only discordance between these two properties occurs in mutant *ts*N1 (complementation group D), which was derived from a subtype B wild-type strain. The P protein of this mutant has the mobility of a subtype A strain (Pringle *et al.*, 1981), but is still recognised by the subtype B specific anti-P protein monoclonal (H. B. Gimenez, personal communication).

Little comparative sequence data are available for different RS virus isolates. Lambden (1985) has determined the complete sequence of the P gene of the Edinburgh strain, an A subtype strain, and shown close homology (97.5%) with the A2 strain: 18 bases in the 726 base coding region were different with only 2 producing a coding change. Sequencing of the F and 22K genes of the RSS-2 strain, another A subtype strain, shows a similar level of homology with the A2 strain for these proteins (97.5% and 97.3% respectively H. N. Baybutt & C. R. Pringle, unpublished data).*

It is not yet clear whether this subtype variation has any biological significance. Temperature-sensitive mutants derived from subtype A and B strains complement efficiently (Gimenez & Pringle, 1978), indicating functional equivalence of gene products. RS virus-specific murine cytotoxic T cells recognised all of a range of human strains representing the two or three antigenic subtypes (Bangham & Askonas, 1986). Qualitative cross-reactivity was apparently complete. These results are similar to the findings with influenza virus, where polyclonal cytotoxic T cells did not distinguish serologically distinct type A strains (Zweerink *et al.*, 1977). It may be that the cross-reactive cytotoxic T cells recognise a conserved internal antigen, rather than the potentially more variable surface antigens, as is the case in influenza virus where the nucleoprotein serves as the major cross-reactive determinant (Townsend & Skehel, 1984; Yewdell *et al.*, 1985). Recombinant vaccinia virus expressing influenza HA protein induces non-cross-reactive cytotoxic T lymphocytes (Bennink *et al.*, 1986), whereas recombinants expressing nucleoprotein induce cross-reactive cytotoxic T lymphocytes (Yewdell *et al.*, 1985). A RS virus vaccine eliciting a cellular immune response would be advantageous because of its inherent cross-reactivity.

It is to be expected that strains of RS virus of differing virulence will also exist and play a role in the epidemiology of RS virus

* Interestingly the F/22K intergenic junction showed markedly less homology (80.4%). Approximately a third of the nucleotide differences in the F gene result in amino acid changes, compared with approximately a fifth and a twentieth respectively in the case of the 22K and P genes.

infection. This is suggested by the results of a study of natural variability using the genetic markers of growth at 39 °C and relative neutralisability (Yurlova, Karpova & Karpukhin, 1986), but this remains to be demonstrated directly.

NON-RESPIRATORY DISEASE

Despite the dramatic rate of progress in determining the molecular nature of paramyxoviruses, many facets of the biology of these viruses remain obscure. Measles virus normally causes a self-limiting acute illness which is followed by life-long immunity. Occasionally a persistent infection develops and subacute sclerosing panencephalitis (SSPE), a fatal neurological disorder, is a rare consequence of this persistence (see ter Meulen & Carter, 1982). Much effort has been expended on comparing isolates of SSPE-associated viruses and standard measles virus strains, but no consistent differences have been observed. The factors promoting persistence *in vitro* have also been investigated extensively. The generation of defective interfering virus by deletion mutation may be involved since persistently infected cultures and SSPE brain tissue release little infectious virus. Persistence is often accompanied by absence or deficiency of the M protein, and some SSPE patients have little or no antibody to this viral protein. Antigenic changes have also been observed in the H protein during persistent infection, and reduced haemadsorption. Recent work has shown that in model systems the M protein deficiency is due to rapid intracellular degradation of this protein and the lack of haemadsorption to failure of processing of the H protein (Young, Heineke & Wechsler, 1985). It has not been clearly established whether these changes are responsible for initiation of persistence or consequences of it. It is likely that the transition from productive infection to the non-productive persistent infection accompanying SSPE is complex and can be achieved by several routes.

Paramyxoviruses are prone to establish persistent infections in cultured cells, and there is some evidence for the persistence of paramyxoviruses *in vivo*. The ubiquity of antibodies to paramyxoviruses such as RS virus and parainfluenza virus type 3 in the adult population, coupled with the often seasonal prevalence of these viruses, are difficult to explain without invoking persistence. For instance Parkinson *et al.* (1980) have documented an outbreak of upper respiratory tract infection due to parainfluenza virus type 3 in a community in Antarctica after a long period of isolation.

At the cellular level Goswami *et al.* (1984*a*) using monoclonal antibodies to the HN and N proteins detected antigens of SV5 in human bone marrows from multiple sclerosis (MS) patients and from apparently healthy individuals. About 60% of bone marrows obtained from MS patients and 25% of control bone marrows contained SV5-related antigens. Also nucleoprotein antigens of parainfluenzaviruses types 1 and 3 were detected using anti-N protein monoclonal antibodies in 25% of bone marrows from MS patients and 50% of control bone marrows. Although cross-reactivity of antibodies to virus proteins with cellular proteins has been recorded [e.g. between the measles virus P protein and vimetin (Fujinami *et al.*, 1983) and between measles virus F protein and stress proteins (Sheshberadaran & Norrby, 1984)] the evidence suggested that paramyxovirus genetic information was present in these cultures. Limited focal spread of the SV5 antigens was observed in the bone marrow cell cultures from some of the MS patients and Vero cells exposed to fluids from antigen-positive cultures expressed SV5 antigens for a limited period. The spread and passage of these antigens suggested that the SV5 genome must persist in some of these cells perhaps in a defective or repressed state. Although spread and passage of antigen were only demonstrable in bone marrow cultures from MS patients, these observations suggest that a high proportion of human bone marrows harbour paramyxovirus antigens. The relationship of persistence to disease processes is less clear. Russell & Goswami (1984) suggested that antibody-mediated antigenic modulation might play a role in the establishment of persistent paramyxovirus infection of bone marrow. An impaired immune response in MS patients could have relaxed this selection pressure allowing the evolution of a less restricted virus population. The evidence for induction of measles virus persistence both *in vitro* and *in vivo* by antibody-induced antigenic modulation has been reviewed (in a previous SGM symposium) by Oldstone & Fujinami (1982).

The pneumovirus PVM normally exists as a latent infection of mice and RS virus regularly causes a persistent infection in EB-virus-transformed human B cell lines (C. R. M. Bangham & A. McMichael, personal communication). In an *in vitro* system described by Pringle *et al.* (1978), phenotypic pseudotransformation of persistently infected BSC-1 cells was observed which was correlated with expression of RS virus antigen at the cell surface (Parry *et al.*, 1979). Another consequence of persistence in this system was a progressive shift in mean chromosomal number of the infected cells. The potential for paramyxovirus involvement in non-respiratory disease is manifest.

RS virus, as well as causing acute respiratory disease in infancy and in old age, has been associated (rarely) with other conditions including bone, heart, neurological and urinary tract diseases. In the case of neurological disease the involvement of RS virus is likely to be as an inducer of a general host response rather than as a result of specific damage (Kennedy *et al.*, 1986). Although viraemia is generally considered to be exceptional, Roberts (1982) observed that RS virus infection of human mononuclear leukocytes *in vitro* resulted in depressed cell-mediated immune responses. The accessory cell function of monocytes and macrophages were unaffected. These observations suggested that RS virus might infect human lymphocytes selectively, resulting in an active suppression of the immune or cell-mediated response to challenge virus. Subsequently Domurat *et al.* (1985) demonstrated, by immunofluorescent staining, that RS virus can infect both human monocyte-macrophages and lymphocytes *in vitro*, indicating that the selective effects on lymphocyte function were not necessarily a result of differential susceptibility. RS virus antigens in leukocytes were not detected until 24 h after infection and reached a maximum at 3–4 days, indicating synthesis rather than retention of input material, although no increase in infectivity was detected. RS virus infection produced a decrease in the number of helper T cells and an increase in the number of suppressor T cells with inversion of the helper:suppressor ratio. Independently it has been observed that patients with bronchiolitis had fewer suppressor phenotype T cells during convalescence than patients with milder manifestations of RS virus infection (Welliver *et al.*, 1984). The F and G surface glycoproteins of RS virus were expressed predominantly by cells with a T suppressor phenotype. Circulating mononuclear leukocytes obtained from children during the acute phase of RS virus infection frequently exhibited RS virus antigen, particularly cells from children under 3 months of age. Expression of RS virus proteins was not detected in healthy children or those experiencing influenza A virus infection.

Therefore initial RS virus infections in children may involve infection of circulating immunocompetant cells, resulting in disturbance of the helper:suppressor lymphocyte balance. Further work is required to establish whether this effect on lymphocyte phenotype contributes to recovery from infection or whether it is responsible for recurrence of RS virus infection by diminishing immunological memory or anamnestic response to recurrent challenge.

Immunosuppression by cyclophosphamide treatment dramatically altered the outcome of RS virus infection in cotton rats (Wong *et al.*, 1985). Shedding of RS virus from the respiratory tract and nasal turbinates continued for up to 6 weeks instead of the normal few days and severe pulmonary pathology developed. Persistent de-generation and regeneration of epithelium was observed. In three out of four cotton rats there was evidence of extrapulmonary infec-tion, although viraemia was not demonstrated. These observations suggest that cell-mediated immunity plays an important role in clear-ance of RS virus from the respiratory tract and in the prevention of dissemination of virus throughout the body.

Many viruses are non-specific modulators of the immune system, enhancing or suppressing the immune response. RS virus is mitogenic for unprimed murine Balb-C spleen cells with both B and T lympho-cytes responding in contrast to most other viruses which are exclu-sively B cell mitogens (Alsheikhly & Norrby, 1986). However, as observed by others, human lymphocytes do not respond in the same way as murine lymphocytes. Consequently, extrapolation from experiments on cell-mediated immunity in animals to the human disease situation requires caution and careful evaluation.

The question of the involvement of paramyxoviruses in Paget's disease of bone, a common disorder in the middle-aged and elderly, illustrates the problems encountered in attempting to establish in-volvement of paramyxovirus infection in the causation of non-respir-atory disease. The aetiology of Paget's disease of bone has not been established, but certain features suggest that it is the result of a slow virus infection (Rebel *et al.*, 1977; Singer, 1980). The lesions are focal and the osteoclasts of patients frequently have nuclear and occasionally cytoplasmic inclusions consisting of tubular filaments that resemble the nuclear capsid of paramyxoviruses. Measurements of the diameter of these filaments (Mills *et al.*, 1981; Howatson & Fournier, 1982) showed that these filaments were more similar to the nucleocapsids of pneumoviruses than of the other paramyxo-viruses. Furthermore Mills *et al.*, (1981) by fluorescent staining were able to detect RS virus antigens in osteoclasts and in cells derived from the lesions maintained *in vitro*. Previously Basle *et al.* (1979) reported instead the presence of measles virus antigen in osteoclasts

and subsequently measles virus mRNA by *in situ* hybridisation using
a cDNA copy of the measles virus nucleocapsid protein gene (Basle
et al., 1986). The discrepancy between these two sets of observations
has not been resolved, although later Mills *et al.* (1982) reported
the simultaneous presence of RS virus and measles virus antigens
in the osteoclasts of six patients. Antigens to the nucleoproteins
of SV5 and parainfluenzavirus type 3 have also been identified using
monoclonal antibodies (Basle *et al.*, 1985; Pringle *et al.*, 1985). It
is possible that the different results reported from different laboratories are a consequence of the sensitivity of the antigens to different
fixation procedures. Serological evidence for (Mills *et al.*, 1981) and
against (Pringle *et al.*, 1985) the association between Paget's disease
and RS virus has been published. A more positive but probably
fortuitous association was observed between diagnosis of Paget's
disease of bone and high anti-PVM serum titres (Pringle & Eglin,
1986). It is possible that paramyxovirus antigens become sequestered
in osteoclasts, or that a crossreactive antigen is present. Antigenic
mimicry is not uncommon; Srinivasappa *et al.* (1986) have reported
that 3.5% of more than 600 monoclonal antibodies against 11 different viruses cross-reacted with host cell antigens, and Norrby, Sheshberadaran & Rafner (1986) have described cross-reactivity of a
measles virus anti-HN monoclonal antibody with the nucleoprotein
of RS virus and Orvell *et al.* (1986) cross-reactivity between parainfluenzavirus type 3 nucleoprotein and the fusion protein of Sendai
virus. The question of the involvement of paramyxoviruses in the
aetiology of Paget's disease of bone has not yet been resolved.

PROSPECTS FOR CONTROL OF DISEASE

The antiviral drug Ribavirin (1-ribofuranosyl-1,2,4-triazole-3-
carboxamide) inhibits RS virus *in vitro* at concentrations as low as
$1 \mu g/ml$, and is effective in reducing the multiplication of RS virus
in the lungs and nasal turbinates of cotton rats. In human patients
ribavirin administered continuously by aerosol exposure is well toler-
ated even by the very young and has ameliorated symptoms and
aided recovery of children experiencing severe influenza virus and
RS virus infections (Knight *et al.*, 1981; Hall *et al.*, 1983*a*, *b*; McClung
et al., 1983; Taber *et al.*, 1983). The level of hospital care required
in administration of the drug and the monitoring of treatment, however, preclude general use of ribavirin. Vaccines remain the only
feasible option in the immediate future.

Effective live and inactivated vaccines have been developed against infection by several paramyxoviruses (measles virus and mumps virus in man and Newcastle disease virus in birds). Measles as an epidemic disease has been eradicated from some countries (notably Czechoslovakia) and all but eradicated from others; flocks of domestic fowl can be protected from infection by Newcastle disease virus by vaccination. Development of vaccines against the viruses which cause acute respiratory diseases in children has been identified as one of the targets of the current WHO Vaccine Development Programme. The viruses ranked in order of importance are respiratory syncytial virus, parainfluenzavirus type 3, followed by the other antigenic types of parainfluenzaviruses.

A series of modified strains of RS virus was developed as live vaccines by passage at low temperature (Friedewald et al., 1968; Kim et al., 1971) or from ts mutants of the A2 strain propagated in bovine embryo kidney cells by Chanock and colleagues (Wright et al., 1971, 1982; Kim et al., 1973; Hodes et al., 1974; Wright et al., 1976), but none of these vaccines were considered suitable for use in children on account of either lack of genetic stability or over-attenuation. Modified strains of the RSS-2 strain of RS virus have been developed similarly in human diploid cells and are being evaluated (E. McKay, C. R. Pringle and D. A. J. Tyrrell, unpublished data). Although potentially better immunogens than non-replicating vaccines, it is unlikely that any of these live modified strains will be developed further until all other approaches have been explored because of the risks inherent in administering live virus to very young children. It is possible that live vaccines could be employed in adult women to boost maternal antibody levels, either alone or in combination with non-replicating vaccine in their offspring.

Parenteral administration of live essentially unmodified RS virus induced neutralising antibody in cotton rats (Prince, Potash & Horswood, 1979) and has been evaluated clinically in children (Bunyak et al., 1978, 1979; Belshe, Van Voris & Mufson, 1982). No disease resulted, but only low levels of antibody were induced and no protection was demonstrable.

Stott et al. (1984) have developed an effective vaccine against bovine RS virus prepared by glutaraldehyde inactivation of persistently infected bovine nasal mucosal cells. On challenge, 11 of 12 calves were protected against infection; protection correlated better with antibody detected by radioimmunoassay than with neutralising antibody. The crude nature of this type of inactivated vaccine (despite its success) precludes its exploitation in human medicine.

Recombinant DNA technology allows new approaches to vaccine development. The immunologically important viral genes can be converted to DNA, inserted into prokaryotic or eukaryotic vectors, amplified and individual gene products expressed singly or in combination in an appropriate host system. Several expression vector-host systems are being developed and evaluated for production of individual paramyxovirus gene products. Already the G protein gene of the A2 strain of RS virus has been recombined into vaccinia virus and the G polypeptide synthesised and transported to the plasma membrane of cells infected with the recombinant vaccinia virus (Ball *et al.*, 1986). Cotton rats have been protected against intranasal RS virus challenge by vaccination with a recombinant virus of this type (Elango *et al.*, 1986). The F polypeptide has been similarly expressed although initially not in a fully functional form, since it reached the plasma membrane uncleaned (Wertz *et al.*, 1985a). In later work regular cleavage has been obtained (G. Wertz, personal communication). It is unlikely that a vaccinia-virus-based gene vector will be used in children, although it could be employed to boost maternal immunity, thereby achieving a measure of disease control by reducing the frequency and severity of disease in newborn children.

Both the F and G proteins of RS virus have been purified by affinity chromatography employing murine anti-F and anti-G monoclonal antibodies respectively. While this is a feasible route to vaccine production, single protein immunogens produced in this way would be uneconomic. Comparison of the nucleotide sequences of different natural strains and the sequencing of variants produced *in vitro* by resistance to neutralisation by specific monoclonal antibodies will eventually define the critical antigenic sites on the G and F proteins of RS virus. Then it will be possible to evaluate polypeptide fragments or synthetic oligopeptides as vaccines. Although synthetic oligopeptides have not proved very effective so far in the control of other enveloped viruses (such as influenza virus and rabies virus), this approach should still be evaluated for the envelope proteins of RS virus because of its potential cost effectiveness.

It is likely that potential vaccines derived by these new approaches will sooner or later become available. The challenge will then be to develop strategies for the evaluation and use of these vaccines in the containment of disease.

Unfortunately the direct derivation of live vaccines remains a stochastic process. Genetic recombination *in vitro* is not known to occur in monomolecular genome negative strand RNA viruses, nor can genetic information be recovered yet from cDNA into negative strand genomes. Solution of these fundamental problems would greatly extend the techniques available for vaccine development and advance the prospects of disease control.

REFERENCES

ALEXANDER, D. J. (1980). Avian paramyxoviruses. *Veterinary Bulletin*, **50**, 737–52.
ALSHEIKHLY, A. R. & NORRBY, E. (1986). *In vitro* stimulation of murine lymphocytes by human respiratory syncytial virus strains. *Journal of General Virology*, in press.
ANDERSON, L. J., HIERHOLZER, J. C., TSOU, C., HENDRY, R. M., FERNIE, B. F., STONE, Y. & MCINTOSH, K. (1985). Antigenic characterisation of respiratory syncytial virus strains with monoclonal antibodies. *The Journal of Infectious Diseases*, **151**, 626–33.
BALL, L. A., YOUNG, K. K. Y., ANDERSON, K., COLLINS, P. L. & WERTZ, G. W. (1986). Expression of the major glycoprotein G of human respiratory syncytial virus from recombinant vaccinia virus vectors. *Proceedings of the National Academy of Sciences, USA*, **83**, 246–50.
BANGHAM, C. R. M. & ASKONAS, B. A. (1986). Murine cytotoxic T-cells specific to respiratory syncytial virus recognise different antigenic sub-types of the virus. *Journal of General Virology*, **67**, 623–9.
BASLE, M., REBEL, A., POUPLARD, A., KOUYOUMDJIAN, S., FILMON, R. & LEPATE-ZOUR, A. (1979). Mise en evidence d'antigens viraux de rougeole dans les osteo-clasts de la maladie osseuse de Paget. *Comptes Rendu de l'Academie des Sciences, Paris*, **289**, 225–8.
BASLE, M. F., FOURNIER, J. G., ROZENBLATT, S., REBEL, A. & BOUTEILLE, M. (1986). Measles virus RNA detected in Paget's bone tissue by *in situ* hybridisation. *Journal of General Virology*, **67**, 907, 907–13.
BASLE, M. F., RUSSELL, W. C., GOSWAMI, K. K. A., REBEL, A., GIRAUDON, P., WILD, F. & FILMON, R. (1985). Paramyxovirus antigens in osteoclasts from Paget's bone tissue detected by monoclonal antibodies. *Journal of General Virology*, **66**, 2103–10.
BELLINI, W. J., ENGLUND, G., ROZENBLATT, S., ARNHEITER, H. & RICHARDSON, C. D. (1985). Measles virus P gene codes for two proteins. *Journal of Virology*, **53**, 908–19.
BELSHE, R. B., BERNSTEIN, J. M. & DANSBY, K. N. (1984). Respiratory syncytial virus. In *Textbook of Human Virology*, pp. 361–83. Littleton, Mass: PSG Publishing Co.
BELSHE, R. B., VAN VORIS, L. P. & MUFSON, M. A. (1982). Parentally administered live respiratory syncytial virus vaccine: results of a field trial. *Journal of Infectious Diseases*, **145**, 311–19.
BENNINK, J. R., YEWDELL, J. W., SMITH, G. L. & MOSS, B. (1986). Recognition of cloned influenza virus haemagglutinin gene products by cytotoxic T lymphocytes. *Journal of Virology*, **57**, 786–91.
BLACK, F. L. (1966). Measles endemicity in insular populations: critical community size and its evolutionary implications. *Journal of Theoretical Biology*, **11**, 207–11.
BLUMBERG, B. M., GIORGI, C., ROSE, K. & KOLAKOFSKY, D. (1985a). Sequence determination of the Sendai virus fusion protein gene. *Journal of General Virology*, **66**, 317–31.

BLUMBERG, B., GIORGI, C., ROUX, L., RAJIN, R., DOWLING, P., CHOLLET, A. & KOLAKOFSKY, D. (1985b). Sequence determination of the Sendai virus HN gene and its comparison to the influenza virus glycoproteins. *Cell*, **41**, 269–78.

BUNYAK, E. B., WEIBEL, R. E., CARLSON, A. J., McLEAN, A. A. & HILLEMAN, M. R. (1979). Further investigations of live respiratory syncytial virus vaccine administered parenterally. *Proceedings of the Society for Experimental Biology and Medicine*, **160**, 272–7.

BUNYAK, E. B., WEIBEL, R. E., McLEAN, A. A. & HILLEMAN, M. R. (1978). Live respiratory syncytial virus vaccine administered parentarally. *Proceedings of the Society for Experimental Biology and Medicine*, **57**, 636–42.

CHAMBERS, P., MILLAR, N. S., BINGHAM, R. W. & EMMERSON, P. T. (1986). Molecular cloning of complementary DNA to Newcastle disease virus, and nucleotide sequence analysis of the junction between the genes encoding the haemagglutinin-neuraminidase and the large protein. *Journal of General Virology*, in press.

CHANOCK, R. M., KIM, H. W., BRANDT, C. & PARROTT, R. H. (1976). Respiratory syncytial virus. In *Viral Infections of Humans, Epidemiology and Control*, ed. A. S. Evans, pp. 365–82. New York: Plenum Press.

CHOPPIN, P. W. & COMPANS, R. W. (1975). Reproduction of paramyxoviruses. In *Comprehensive Virology*, ed. H. Fraenkel-Conrat & R. R. Wagner (Volume 4), pp. 95–178. New York: Plenum Press.

CHOPPIN, P. W. & SCHEID, A. (1980). The role of viral glycoproteins in adsorption, penetration and pathogenicity of viruses. *Reviews of Infectious Diseases*, **2**, 40–61.

COLLINS, P. L., ANDERSON, K., LANGER, S. J. & WERTZ, G. W. (1985). Correct sequence for the major nucleocapsid protein mRNA of respiratory syncytial virus. *Virology*, **146**, 69–77.

COLLINS, P. L., HUANG, Y. T. & WERTZ, G. W. (1984a). Identification of a tenth mRNA of respiratory syncytial virus and assignment of polypeptides to the 10 viral genes. *Journal of Virology*, **49**, 572–8.

COLLINS, P. L., HUANG, Y. T. & WERTZ, G. W. (1984b). Nucleotide sequencing of the gene encoding the fusion (F) glycoproteins of human respiratory syncytial virus. *Proceedings of the National Academy of Sciences, USA*, **81**, 7683–7.

COLLINS, P. L. & WERTZ, G. W. (1983). cDNA cloning and transcriptional mapping of nine polyadenylated RNAs encoded by the genome of human respiratory syncytial virus. *Proceedings of the Academy of Sciences, USA*, **80**, 3208–12.

COLLINS, P. L. & WERTZ, G. W. (1985a). Nucleotide sequences of the 1B and 1C nonstructural protein mRNAs of human respiratory syncytial virus. *Virology*, **143**, 442–51.

COLLINS, P. L. & WERTZ, G. W. (1985b). Gene products and genome organisation of human respiratory syncytial (RS) virus. In: *Modern Approaches to Vaccines*, ed. R. A. Lerner, R. M. Chanock & F. Brown. Cold Spring Harbor Laboratory, New York.

COLLINS, P. L., WERTZ, G. W. & BALL, L. A. (1982). Coding assignments of the five smaller mRNAs of Newcastle disease virus. *Journal of Virology*, **43**, 1024–31.

DOMURAT, F., ROBERTS, N. J. JR., WALSH, E. E. & DAGAN, R. (1985). Respiratory syncytial virus infection of human mononuclear leukocytes *in vitro* and *in vivo*. *The Journal of Infectious Diseases*, **152**, 895–902.

DOWLING, P. C., BLUMBERG, B. M., MENONNA, J., ADAMUS, J. E., COOK, P., CROWLEY, J. C., KOLAKOFSKY, D. & COOK, S. D. (1986). Transcriptional map of the measles virus genome. *Journal of General Virology*, **67**, 1987–92.

EDWARDS, K. M., SNYDER, P. N. & WRIGHT, P. F. (1986). Complement activation by respiratory syncytial virus-infected cells. *Archives of Virology*, **88**, 49–56.

ELANGO, N., PRINCE, G. A., MURPHY, B. R., VENKATESAN, S., CHANOCK, R. M.

& Moss, B. (1986). Resistance to human respiratory syncytial virus (RSV) infection induced by immunization of cotton rats with a recombinant vaccinia virus expressing the RSV G glycoproteins. *Proceedings of the National Academy of Sciences, USA*, **83**, 1906–10.

FAULKNER, G. P., FOLLETT, E. A. C., SHIRODARIA, P. V. & PRINGLE, C. R. (1976). Respiratory syncytial virus *ts* mutants and nuclear immunofluorescence. *Journal of Virology*, **20**, 487–500.

FERNIE, B. F., DAPOLITO, G., COTE, P. J. JR & GERIN, J. L. (1985*a*). Kinetics of synthesis of respiratory syncytial virus glycoproteins. *Journal of General Virology*, **66**, 1983–90.

FERNIE, B. F., DAPOLITO, G., COTE, P. J. JR & GERIN, J. L. (1985*b*). Maturation of the large glycoprotein (VP84) of respiratory syncytial virus. Virus Research Supplement 1, 12.

FERNIE, B. F. & GERIN, J. (1982) Immunochemical identification of viral and non-viral proteins of respiratory syncytial virus. *Infection and Immunity*, **37**, 243–9.

FRIEDEWALD, W. T., FORSYTH, B. R., SMITH, C. B., GHARPURE, M. A. & CHANOCK, R. M. (1968). Low temperature grown RS virus in adult volunteers. *Journal of the American Medical Association*, **204**, 690–4.

FUJINAMI, R. S., OLDSTONE, M. B. A., WROBLEWSKA, Z., FRANKEL, M. E. & KOPROWSKI, H. (1983). Molecular mimicry in virus infection: cross-reaction of measles virus phosphoprotein or of herpes simplex virus protein with human intermediate filaments. *Proceedings of the Academy of Sciences, USA*, **80**, 2346–50.

GIMENEZ, H. B., CASH, P. & MELVIN, W. T. (1984). Monoclonal antibodies to human respiratory syncytial virus and their use in comparison of different virus isolates. *Journal of General Virology*, **65**, 963–71.

GIMENEZ, H. B., HARDMANN, N., KEIR, H. M. & CASH, P. (1986). Antigenic variation between human respiratory syncytial virus isolates. *Journal of General Virology*, **67**, 863–70.

GIMENEZ, H. B. & PRINGLE, C. R. (1978). Seven complementation groups of respiratory syncytial virus temperature-sensitive mutants. *Journal of Virology*, **27**, 459–64.

GIORGI, C., BLUMBERG, B. M. & KOLAKOFSKY, D. (1983). Sendai virus contains overlapping genes expressed from a single mRNA. *Cell*, **35**, 829–36.

GOSWAMI, K. K. A. & RUSSELL, W. C. (1982). A comparison of paramyxoviruses by immunoprecipitation. *Journal of General Virology*, **60**, 177–83.

GOSWAMI, K. K. A., CAMERON, K. R., RUSSELL, W. C., LANGE, L. S. & MITCHELL, D. N. (1984*a*). Evidence for the persistence of paramyxoviruses in human bone marrows. *Journal of General Virology*, **65**, 1881–88.

GOSWAMI, K. K. A., LANGE, L. S., MITCHELL, D. N., CAMERON, K. R. & RUSSELL, W. C. (1984*b*). Does simian virus 5 infect humans? *Journal of General Virology*, **65**, 1295–303.

GRUBER, C. & LEVINE, S. (1985). Respiratory syncytial virus polypeptides. V. The kinetics of glycoprotein synthesis. *Journal of General Virology*, **66**, 1241–7.

GUPTA, K. C. & KINGSBURY, D. W. (1982). Conserved polyadenylation signals in two negative strand RNA virus families. *Virology*, **120**, 518–23.

GUPTA, C. & KINGSBURY, D. W. (1984). Complete sequences of the intergenic and mRNA start signals in the Sendai virus genome: homologies with the genome of vesicular stomatitis virus. *Nucleic Acids Research*, **12**, 3829–41.

HALL, C. B. (1980). Prevention of infections with respiratory syncytial virus: the hopes and hurdles ahead. *Reviews of Infectious Disease*, **2**, 384–92.

HALL, C. B., MCBRIDE, J. T., WALSH, E. E., BELL, D. M., GALA, C. L., HILDRETH, S., TEN EYCK, L. G. & HALL, W. J. (1983). Aerosolized ribavirin treatment of infants with respiratory syncytial viral infection: a randomised double-blind study. *New England Journal of Medicine*, **308**, 1443–7.

HALL, C. B., WALSH, E. E., HRUSKA, J. F., BETTS, R. F. & HALL, W. J. (1983*a*). Ribavirin treatment of experimental respiratory syncytial viral infection: a controlled double-blind study in young adults. *Journal of the American Medical Association*, **249**, 2666–70.

HEMMING, V. G., PRINCE, G. A., HORSWOOD, R. L., LONDON, W. T., MURPHY, B. R., WALSH, E. E., FUCHER, G. W., WEISMAN, L. E., BARON, P. A. & CHANOCK, R. M. (1985). Studies of passive immunotherapy for infections of respiratory syncytial virus in the respiratory tract of a primate model. *Journal of Infectious Diseases*, **152**, 1083–7.

HENDRY, R. M., GODFREY, E., ANDERSON, L. J., FERNIE, B. F. & McINTOSH, K. (1985). Quantification of respiratory syncytial virus polypeptides in nasal secretions by monoclonal antibodies. *Journal of General Virology*, **66**, 1705–14.

HERMAN, R. C. (1986). Internal initiation of translation on the vesicular stomatitis virus phosphoprotein mRNA yields a second protein. *Journal of Virology*, **58**, 797–804.

HERRLER, G. & COMPANS, R. W. (1982). Synthesis of mumps virus polypeptides in infected Vero cells. *Virology*, **100**, 433–49.

HIEBERT, S. W., PATERSON, R. G. & LAMB, R. A. (1985). Identification and predicted sequence of a previously unrecognised small hydrophobic protein, SH, of the paramyxovirus simian virus 5. *Journal of Virology*, **55**, 744–51.

HODES, D. S., KIM, H. W., PARROTT, R. H., CAMARGO, E. & CHANOCK, R. M. (1974). Genetic alteration in a temperature-sensitive mutant of respiratory syncytial virus after replication *in vivo*. *Proceedings of the Society of Experimental Biology and Medicine*, **145**, 1158–64.

HORSFALL, F. L. & HAHN, R. G. (1939). A pneumonia virus of Swiss mice. *Proceedings of the Society for Experimental Biological Medicine*, **40**, 684–6.

HORSFALL, F. L. & HAHN, R. G. (1940). A latent virus in normal mice capable of producing pneumonia in its natural host. *Journal of Experimental Medicine*, **71**, 391–408.

HOWATSON, A. F. & FOURNIER, V. L. (1982). Microfilaments associated with Paget's disease of bone: comparison with nucleocapsids of measles virus and respiratory syncytial virus. *Intervirology*, **18**, 150–9.

HSU, M. C., SCHEID, A. & CHOPPIN, P. W. (1985). Protease activation mutants of Sendai virus: single nucleotide change in the F protein gene alter fusion activation – host range. *Virus Research Supplement I*, 2.

HUDSON, L. D., CONDRA, C. & LAZZARINI, R. A. (1986). Cloning and expression of a viral phosphoprotein: structure suggests NS may function by mimicking an RNA template. *Journal of General Virology*, **67**, 1571–9.

KAUL, T. N., FADEN, H., BAKER, R, & OGRA, P. L. (1984). Virus-induced complement activation and neutrophil-mediated cytotoxicity against respiratory syncytial virus (RSV). *Clinical and Experimental Immunology*, **56**, 501–8.

KAUL, T. N., WELLIVER, R. C. & OGRA, P. L. (1983). Development of antibody-dependent cell-mediated cytotoxicity in the respiratory tract after natural infection with respiratory syncytial virus. *Infection and Immunity*, **37**, 492–8.

KENNEDY, C. R., ROBINSON, R. O., VALMAN, H. B., CHRZANOWSKA, K., TYRRELL, D. A. J. & WEBSTER, A. D. B. (1986). A major role for viruses in acute childhood encephalopathy. *Lancet*, **1**, 989–91.

KIM, H. W., ARROBIO, J., BRANDT, C. D., WRIGHT, P., HODES, D., CHANOCK, R. M. & PARROTT, R. H. (1973). Safety and antigenicity of temperature-sensitive mutant respiratory syncytial virus (RSV) in infants and children. *Pediatrics*, **52**, 56–63.

KIM, H. W., ARROBIO, J. O., PYLES, G., BRANDT, C. D., CAMARGO, E., CHANOCK, R. M. & PARROTT, R. H. (1971). Clinical and immunological response of infants and children to administration of low-temperature adapted respiratory syncytial virus. *Pediatrics*, **48**, 745–55.

KINGSBURY, D. W., BRATT, M. A., CHOPPIN, P. W., HANSON, R. P., HOSAKA, Y., TER MEULEN, NORRBY, E., PLOWRIGHT, W., ROTT, R. & WUNNER, W. H. (1978). Paramyxoviridae. *Intervirology*, **10**, 137–52.

KNIGHT, V., McCLUNG, H. W., WILSON, S. Z., WATERS, B. K., QUARLES, J. M., CAMERON, R. W., GREGGS, S. E., ZERWAS, J. M. & COUCH, R. B. (1981). Ribavirin small-particle aerosol treatment of influenza. *Lancet*, 945–49.

KURATH, G., AHREN, K. G., PEARSON, G. D. & LEONG, J. C. (1985). Molecular cloning of the six mRNA of infectious hematopoietic necrosis virus, a fish rhabdovirus, and gene order determination of R-loop mapping. *Journal of Virology*, **53**, 469–76.

KURATH, G. & LEONG, J. C. (1985). Characterisation of infectious hematopoietic necrosis virus mRNA species reveals a nonvirion rhabdovirus protein. *Journal of Virology*, **53**, 462–8.

KURILLA, M. G., STONE, H. O. & KEENE, J. O. (1985). RNA sequence and transcriptional properties of the 3' end of the Newcastle disease virus genome. *Virology*, **145**, 203–12.

LAMBDEN, P. R. (1985). Nucleotide sequence of the respiratory syncytial virus phosphoprotein gene. *Journal of General Virology*, **66**, 1607–12.

LUNDGREN, D. L., MAGNUSON, M. G. & CLAPPER, W. E. (1969). A serological survey in dogs for antibody to human respiratory viruses. *Laboratory Animal Care*, **19**, 352–9.

McCLUNG, H. W., KNIGHT, V., GILBERT, B. E., WILSON, S. Z., QUARLES, J. M. & DIVINE, G. W. (1983). Ribavirin aerosol treatment of influenza B virus infection. *Journal of the American Medical Association*, **249**, 2671–4.

MERZ, D. C., SCHEID, A. & CHOPPIN, P. W. (1980). Importance of antibodies to the fusion glycoprotein of paramyxoviruses in the prevention of spread of infection. *Journal of Experimental Medicine*, **151**, 275–88.

MILLAR, N. S., CHAMBERS, P. & EMMERSON, P. T. (1986). Nucleotide sequence analysis of the haemagglutinin-neuraminidase gene of Newcastle disease virus. *Journal of General Virology*, **67**, 1917–27.

MILLS, B. G., SINGER, F. R., WEINER, L. P. & HOLST, P. A. (1981). Immunohistological demonstration of respiratory syncytial virus antigens in Paget's disease of bone. *Proceedings of the National Academy of Sciences, USA*, **78**, 1209–13.

MILLS, B. G., STABILE, E., HOLST, P. A. & GRAHAM, C. (1982). Antigens of two different viruses in Paget's disease of bone. *Journal of Dental Research*, **61**, 347–51.

MOTOYAMA, E. K. (1977). Pulmonary mechanics during early postnatal years. *Pediatric Research*, **11**, 220–3.

MUFSON, M. A., ORVELL, C., RAFNAR, B. & NORRBY, E. (1985). Two distinct subtypes of human respiratory syncytial virus. *Journal of General Virology*, **66**, 2111–24.

NAGAI, Y. & KLENK, H.-D. (1977). Activation of precursors to both glycoproteins of NDV by proteolytic cleavage. *Virology*, **77**, 125–34.

NAGAI, Y., KLENK, H.-D. & ROTT, R. (1976). Proteolytic cleavage of the viral glycoproteins and its significance for the virulence of Newcastle disease virus. *Virology*, **72**, 494–508.

NORRBY, E. & GOLLMAR, Y. (1975). Identification of measles virus-specific hemolysis-inhibiting antibodies separate from hemagglutination-inhibiting antibodies. *Infection and Immunity*, **11**, 231–9.

NORRBY, E., SHESHBERADARAN, H., McCULLOUGH, K. C., CARPENTER, W. C. & ORVELL, C. (1985). Is rinderpest virus the archevirus of the morbillivirus genus? *Intervirology*, **23**, 228–32.

NORRBY, E., SHESHBERADARAN, H. & RAFNER, B. (1986). Antigen mimicry involving measles virus hemagglutinin and human respiratory syncytial virus nucleoprotein. *Journal of Virology*, **57**, 394–6.

OLDSTONE, M. B. A. & FUJINAMI, R. S. (1982). Virus persistence and avoidance of immune surveillance: how measles viruses can be induced to persist in cells, escape immune assault and injure tissues. In *Virus Persistence, SGM Symposium*, **33**, ed. B. W. J. Mahy, A. C. Minson & G. K. Darby, pp. 185–202. Cambridge University Press.

ORVELL, C., RYDBECK, R. & LOVE, A. (1986). Immunological relationships between mumps virus and parainfluenza viruses studied with monoclonal antibodies. *Journal of General Virology*, **67**, 1929–39.

PALESE, P., TOBITA, K., UEDA, M. & COMPANS, R. N. (1974). Characterisation of temperature-sensitive influenza virus mutants defective in neuraminidase. *Virology*, **61**, 397–410.

PARKINSON, A. J., MUCHMORE, H. G., McCONNELL, T. A., SCOTT, L. V. & MILES, J. A. R. (1980). Serologic evidence for parainfluenza virus infection during isolation at South Pole Station, Antarctica, *American Journal of Epidemiology*, **112**, 334–40.

PARROTT, R. H., KIM, H. W., ARROBIO, J. O., HODES, D. S., MURPHY, B. R., BRANDT, C. D., CAMARGO, E. & CHANOCK, R. M. (1973). Epidemiology of respiratory syncytial virus infection in Washington DC. II. Infection and disease with respect to age; immunologic status, race and sex. *American Journal of Epidemiology*, **98**, 289–300.

PARRY, J. E., SHIRODARIA, P. V. & PRINGLE, C. R. (1979). Pneumoviruses: the cell surface of lytically and persistently infected cells. *Journal of General Virology*, **44**, 479–91.

PATERSON, R. G., HARRIS, T. J. R. & LAMB, R. A. (1984a). Analysis and gene assignment of mRNAs of a paramyxovirus, simian virus 5. *Virology*, **138**, 310–23.

PATERSON, R. G., HARRIS, T. J. R. & LAMB, R. A. (1984b). Fusion protein of the paramyxovirus simian virus 5: nucleotide sequence of mRNA predicts a highly hydrophobic glycoprotein. *Proceedings of the National Academy of Sciences, USA*, **81**, 6706–10.

PELUSO, R. W., LAMB, R. A. & CHOPPIN, R. W. (1977). Polypeptide synthesis in simian virus 5-infected cells. *Journal of Virology*, **23**, 177–87.

PENNINGTON, T. H. & PRINGLE, C. R. (1978). Negative strand viruses in enucleate cells. In *Negative Strand Viruses and the Host Cell*, ed. B. W. J. Mahy & R. D. Barry, pp. 451–64. New York: Academic Press.

PIRIE, H. M., PETRIE, L., PRINGLE, C. R., ALLAN, E. M. & KENNEDY, G. J. (1981). Acute fatal pneumonia in calves due to respiratory syncytial virus. *Veterinary Record*, **108**, 411–16.

PRINCE, G. A., HEMMING, V. G., HORSWOOD, R. L. & CHANOCK, R. M. (1985a). Immunoprophylaxis and immunotherapy of respiratory syncytial virus infection in the cotton rat. *Virus Research*, **3**, 193–206.

PRINCE, G. A., HORSWOOD, R. L., CAMARGO, E., KOENIG, D. & CHANOCK, R. M. (1983). Mechanisms of immunity to respiratory syncytial virus in cotton rats. *Infection and Immunity*, **42**, 81–7.

PRINCE, G. A., HORSWOOD, R. L. & CHANOCK, R. M. (1985b). Quantitative aspects of passive immunity to respiratory syncytial virus infection in infant cotton rats. *Journal of Virology*, **55**, 517–20.

PRINCE, G. A., JENSON, A. B., HEMMING, V. G., MURPHY, B. R., WALSH, E. E.,

HORSWOOD, R. L. & CHANOCK, R. M. (1986). Enhancement of respiratory syncytial virus pulmonary pathology in cotton rats by prior intramuscular inoculation of formalin-inactivated virus. *Journal of Virology*, **57**, 721–8.

PRINCE, G. A., POTASH, L. & HORSWOOD, R. L. (1979). Intramuscular inoculation of live respiratory syncytial virus induces immunity in cotton rats. *Infection and Immunity*, **23**, 723–8.

PRINGLE, C. R. (1977). Enucleation as a technique in the study of virus–host interactions. *Current Topics in Microbiology and Immunology*, **76**, 49–82.

PRINGLE, C. R. & CROSS, A. (1978). Neutralization of respiratory syncytial virus by cat serum. *Nature*, **276**, 501–2.

PRINGLE, C. R. & EGLIN, R. P. (1986). Murine pneumonia virus: seroepidemiological evidence of widespread human infection. *Journal of General Virology*, **67**, 975–82.

PRINGLE, C. R. & PARRY, J. E. (1980). Location and quantitation of antigen on the surface of virus-infected cells by specific bacterial adherence and scanning electron microscopy. *Journal of Virological Methods*, **1**, 61–75.

PRINGLE, C. R., SHIRODARIA, P. V., CASH, P., CHISWELL, D. J. & MALLOY, P. (1978). Initiation and maintenance of persistent infection by respiratory syncytial virus. *Journal of Virology*, **28**, 199–211.

PRINGLE, C. R., SHIRODARIS, P. V., GIMENEZ, H. B. & LEVINE, S. (1981). Antigen and polypeptide synthesis by temperature sensitive mutants of respiratory syncytial virus. *Journal of General Virology*, **54**, 173–83.

PRINGLE, C. R., WILKIE, M. L. & ELLIOTT, R. M. (1985). A survey of respiratory syncytial virus and parainfluenza virus type 3 neutralising and precipitating antibodies in relation to Paget's disease. *Journal of Medical Virology*, **17**, 377–86.

RAY, R. & COMPANS, R. W. (1986). Monoclonal antibodies reveal extensive antigenic differences between the hemagglutinin-neuraminidase glycoproteins of human and bovine parainfluenza 3 viruses. *Virology*, **148**, 232–6.

REANNEY, D. C. (1984). The molecular evolution of viruses. In *The Microbe*, 1984, I: *Viruses*, ed. B. W. J. Mahy & J. R. Pattison, pp. 175–96. Cambridge University Press.

REBEL, A., MALKANI, K., BASLE, M. & BREGEON, CH. (1977). Is Paget's disease of bone a viral infection? *Calcified Tissue Research*, **22** suppl., 283–6.

RICHARDSON-WYATT, L. S., BELSHE, R. B., LONDON, W. T., SLY, D. L., CAMARGO, E. & CHANOCK, R. M. (1981). Respiratory syncytial virus antibodies in non-human primates and domestic animals. *Laboratory Animal Science*, **31**, 413–15.

RIMA, B. K., BACZKO, K., CLARK, D. K., CURRAN, M. D., MARTIN, S. J., BILLETER, M. & TER MEULEN, V. (1986). Characterisation of clones for the sixth (L) gene and a transcriptional map for morbilliviruses. *Journal of General Virology*, **67**, 1971–8.

ROBERTS, N. J. JR (1982). Different effects of influenza virus, respiratory syncytial virus, and Sendai virus on human lymphocytes and macrophages. *Infection and Immunity*, **35**, 1142–6.

RUSSELL, W. C. & GOSWAMI, K. K. A. (1984). Antigenic relationships in the paramyxoviridae – implications for persistent infections in the central nervous system. In *Viruses and Demyelinating Diseases*, ed. C. A. Mims, M. L. Cuzner & R. E. Kelly, pp. 89–99. New York: Academic Press.

SATAKE, M., COLIGAN, J. E., ELANGO, N., NORRBY, E. & VENKATESAN, S. (1985). Respiratory syncytial virus envelope glycoprotein (G) has a novel structure. *Nucleic Acids Research*, **13**, 7795–812.

SATAKE, M. & VENKATESAN, S. (1984). Nucleotide sequence of the gene encoding respiratory syncytial virus matrix protein. *Journal of Virology*, **50**, 92–9.

SCHUBERT, M., HARMISON, G. G. & MEIER, E. (1984). Primary structure of the vesicular stomatitis virus polymerase (L) gene: evidence for a high frequency of mutations. *Journal of Virology*, **51**, 505.

SCHUY, W., GARTEN, W., LINDER, D. & KLENK, H-D. (1984). The carboxy-terminus of the haemagglutinin-neuraminidase of Newcastle disease virus is exposed at the surface of the viral envelope. *Virus Research*, **1**, 415–26.

SCOTT, R., SCOTT, M. & TOMS, G. L. (1981). Cellular and antibody responses to respiratory syncytial (RS) virus in human colostrum, maternal blood and cord blood. *Journal of Medical Virology*, **8**, 55–56.

SCOTT, R., SCOTT, M. & TOMS, G. L. (1985). Cellular reactivity to respiratory syncytial virus in human colostrum and breast milk. *Journal of Medical Virology*, **17**, 83–93.

SHESHBERADARAN, H. & NORRBY, E. (1984). Three monoclonal antibodies against measles virus F protein cross-react with cellular stress proteins. *Journal of Virology*, **52**, 995–9.

SHIODA, T., HIDAKA, Y., KANDA, T., SHIBUTA, H., NAMOTO, A. & IWASAKI, K. (1983). Sequence of 3687 nucleotides from the 3' end of Sendai virus genomic RNA and the predicted amino acid sequences of viral NP, P and C proteins. *Nucleic Acids Research*, **11**, 7317–31.

SHIODA. T., IWASAKI, K. & SHIBUTA, H. (1986). Determination of the complete nucleotide sequence of the Sendai virus genome RNA and the predicted amino acid sequences of the F, HN and L proteins. *Nucleic Acids Research*, **14**, 1545–63.

SINGER, F. R. (1980). Paget's disease of bone: a slow virus infection? *Calcified Tissue International*, **31**, 185–7.

SPRING, S. B. & TOLPIN, M. D. (1983). Enzymatic cleavage of a glycoprotein of respiratory syncytial virus. *Archives of Virology*, **76**, 359–63.

SRINIVASAPPA, J., SAEGUSA, J., PRABHAKAR, B. S., GENTRY, M. K., BUCHMEIER, M. J., WIKTOR, T. J., KOPROWSKI, H., OLDSTONE, M. B. A. & NOTKINS, A. L. (1986). Molecular mimicry: frequency of reactivity of monoclonal antibodies with normal tissues. *Journal of Virology*, **57**, 397–401.

STOTT, E. J. & TAYLOR, G. (1985). Respiratory syncytial virus: brief review. *Archives of Virology*, **84**, 1–52.

STOTT, E. J., THOMAS, L. H., TAYLOR, G., COLLINS, A. P., JEBBETT, J. & CROUCH, S. (1984). A comparison of three vaccines against respiratory syncytial virus in calves. *Journal of Hygiene*, **93**, 251–61.

SUFFIN, S. C., PRINCE, G. A., MUCK, K. B. & PORTER, D. D. (1979). Immunoprophylaxis of respiratory syncytial virus infection in the infant ferret. *Journal of Immunology*, **123**, 10–13.

TABER, L. H., KNIGHT, V., GILBERT, B. E., McCLUNG, H. W., WILSON, S. Z., NORTON, H. J., THURSON, J. M., GORDON, W. H., ATMAR, R. L. & SCHLAUDT, W. R. (1983). Ribavirin aerosol treatment of bronchiolitis associated with respiratory syncytial virus infection in infants. *Pediatrics*, **72**, 613–18.

TASHIRO, M. & HOMMA, M. (1983). Pneumotropism of Sendai virus in relation to protease mediated activation in mouse lungs. *Infection and Immunity*, **39**, 879–88.

TAYLOR, G., STOTT, E. J., BEW, M., FERNIE, B. F. & COTE, P. J. (1983). Monoclonal antibodies protect against respiratory syncytial virus. *Lancet*, **2**, 976.

TAYLOR, G., STOTT, E. J., BEW, M., FERNIE, B. F., COTE, P. J., COLLINS, A. P., HUGHES, M. & JEBBETT, J. (1984). Monoclonal antibodies protect against respiratory syncytial virus infection in mice. *Immunology*, **52**, 137–42.

TER MEULEN, V. & CARTER, M. J. (1982). Morbillivirus persistent infections in animals and man, In *Virus Persistence*, 33rd SGM Symposium, ed. B. W. J. Mahy, A. C. Minson & G. K. Darby, pp. 97–132. Cambridge University Press.

TORDO, N., POCH, O., KEITH, G. & ERMINE, A. (1985). Nucleotide sequence of the rabies Pasteur virus genome. *Virus Research Supplement*, **1**, 94.

TOWNSEND, A. R. M. & SKEHEL, J. J. (1984). The influenza A virus nucleoprotein gene controls the induction of both subtype specific and cross-reactive cytotoxic T cells. *Journal of Experimental Medicine*, **160**, 552–63.

TRUDEL, M., NADON, F., SEGUIN, C., GHOUBRIL, S., PAYMENT, P. & TREPANIER, P. (1986). Immunovirological studies on human respiratory syncytial virus structure proteins. *Canadian Journal of Microbiology*, **32**, 15–21.

VAINIONPAA, R., MEURMAN, O. & SARKINEN, H. (1985). Antibody response to respiratory syncytial virus structural proteins in children with acute respiratory syncytial virus infection. *Journal of Virology*, **53**, 976–9.

VARSANYI, T. M., JORNVALL, H. & NORRBY, E. (1985). Isolation and characterisation of the measles virus F1 polypeptide: comparison with other paramyxovirus fusion proteins. *Virology*, **147**, 110–17.

WALSH, E. E., BRANDRISS, M. W. & SCHLESINGER, J. J. (1985*a*). Purification and characterisation of the respiratory syncytial virus fusion protein. *Journal of General Virology*, **66**, 409–15.

WALSH, E. E. & HRUSKA, J. F. (1983). Monoclonal antibodies to respiratory syncytial virus proteins: identification of the fusion protein. *Journal of Virology*, **47**, 171–7.

WALSH, E. E., SCHLESINGER, J. J. & BRANDRISS, M. W. (1984*b*). Purification and characterisation of GP90, one of the envelope glycoproteins of respiratory syncytial virus. *Journal of General Virology*, **65**, 761–7.

WALSH, E. E., SCHLESINGER, J. J. & BRANDRISS, M. W. (1984*c*). Protection from respiratory syncytial virus infection in cotton rats by passive transfer of monoclonal antibodies. *Infection and Immunity*, **43**, 756–8.

WARD, K. A., LAMBDEN, P. R., OGLIVIE, M. M. & WATT, P. J. (1983). Antibodies to respiratory syncytial virus polypeptides and their significance in human infection. *Journal of General Virology*, **64**, 1867–76.

WELLIVER, R. C., KAUL, T. N., SUN, M. & OGRA, P. L. (1984). Defective regulation of immune responses in respiratory syncytial virus infection. *Journal of Immunology*, **33**, 1925–30.

WERTZ, G., ANDERSON, K., YOUNG, K. & BALL, A. (1985*a*). Expression of the fusion (F) protein of human respiratory syncytial virus from recombinant vaccinia virus vectors. *Virus Research Supplement*, **1**, 31.

WERTZ, G. W., COLLINS, P. L., HUANG, Y., GRUBER, G., LEVINE, S. & BALL, L. A. (1985*b*). Nucleotide sequence of the G protein gene of human respiratory syncytial virus reveals an unusual type of viral membrane protein. *Proceedings of the National Academy of Sciences, USA*, **82**, 4075–9.

WONG, D. T., ROSENBAND, M., HOVEY, K. & OGRA, P. L. (1985). Respiratory syncytial virus infection in immunosuppressed animals: Implications in human infection. *Journal of Medical Virology*, **17**, 359–70.

WRIGHT, P. F., BELSHE, R. B., KIM, H. W., VAN VORIS, L. P. & CHANOCK, R. M. (1982). Administration of a highly attenuated live respiratory syncytial virus vaccine to adults and children. *Infection and Immunity*, **37**, 397–400.

WRIGHT, P. F., MILLS, J. & CHANOCK, R. M. (1971). Evaluation of a temperature-sensitive mutant of respiratory syncytial virus in adults. *Journal of Infectious Diseases*, **124**, 505–11.

WRIGHT, P. F., SHINOZAKI, T., FLEET, W., SELL, S. H., THOMPSON, J. & KARZON, D. T. (1976). Evaluation of live attenuated respiratory syncytial virus vaccine in infants. *Journal of Pediatrics*, **88**, 931–6.

YEWDELL, J. W., BENNINK, J. R., SMITH, G. L. & MOSS, B. (1985). Influenza A virus nucleoprotein is a major target antigen for cross-reactive anti-influenza A virus cytotoxic T lymphocytes. *Proceedings of the National Academy of Sciences, USA*, **82**, 1785–9.

YOUNG, K. K. Y., HEINEKE, B. E. & WECHSLER, S. L. (1985). M protein instability and lack of H protein processing associated with nonproductive persistent infection of HeLa cells by measles virus. *Virology*, **143**, 536–45.

YURLOVA, T. I., KARPOVA, L. S. & KARPUKHIN, G. I. (1986). Dynamics of some genetic marker changes of respiratory syncytial virus strains circulating among nursery-school children. *Acta Virologica*, **30**, 45–50.

ZWEERINK, H. J., ASKONAS, B. A., MILLICAN, D., COURTNEIDGE, S. A. & SKEHEL, J. J. (1977). Cytotoxic T cells to type A influenza virus: viral haemagglutinin induces A-strain specificity while infected cells confer cross-reactivity. *European Journal of Immunology*, **7**, 630–5.

RABIES VIRUS AND ITS INTERACTION WITH THE CNS

ANNE FLAMAND*, PATRICE COULON*,
CHRISTOPHE PREHAUD*, ADAMA DIALLO*,
PAUL KUCERA† AND MICHEL DOLIVO†

*Laboratoire de Génétique des Virus, CNRS, 91190 Gif sur Yvette,
France
†Institut de Physiologie de la Faculté de Médecine, Université de
Lausanne, Ch-1011 Lausanne, Switzerland

Rabies disease is endemic over most of the world but human cases are almost exclusively found in developing countries. Most of the victims have been bitten by dogs. Virtually all wild and domestic mammals, such as dogs, cats, foxes, mongooses, vampire bats and others can be infected. Although it is theoretically possible to vaccinate domestic animals, field rabies will be almost impossible to eradicate. Therefore, it is still crucial to understand the molecular basis of rabies virulence. In addition the well-known neurotropism of the virus is a fascinating problem and careful investigations may provide valuable information on the function and properties of the CNS.

The infectious agent is a lyssavirus, which belongs to the rhabdovirus family. It is a bullet shaped, negative stranded RNA virus. The genome which codes for five proteins, has a molecular weight of 4.6×10^6 daltons (Sokol et al., 1969). It is tightly wrapped in a protein structure, the nucleocapsid, composed of three viral proteins: N, L and M1, surrounded by a lipid bilayer including viral proteins G and M2 and some contaminants of cellular origin. G is a glycoprotein, which protrudes as spikes visible at the surface of the virion. There are respectively around 1750, 900, 100, 1600 and 1900 molecules of N, M1, L, M2 and G proteins per virion (Madore & England, 1977; Dietzschold, Cox & Schneider, 1979). Studies performed on infected cell cultures have provided useful information on the characteristics of the rabies cycle which could possibly be generalized to the animal. Such results are summarized in the following paragraph, before describing what is known on the subject of this review.

THE RABIES CYCLE

What is known about the rabies cycle indicates that it shares general characteristics with other rhabdoviruses. Therefore, it can probably be described as follows, although several aspects have not been directly proven.

The first step is the penetration of the virus which certainly involves interactions between the viral glycoprotein and some receptor molecules present on external membranes of target cells. Whether the delivery of the nucleocapsid into the cytoplasm occurs by fusion of viral and cellular membranes or by pinocytosis of the virion in coated vesicles is still controversial.

Once the nucleocapsid is in the cytoplasm, primary transcription immediately follows. The transcriptase present in this structure (probably as an L, M1 complex) starts from the 3' end and sequentially transcribes the five viral genes according to their location on the genome (N M1 M2 G L) (Flamand & Delagneau, 1978). It is generally believed that the transcriptase stops at the end of each messenger and then reinitiates. Viral genes do not contain introns and the corresponding messengers, besides being methylated and polyadenylated (Ermine, 1979; Holloway & Obijeski, 1980), are not further processed and are immediately translated.

Replication follows the appearance of the viral proteins in sufficient amounts to allow the encapsidation of nascent molecules, giving rise to complete and encapsidated positive stranded copies of the genome. The second step of the replication consists in copying positive intermediates into complete negative molecules, which are also immediately encapsidated. The nascent chain and the template do not remain double stranded for more than a few nucleotides. Nucleocapsids accumulate in the cytoplasm in aggregates known as Negri's bodies.

G protein is probably delivered through internal membranes to the cell surface. It is glycosylated and associates as trimers (Wunner *et al.*, 1985). Viral maturation occurs when nucleocapsids come in contact with G-containing portions of the membrane. The role of M2 is not clear at the moment, but it is postulated that it could stand on the cytoplasmic side of the membrane and act as a linker between the nucleocapsid and the glycoproteins. As first described *in vitro* by Iwasaki, Wiktor & Koprowski (1973) rabies budding seems to occur on external and internal membranes which differentiates it from vesicular stomatitis virus (VSV).

The viral cycle has a mean duration of 24 hours in every cellular system so far investigated, including BHK 21 cells and its derivatives (BSR or CER cells), chick embryo fibroblasts, neuroblastoma cells and brain primary cultures from mouse or rat embryos. At this time, maximal viral production is observed and the cells contain viral inclusions, which can be easily visualized with fluorescent antinucleocapsid antibodies. Fluorescence is barely detectable before 15–18 hours after infection.

Compared to VSV, rabies has a slower cytopathic effect on the host cell. DNA and protein syntheses continue normally for 36 or 48 hours, and then decrease while cells round up and detach. After three days or so very few surviving cells are observed, which indicates that cellular metabolism has been impaired by the infectious process. The glycoprotein which integrates into external membranes and may change their intrinsic properties, could very well be responsible for such a deleterious effect. A partial answer to this question may be obtained from studies on some of our temperature-sensitive mutants which are affected in the glycoprotein gene. These do not kill the host cell at non-permissive temperature, although viral protein synthesis is apparently normal.

In the case of VSV, it has been postulated that the cytopathic effect is due to the leader RNA which is a 50 to 53 bases long copy of the 3' end of the genome (Giorgi, Blumberg & Kolakofski, 1983). The leader enters the nucleus and probably inhibits RNA polymerases II and III (McGowan, Emerson & Wagner, 1982; Grinnell & Wagner, 1984, 1985). The synthesis of a rabies leader RNA molecule and its migration into the nucleus has also been demonstrated (Kurilla et al., 1984). Since it does not affect cellular synthesis for at least 36 hours, its role remains obscure.

From the above, it is clear that the rabies cycle is predominantly cytoplasmic. There is no DNA intermediate in the replication process. Transcription and replication occur on a complex structure, the nucleocapsid. Naked RNA cannot be transcribed and is not infectious (Sokol et al., 1969). These facts have to be kept in mind to interpret results obtained in animals correctly.

THE RABIES DISEASE

In wild populations, the characteristics of the disease vary from animal to animal. In laboratory conditions, where the route of inoculation, the viral strain and the dose are controlled, the symptoms are very reproducible although dependent on the species. For instance, dogs most often develop the furious form of rabies while mice are paralyzed.

Rabbits, mice, rats, hamsters, foxes, dogs and cats have been chosen by laboratory workers, either because they are easy to handle and give reproducible results, or because they are the most important vectors of the disease in the field.

Many strains of rabies virus are available for experimentation. Some of them are the so-called 'fixed' strains which derived from a field isolate after several thousand successive passages in embryonated eggs (Flury LEP, HEP and Kelef) or in rabbit and/or mouse brain (SAD, ERA, CVS). Others are the 'field' or 'street' isolates with less than ten passages in mice or cell cultures (for instance the GS isolate which was obtained from a rabid fox (Dubreuil *et al.*, 1979)).

It is generally stated that the characteristics of the disease induced by fixed or field strains are different. This point is debatable: it is clear from the compilation of the published data that a strain is adapted to the species from which it has been isolated. For instance a strain isolated from a fox is far less pathogenic for dogs and cats than for foxes. This is an important characteristic which probably limits the horizontal extension of the epidemic. This should also be considered when choosing an experimental model: as a rule the strain should be adapted to the species which is to be studied. It is quite possible that the characteristics of the disease induced by CVS in mice or by GS in foxes would not be so different, despite the fact that CVS and GS give a quite different picture in mice.

Several fixed strains develop plaques on appropriate cell cultures under agarose or form foci of infected cells which can be revealed with fluorescent antinucleocapsid antibodies. The titre of the viral suspension can therefore be expressed in plaque forming units (PFU) or focus forming units (FFU). The virus can also be titrated by intracerebral inoculation of adult mice, leading to the definition of a titre in lethal dose 50 (LD 50), i.e. the dose of virus which kills 50% of the animals. When the comparison is possible, it is clear that 1 PFU or FFU is sufficient to kill a mouse by this route of inoculation. The infectiousness depends on the route of inoculation. The intramuscular route is a thousand times less efficient than the intracerebral route, while oral, peritoneal or intravenous administration of the virus requires 10^5 to 10^6 more infectious particles to kill. In suckling mice it is possible to titre some strains which are apathogenic for adult animals, for instance HEP, Kelef or avirulent mutants (see next chapter). Young animals are susceptible to all known strains of rabies virus, either because they lack an efficient immune response or because their immature neurones are permissive for strains otherwise restricted in mature neurones.

What are the general features of the disease? A considerable amount of data has been accumulated over many years by several groups of workers and has been reviewed by Murphy (1977), who contributed largely to the field. Further progress was delayed for several years due to the lack of precise information on the virus cycle *in vitro* (a gap which has now been filled), on the nature of the central and peripheral nervous system, and on the molecular mechanism of the interactions between the virus and the nerve endings. Current knowledge will be briefly summarized while more recent contributions from our laboratory will be discussed in more detail.

After intramuscular inoculation, the virus enters peripheral nerves, travels to the spinal cord and then to the brain. There it disseminates rapidly and within a few days it is widely distributed throughout the central and peripheral nervous systems. The infection induces a generalized encephalitis with drastic dysfunction of the neurones leading to paralysis and death. At the end of the infectious process the presence of the virus can be detected in several extra-neural tissues such as corneal cells, salivary gland epithelium, Auerbach and Meissner plexuses in many levels of the intestine, some chromaffin cells of the adrenal gland, a few exocrin acini of the pancreas, brown adipose tissue and myocardium (Murphy *et al.*, 1973). Most of these tissues are innervated by autonomic fibres.

The length of the incubation period depends on several factors, among which are (i) the size of the animal, probably because of the average distance the virus has to travel to enter and invade the brain, (ii) the site of inoculation, and (iii) the dose of virus.

In mice and rats we have observed an extensive loss of weight starting two or three days before death. For instance, an average loss of 2 g per animal per day (i.e. around 10% of the total weight) is currently observed with mice (Fig. 1). So far, we have no explanation for such a considerable loss of weight.

A dual role of the immune system is usually postulated (Smith *et al.*, 1982). Genetically or chemically induced immuno-deficient

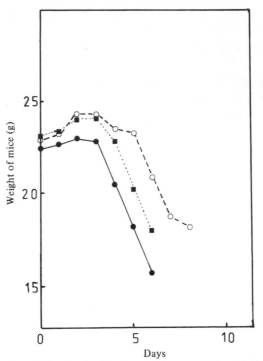

Fig. 1. Change in the weight of mice inoculated intramuscularly with CVS. Groups of five mice were inoculated on day zero in the masseter muscle with 2×10^6 (●——●), 2×10^5 (■······■) or 2×10^4 (○−−○) PFU of the CVS strain of rabies virus. The mice were weighed daily and their average weight calculated.

mice are more susceptible to rabies infection but they die later and with different symptoms from normal animals (Stochman *et al.*, 1973; Kaplan, Wiktor & Koprowski, 1975; Iwasaki, 1978; Miller *et al.*, 1978; Prabhakar, Fischman & Nathanson, 1981). Although undoubtedly protective at the beginning of the infection, the immune system seems to accelerate the onset of the fatal issue once the brain is invaded. The presence of complement and circulating antibodies could very well be deleterious for infected tissues. It has also been postulated that circulating antibodies are responsible for the early death phenomenon (Sikes *et al.*, 1971; Tignor *et al.*, 1974; Blancou, Andral & Andral, 1980; Prabhaker & Nathanson, 1981).

PENETRATION IN THE PERIPHERAL AND CENTRAL NERVOUS SYSTEM

Penetration of the virus has been followed after inoculation in the anterior chamber of the eye of adult rats as already described by

Dolivo (1980). This route has been chosen because it is delimited and because innervation of the eye is relatively well known. The propagation of the virus has been followed with fluorescent anti-nucleocapsid antibodies on serial sections of the eye, retrobulbar tissue containing the ciliary ganglion and brain being removed daily from infected animals and rapidly frozen.

Bright viral fluorescence was detected in a few cellular bodies of the ciliary ganglion as early as 24 hours after inoculation (Fig. 2a). Since it takes between 15 and 18 hours to detect viral fluorescence

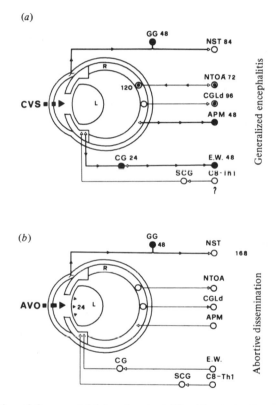

Fig. 2. Propagation of the parental (a) and mutant (b) rabies strains through the trigeminal (top), visual (centre), and autonomic (bottom) interconnections between the eye and brain. 4×10^5 PFU of CVS or avirulent mutant were injected in the anterior chamber of the eye. Symbols: open arrows, direction of synaptic transmission; closed arrows, direction of propagation of the virus; circles, peripheral and central neuronal somata infected primarily (closed), secondarily (dots), and not infected (open) at each interval of time, indicated in hours after inoculation. Abbreviations: CG, ciliary ganglion; CGLd, lateral geniculate body (dorsal part); C8-Th1, spinal preganglionary sympathetic neurons; GG, trigeminal ganglion of Gasser; L, eye lens; NTOA, terminal nuclei of the accessory optic system; NST, terminal trigeminal sensory nucleus; R, retina; SCG, superior cervical sympathetic ganglion; APM, pretectalis medialis area; EW, Edinger Westphal nucleus. Numbers indicate the time when viral fluorescence was first detected. Reprinted with permission from Kucera et al. (1985).

in infected cells (cf. first paragraph), this observation strongly suggests that the virus directly entered the corresponding parasympathetic nerve endings without intermediate cycle of multiplication in the muscle cells which they innervate (iris and ciliary bodies). The fact that the intensity of fluorescence in the infected cell bodies was similar also suggested that they were simultaneously infected. This all-or-none situation suggests that an infectious virus becomes rapidly unavailable at the site of inoculation either because it is trapped by surrounding cells or because it is inactivated in the aqueous humour. A similar situation has been found after intramuscular inoculation into the hind limbs. We did not find any viral multiplication in muscle cells at the site of inoculation (unpublished results), an observation which differs from that of Murphy & Bauer (1974) and Charlton & Casey (1979). Whether the discrepancy is due to the fact that these authors were working on new-born hamsters or adult skunks while we are studying adult mice is not known.

From results obtained in vitro one can postulate what happens once the nucleocapsid has entered the nerve endings and is passively transported to the cell body by the retrograde axoplasmic flow. The primary transcription for which only the presence of cellular ribonucleotides is necessary can probably start immediately but viral replication cannot proceed since the axon lacks the protein synthesizing machinery.

If this is true, the appearance of viral fluorescence in any cellular body is delayed by the time the viral nucleocapsid takes to travel along the axon. Since the speed of the transport by the retrograde axoplasmic flow has been estimated to be around 0.5 mm/h for the virus (Kucera et al., 1985), this delay could be calculated knowing the length of the axon. This is the reason why we consider that neurones from the Gasser ganglion (GG) and pretectalis medialis area (APM) which were found fluorescent only 48 hours after inoculation, were also primarily infected. The number of fluorescent cell bodies in these two nuclei was rather small, from ten to several tens; again fluorescence was comparable from cell to cell which confirms a synchronous evolution of viral infection.

As a second cycle of infection proceeded, the virus migrated from the ciliary ganglion to the Edinger Westphal nucleus (EW) where bright fluorescence could be detected 24 hours later, i.e. 48 hours after inoculation. From the EW and the APM nuclei, the virus continued to propagate into functionally related, more distant neurones. At 96 hours, practically all cerebral neurones were infected. Glial elements surrounding infected neurones did not become fluorescent. Reinfection of neighbouring neurones probably occurs at the synapses through viral maturation and readsorption of mature particles.

No fluorescence was ever detected in the superior cervical sympathetic ganglion, confirming the finding of Dolivo, Kucera & Bommeli (1982) and Tsiang, Derer & Taxi (1983) that the sympathetic route was not permissive for the virus. The virus also did not primarily infect the optic nerve.

It has been proposed that the entry of rabies virus could be mediated by the nicotinic acetylcholine receptor (Lentz et al., 1982). Our results suggest a more complex situation. Among the intraocular endings infected by the virus, some are indeed cholinergic but of a muscarinic type of transmission, thus involving a different receptor. In addition the virus is able to penetrate two other systems of nerve fibres, one efferent and one afferent. The nature of transmitters or membrane molecules in the pretectal fibres to the retina or in the trigeminal nerve remains to be determined. It could be speculated, of course, that these two systems might also be cholinergic. Alternatively, the virus could interact with other membrane components present only in some species of neurons or other cells.

In vitro studies suggest that rabies virus may attach to a variety of molecules present at the cell surface, such as phospho- or glycolipids (Superti, Derer & Tsiang, 1984). Other studies indicate that the cellular receptors for rabies are saturable (Wünner, Reagan & Koprowski, 1984). Pathogenic and apathogenic viruses differ in their ability to infect neuroblastoma cells in vitro (Dietzschold et al., 1985). One explanation is that the two viruses attach to different species of cell surface receptors. Despite the fact that in vitro models do not necessarily reflect the complexity of the in vivo situation, one can probably assume that some light will come from a constant 'va et vient' between in vivo and in vitro systems.

Whatever the nature of the putative receptor, penetration is likely to involve its recognition by the viral glycoprotein. Since the first demonstration that it was possible to select avirulent mutants among survivors to neutralization with appropriate monoclonal antibodies (Coulon et al., 1982), it has been found that these mutants differ from the parental strain by a single point mutation on the glycoprotein involving an arginine in position 333 (Dietzschold et al., 1983; Seif et al., 1985). Arginine 333 is located within antigenic site III of the viral glycoprotein (Lafon, Wiktor & Macfarlan, 1983). A careful investigation of the properties of 58 antigenic mutants affected in site III has been undertaken by Seif *et al.* (1985). Mutants have been classified into five groups on the basis of the pattern of resistance to the four monoclonal antibodies which define this site. An amino acid substitution in the glycoprotein has been identified in each of 16 representative mutants as either lysine 330, arginine 333, asparagine 336, or isoleucine 338 (Fig. 3). Only substitutions

```
K F P I Y T I P D K L G P W S P I D I H H L R C P N N L V V E D E G C I N L S G F S Y M E L K V G Y      50

I S A I K V N G F T C T G V V T E A E T Y T L F V G Y V T T T F K R K H F R P T P D A C R A A Y N W      100

K M A G D P R Y E E S L Q N P Y P D Y H W L R T V R T T K E S L I I I S P S V T D L D P Y D K S L H      150

S R V F P S G K C S G I T V S S T Y C S T N H D Y T I W M P E N P R P G T P C D I F T N S R G K R A      200

S N G N K T C G F V D E R G L Y K S L K G A C R L K L C G V L G L R L M D G T W V A M Q T S D E T K      250

W C S P D Q L V N L H D F R S D E I E H L V V E E L V K K R E E C L D T L E S I M T T K S V S F R R      300

L S H L R K L V P G F G K A Y T I F N K T L M E A D A H Y R S V R T W N E T I P S K G C L K V G G R      350

C H P H V N G V F F N G I I L G P D D R V L I P E M Q S S L L R Q H M E L L E S S V I P L M H P L A      400

D P S T V F K E G D E A E D F V E V H L P D V Y K Q I S G V D L G L P N W G K(Y V L M T A G A M I G      450

L V L I F S L M T W C)R R A N R P E S K Q R S F G G T G G N V S V T S Q S G K V I P S W E S Y K S G      500

G E I R L                                                                                                505
```

Fig. 3. The amino acid composition of CVS and mutant mature glycoproteins. Amino acid residues are symbolized as follows: A, Ala; C, Cys; D, Asp; E, Glu; F, Phe; G, Gly; H. His; I, Ile; K, Lys; L, Leu; M, Met; N, Asn; P, Pro; Q, Gln; R, Arg; S, Ser; T, Thr; V, Val; W, Trp; Y, Tyr. The open circles designate the two putative glycosylation sites (NKT). The parentheses enclose the putative membrane-anchoring region. The sequence is identical to that published by Yelverton *et al.* (1983) except for amino acids 36 and 122 where those authors found T and V instead of I and L. Symbols designate amino acid residues which are substituted in antigenic mutants of site II (▲), or III (●) or in temperature sensitive mutant tsG1 (■).

at arginine 333 affect pathogenicity. Whether another amino acid could be substituted in position 333 without abolishing pathogenicity for adult animals is under investigation. The answer to this question should help to delineate the nature of the interaction between the virus and the receptor(s).

What are the consequences of a substitution of arginine 333 on the penetration of the virus into the peripheric nervous system? After intraocular inoculation of this mutant virus, a few neurones of the trigeminal Gasser ganglion became fluorescent within 48 hours (Fig. 2*b*). The terminal sensory nucleus did not seem to be infected, as already found for CVS, but a few fluorescent neurones were found later in the cerebellum and basal ganglion. These were probably invaded from the afferent trigeminal fibres. No other route of entry seems to be permissive for the mutant. The residual infection cleared up within three weeks and animals survived the inoculation. From considerations on the role of the glycoprotein in the viral cycle, one can assume that the lack of multiplication of the mutant in neurones, which are permissive for the CVS parental strain, is due to a lack of penetration, probably because the mutated glycoprotein no longer recognizes receptors of parasympathetic and retinopetal fibres. It must still, however, be able to recognize those of trigeminal fibres suggesting that these receptors are different. But this route of entry seems to be abortive for the virus.

Two other unexpected results for which there is no explanation have been obtained with our avirulent mutants; (i) they are able to infect the lens epithelium, which is not the case for CVS (Coulon *et al.*, 1983), and (ii) inoculation of high doses of the avirulent virus induces a transient loss of weight, similar to that observed with CVS, except that after six or seven days the animals recover while CVS infected animals die (Fig. 4).

Are there other regions of the glycoprotein which are implicated in virulence? The answer seems to be yes. For instance, several mutants affected in the major antigenic site of the glycoprotein (site II) have a reduced pathogenicity. A detailed analysis of this site has been undertaken in our laboratory. Preliminary results indicate that, different from site III, it is a non-linear complex structure, since we have several independent mutants which show amino acid substitution at positions 40, 42 or 198 (Fig. 3). Interestingly enough, amino acid 198 is located in the region of the glycoprotein which shows sequence homology with the snake toxins known as acetylcholine competitors (Lentz *et al.*, 1985). Corresponding mutants are slightly less pathogenic than the parental CVS (it takes ten times more infectious particles to kill adult mice after intramuscular inoculation). Although far less spectacular than mutants affected at arginine 333, these mutants certainly deserve further experimentation. They could reveal a different kind of interaction with the host and they could help to clarify the question of the relation between acetylcholine receptors and rabies.

Besides antigenic mutants, which are of considerable interest as a tool to delineate the multiple aspects of viral pathogenicity, we have isolated one temperature-sensitive mutant, TsG1, affected in the glycoprotein, which has lost its pathogenicity after intramuscular inoculation. The mutation in this case has been located at amino acid 132 (leucine → phenylalanine) (Fig. 3). As mentioned in the first section, this mutant has no cytopathic effect on infected cells at non-permissive or semi-permissive temperature. Its penetration

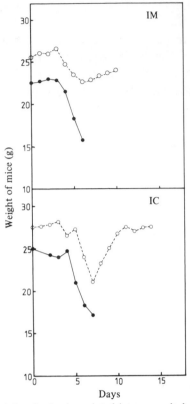

Fig 4. Change in the weight of mice inoculated intramuscularly or intracerebrally with the parental CVS or an avirulent mutant. A series of five mice were inoculated in the masseter muscle (IM) or intracerebrally (IC) with 4×10^6 PFU of CVS (●—●) or the avirulent mutant (○··○). The mice were weighed daily and their average weight calculated.

and propagation in the CNS will be the subject of future studies. It would be very interesting to have a mutant which was invasive without impairing the activity of the neurones.

Such studies will hopefully draw some light on certain ignored aspects of the rabies pathogenesis. They could also provide information on the nature of the molecules present on various types of nerve endings. In addition, investigation of unexplored aspects of the disease, such as loss of weight or penetration into the lens, may address some fundamental questions of physiology.

REFERENCES

BLANCOU, J., ANDRAL, B. & ANDRAL, L. (1980). A model in mice for the study of the early death phenomenon after vaccination and challenge with rabies virus. *Journal of General Virology*, **50**, 433–5.

CHARLTON, K. M. & CASEY, G. A. (1979). Experimental rabies in skunks. Immuno-fluorescence light and electron microscopic studies. *Laboratory Investigation*, **41**, 36–44.

COULON, P., ROLLIN, P., AUBERT, M., DOLIVO, M., KUCERA, P., PORTNOI, D., KITA, M. & FLAMAND, A. (1983). Molecular approach to virulence: isolation and characterization of avirulent mutants of rabies virus. In *Mechanisms of Viral Pathogenesis: From Gene to Pathogen*, ed. A. Kohn & P. Fuchs, pp. 201–16. Boston: M. Nijhoff.

COULON, P., ROLLIN, P., AUBERT, M. & FLAMAND, A. (1982). Molecular basis of rabies virus virulence. I. Selection of avirulent mutants of the CVS strain with anti-G monoclonal antibodies. *Journal of General Virology*, **61**, 97–100.

DIETZSCHOLD, B., COX, J. H. & SCHNEIDER, L. G. (1979). Rabies virus strains: a comparison study by polypeptide analysis of vaccine strains with different patho-genic patterns. *Virology*, **98**, 63–75.

DIETZSCHOLD, B., WIKTOR, T. J., TROJANOWSKI, J., RODERICK, I., MACFARLAN, R. I., WUNNER, W. H., TORRES-ANJEL, M. J. & KOPROWSKI, H. (1985). Differ-ences in cell-to-cell spread of pathogenic and apathogenic rabies virus *in vivo* and *in vitro*. *Journal of Virology*, **56**, 12–18.

DIETZSCHOLD, B., WUNNER, W. H., WIKTOR, T. J., LOPES, A. D., LAFON, M., SMITH, C. L. & KOPROWSKI, H. (1983). Characterization of an antigenic determi-nant of the glycoprotein that correlates with pathogenicity of rabies virus. *Pro-ceedings of the National Academy of Sciences, USA*, **80**, 70–4.

DOLIVO, M. (1980). A neurobiological approach to neurotropic viruses. *Trends in Neuroscience*, **3**, 149–52.

DOLIVO, M., KUCERA, P. & BOMMELI, W. (1982). Etude de la progression du virus rabique dans le système visuel du rat. *Comparative Immunology and Microbiology of Infectious Diseases*, **5**, 67–9.

DUBREUIL, M., ANDRAL, L., AUBERT, M. & BLANCOU, J. (1979). The oral vaccina-tion of foxes against rabies. An experimental study. *Annales de la Recherche vétérinaire*, **10**, 9–21.

ERMINE, A. (1979). Polyadenylation des ARN messagers du virus rabique. *Annales de Microbiologie (Institut Pasteur)*, **130B**, 227–34.

FLAMAND, A. & DELAGNEAU, J. F. (1978). Transcriptional mapping of rabies virus *in vivo*. *Journal of Virology*, **28**, 518–23.

GIORGI, C., BLUMBERG, B. & KOLAKOFSKI, D. (1983). Sequence determination of the (+) leader RNA regions of the vesicular stomatitis virus Chandipura, Cocal and Piry serotypes genomes. *Journal of Virology*, **46**, 125–30.

GRINNELL, B. W. & WAGNER, R. R. (1984). Nucleotide sequence and secondary structure of VSV leader RNA and homologous DNA involved in inhibition of DNA-dependent transcription. *Cell*, **36**, 533–43.

GRINNELL, B. W. & WAGNER, R. R. (1985). Inhibition of DNA-dependent transcrip-tion by the leader RNA of vesicular stomatitis virus: role of specific nucleotide sequences and cell protein binding. *Molecular and Cellular Biology*, **5**, 2502–13.

HOLLOWAY, B. P. & OBIJESKI, J. F. (1980). Rabies virus-induced RNA synthesis in BHK21 cells. *Journal of General Virology*, **49**, 181–95.

IWASAKI, Y. (1978). Experimental virus infections in nude mice. In *The Nude Mouse in Experimental and Clinical Research*, ed. J. Fogh & B. Giovanella, pp. 457–75. New York: Academic Press.

IWASAKI, Y., WIKTOR, T. J. & KOPROWSKI, H. (1973). Early events of rabies virus replication in tissue cultures. An electron microscopy study. *Laboratory Investiga-tion*, **28**, 142–8.

KAPLAN, M. M., WIKTOR, T. J. & KOPROWSKI, H. (1975). Pathogenesis of rabies in immunodeficient mice. *Journal of Immunology*, **114**, 1761–5.

Kucera, P., Dolivo, M., Coulon, P. & Flamand, A. (1985). Pathways of the early propagation of virulent and avirulent rabies strains from the eye to the brain. *Journal of Virology*, **55**, 158–62.

Kurilla, M. G., Cabradilla, C. D., Holloway, B. P. & Keene, J. D. (1984). Nucleotide sequence and host La protein interactions of rabies virus leader RNA. *Journal of Virology*, **50**, 773–8.

Lafon, M., Wiktor, T. J. & Macfarlan, R. I. (1983). Antigenic sites on the CVS rabies virus glycoprotein: analysis with monoclonal antibodies. *Journal of General Virology*, **64**, 843–51.

Lentz, T. L., Burrage, T. G., Smith, A. L., Crick, J. & Tignor, G. H. (1982). Is the acetylcholine receptor a rabies virus receptor? *Science*, **215**, 182–4.

Lentz, T. L., Wilson, P. T., Hawrot, E. & Speicher, D. W. (1985). Amino acid sequence similarity between rabies virus glycoprotein and snake curaremimetic neurotoxins. *Science*, **226**, 847–8.

McGowan, J. J., Emerson, S. U. & Wagner, R. R. (1982). The plus-strand leader RNA of VSV inhibits DNA-dependent transcription of adenovirus and SV40 genes in a soluble whole cell extract. *Cell*, **28**, 325–33.

Madore, H. P. & England, J. M. (1977). Rabies virus protein synthesis in infected BHK21 cells. *Journal of Virology*, **22**, 102–12.

Miller, A., Morse, H. C., Winkelstein, J. & Nathanson, N. (1978). The role of antibody in recovery from experimental rabies. I. Effect of depletion of B and T cells. *Journal of Immunology*, **121**, 321–6.

Murphy, F. A. (1977). Rabies pathogenesis. *Archives of Virology*, **54**, 279–97.

Murphy, F. A. & Bauer, S. P. (1974). Early street rabies virus infection in striated muscle and later progression to the central nervous system. *Intervirology*, **3**, 256–68.

Murphy, F. A., Harrison, A. K., Winn, W. C. & Bauer, S. P. (1973). Comparative pathogenesis of rabies and rabies-like viruses. Infection of the central nervous system and centrifugal spread of virus to peripheral tissues. *Laboratory Investigation*, **29**, 1–6.

Prabhakar, B. S., Fischman, H. R. & Nathanson, N. (1981). Recovery from experimental rabies by adoptive transfer of immune cells. *Journal of General Virology*, **56**, 25–31.

Prabhakar, B. S. & Nathanson, N. (1981). Acute rabies death mediated by antibody. *Nature (London)*, **290**, 590–1.

Seif, I., Coulon, P., Rollin, P. & Flamand, A. (1985). Rabies virulence: effect on pathogenicity and sequence characterization of rabies virus mutations affecting the antigenic site III of the glycoprotein. *Journal of Virology*, **53**, 926–34.

Sikes, R. K., Cleary, W. F., Koprowski, H., Wiktor, T. J. & Kaplan, M. M. (1971). Effective protection of monkeys against death from street virus by post-exposure administration of tissue culture rabies vaccine. *Bulletin of WHO*, **45**, 1–11.

Smith, J., McClelland, C. L., Reid, F. L. & Baer, G. M. (1982). Dual role of the immune response in street rabies virus infection of mice. *Infection and Immunity*, **35**, 213–21.

Sokol, F., Schlumberger, H. D., Wiktor, T. J. & Koprowski, H. (1969). Biochemical and biophysical studies on the nucleocapsid and on the RNA of rabies virus. *Virology*, **38**, 651–65.

Stockman, G. D., Heem, L. R., South, M. A. & Trentin, J. J. (1973). Differential effects of cyclophosphamide on the B and T cell compartments of adult mice. *Journal of Immunology*, **110**, 277–82.

Superti, F., Derer, M. & Tsiang, H. (1984). Mechanism of rabies virus entry into CER cells. *Journal of General Virology*, **65**, 781–9.

TIGNOR, G. H., SHOPE, R. E., GERSHON, R. K. & WAKSMAN, B. H. (1974). Immuno-pathologic aspects of infection with Lagos bat virus of the rabies serogroup. *Journal of Immunology*, **112**, 260–5.

TSIANG, H., DERER, M. & TAXI, J. (1983). An *in vivo* and *in vitro* study of rabies virus infection of the rat superior vervical ganglia. *Archives of Virology*, **76**, 231–43.

WUNNER, W. H., DIETZSCHOLD, B., SMITH, C. L., LAFON, M. & GOLUB, E. (1985). Antigenic variants of CVS rabies virus with altered glycosylation sites. *Virology*, **140**, 1–12.

WUNNER, W. H., REAGAN, K. J. & KOPROWSKI, H. (1984). Characterization of saturable binding sites for rabies virus. *Journal of Virology*, **50**, 691–7.

YELVERTON, E., NORTON, S., OBIJESKI, J. F. & GOEDDEL, D. V. (1983) Rabies virus glycoprotein analogs: biosynthesis in *Escherichia coli*. *Science*, **219**, 614–20.

GENETIC STUDIES OF THE ANTIGENICITY AND THE ATTENUATION PHENOTYPE OF POLIOVIRUS

AKIO NOMOTO* AND ECKARD WIMMER†

*Department of Microbiology, Faculty of Medicine
University of Tokyo, Tokyo, Japan
†Department of Microbiology, School of Medicine
SUNY at Stony Brook, Stony Brook, NY, USA

INTRODUCTION

Poliomyelitis, an acute disease of the central nervous system caused by poliovirus, remains a serious health problem throughout the world. Estimates of the World Health Organization indicate that the incidence of persisting paralytic poliomyelitis in developing countries in 1983 was 400 000 cases, but the actual number may have been much larger. Only in countries where the existing vaccines are regularly administered is the disease a rarity, and, as we will see, it is usually the result of immunization with the live, oral vaccine. However, occasional outbreaks of poliomyelitis even in developed countries, albeit rare and limited in number, are evidence of the possibility of invasion by neurovirulent strains. Such an outbreak occurred in Finland in 1984/1985 (Centers of Disease Control (USA), 1986a) and was possibly due to a gradual decline in vaccination coverage and/or to a lower potency of a component of the vaccine used in preceding years.

Two vaccines are currently being used to control poliomyelitis. One is inactivated virus (Salk, 1960); the other, attenuated virus strains of which those of A. B. Sabin were found to be most effective (Melnick, 1984; Sabin & Boulger, 1973). The virtues of both vaccines have been widely recognized and will not be discussed here (see Horstmann, Quinn & Robbins, 1984). We shall review, however, the antigenic properties of poliovirus in regard to the induction of neutralizing antibodies by virions and the recent studies aimed at understanding the molecular basis of the attenuation phenotype.

Poliovirus belongs to the family of Picornaviridae, one of the largest groups of human pathogens. Members of two genera of the family, the human enteroviruses (72 serotypes) and the human rhinoviruses (HRV; over 100 serotypes), cause a bewildering array of disease syndromes, the molecular basis of which is unknown. An important determinant in the development of disease may be the characteristic tissue tropism of different picornaviruses, which appears to be governed by the accessibility of specific cellular receptors. Virulence, on the other hand, may be determined by genetic traits of the viral strain. Relatively little is known about receptors, tissue tropism and disease, but this fascinating field is rapidly developing (Crowell *et al.*, 1985; Tomassini & Colonno, 1986; Mendelsohn *et al.*, 1986).

Picornaviruses are naked (non-enveloped) particles composed of four capsid proteins (VP1, VP2, VP3 and VP4) and a protein-linked, polyadenylybated, single-stranded RNA (7500 N) of plus-strand polarity. The chemical (Kitamura *et al.*, 1981) and crystal structures have been elucidated for HRV and poliovirus (Rossmann *et al.*, 1985; Hogle *et al.*, 1985). Since the genetic map position and the amino-acid sequences of all known gene products of these biological agents is also known, picornaviruses are among the best characterized viruses (Kitamura *et al.*, 1981; Rueckert, 1985).

Poliovirus, which does not share its cellular receptor with any other known picornavirus, occurs in three serotypes. The nucleotide sequences of the genomes of the different poliovirus types and also of the three attenuated Sabin strains have been determined (Kitamura *et al.*, 1981; Toyoda *et al.*, 1984; Stanway *et al.*, 1984; La Monica, Meriam & Racaniello, 1986, and ref. therein). Amino- and carboxy-terminal sequence analyses of the known viral proteins (Semler *et al.*, 1981a, b; Larsen *et al.*, 1982; Dorner *et al.*, 1982; Emini *et al.*, 1982) have led to a precise genetic map shown for polio 1 in Fig. 1. All virus-specific proteins arise from proteolytic cleavage of the polyprotein (Nicklin *et al.*, 1986). The capsid polypeptides map at the N-terminus (P1 region); the rest of the polyprotein are the replication proteins (P2 and P3 region) of which 2A and 3C are proteinases and 3D is the RNA polymerase (Kuhn & Wimmer, 1986). The open reading frame for the polyprotein is preceded by a long, untranslated region of 742 nucleotides (Dorner *et al.*, 1982) whose function is not known. As we shall see, however, the primary sequence of this region influences the attenuation phenotype of poliovirus remarkably.

An important development in the molecular biology of poliovirus was the construction of cDNA clones which, upon transfection of mammalian cells, produce authentic poliovirus (Racaniello & Baltimore, 1981; Semler, Dorner & Wimmer, 1984; Omata *et al.*, 1984).

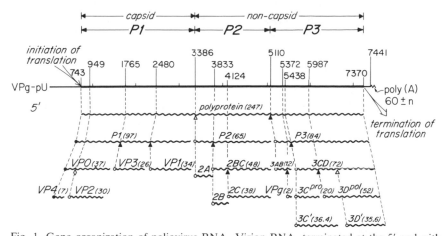

Fig. 1. Gene organization of poliovirus RNA. Virion RNA, terminated at the 5' end with the genome-linked protein VPg and at the 3' end with poly(A), is shown as a solid line, the translated region being more pronounced than the non-coding regions. The numbers above the virion RNA refer to the first nucleotide of the codon specifying the N-terminal amino acid for the viral specific proteins. The coding region has been divided into three regions (P1, P2, P3), corresponding to rapid cleavages of the polyprotein. The newly adopted nomenclature of polypeptides is according to Rueckert & Wimmer (1984). Numbers in parentheses are calculated molecular weights. Open circles indicate that the terminal amino acid has been experimentally determined. The N-termini are glycine in all cases except for VP2 where it is serine. The C-terminal amino acid of 3D is phenylalanine. Closed circles indicate that the N-termini are known to be blocked. Closed triangles: Gln–Gly pairs that are cleaved during proteolytic processing of a polypeptide by the virus-encoded proteinase 3C. Open triangles: Tyr–Gly pairs cleaved by viral proteinase 2A. Open diamond: Asn–Ser pair cleaved only during morphogenesis. Polypeptides 3C' and 3D' are products of an alternative cleavage, the biological significance of which is unknown. (Modified after Kitamura et al., 1981)

Such cDNA clones were recently engineered into suitable transcription vectors such that virtually unlimited amounts of highly infectious poliovirus RNA can be synthesized *in vitro* with coliphage T7 RNA polymerase (van der Werf *et al.*, 1986).

ANTIGENIC STRUCTURE OF POLIOVIRUS

Poliovirus elicits a strong immune response in mammals. Antibodies directed to the surface of the virion may bind to the particle without any influence on infectivity, or they may bind *and* neutralize the virus. We will call the determinants involved in neutralization 'neutralization antigenic sites' (N-Ag). N-Ags of poliovirus (also termed D-antigen) are sensitive to denaturation. Incubation of poliovirus at elevated temperature (56 °C) or at elevated pH leads to the loss of D-antigen; instead, the particle assumes a structure expressing 'H' (or 'C')-antigen specific determinants that will neither elicit nor bind neutralizing antibodies (for a review, see Emini, Jameson & Wimmer, 1985).

Mapping of the N-Ags to specific regions of the capsid polypeptides of poliovirus and to the tertiary folds of the crystal structure of the virion has been achieved during the last four years, and the trail of this investigation proved full of surprises (reviewed by Wimmer, Emini & Diamond, 1986). As with other viral systems, notably influenza virus, the most reliable approach appeared to be the sequence analysis of genomic RNA of neutralization escape mutants that had been selected by neutralizing monoclonal antibodies (N-mcAb). Large panels of the N-mcAbs to polio 1 and polio 3 were isolated by the standard procedure of immunizing BALB/c mice and fusing the mouse spleen cells to non-secretor mouse myeloma cells (Köhler & Milstein, 1975). Whereas the N-mcAbs to polio 3 fell into one group which recognized predominantly a single N-Ag with overlapping neutralization epitopes (N-Ep) (Minor et al., 1983, 1985, and ref. therein), the N-mcAbs to polio 1 could be divided into six groups corresponding to N-Eps which could function independently (Emini et al., 1983; Crainic et al., 1983). Mutations of the non-neutralizable variants of polio 3, selected with N-mcAbs, nearly all mapped to amino acids 90–103 in VP1 (Minor et al., 1985; and ref. therein) a region that we will call N-AgI (Wimmer et al., 1986). On the other hand, none of the mutations in neutralization-resistant variants of polio 1, selected by the corresponding N-mcAbs, mapped to N-AgI (Diamond et al., 1985). Instead, the mutations in polio 1 seemed to be scattered about in all three large capsid polypeptides (Fig. 2A). An exceptional N-mcAb (C3) to polio 1 was isolated by Blondel et al. (1983) using heat-inactivated virus; this antibody can recognize denatured VP1 of polio 1 and was found to bind to N-AgI (van der Werf et al., 1983). Accordingly, N-mcAb C3 selected neutralization-resistant mutants which map to the region 93–103 of VP1 (Blondel et al., 1986) (filled box in Fig. 2A). Thus N-AgI can function in polio 1 as a neutralization antigenic site.

The solution of the crystal structure of HRV 14 and polio 1 (Rossmann et al., 1985; Hogle et al., 1985) then revealed that all mutations found by Diamond and his colleagues in the polio 1 capsid proteins are located at the surface of the virion and cluster into three distinct sites (filled, hatched and stippled areas in Fig. 2B). These regions probably represent distinct antibody binding sites. The filled areas in Fig. 2B are the N-AgI sites. The mutations at 222–224 in VP1, the hatched box in VP2 and the mutations at 270 in VP2 (Fig. 2A) comprise N-AgII (hatched regions in Fig. 2B), and the stippled box and mutations in VP3 (Fig. 2A) comprise N-AgIII (stippled regions in Fig. 2B). Because of the properties of the N-mcAbs used to select the neutralization escape mutants, however, we favour the possibility that the $T > M$ mutation belongs to N-AgIII (Wimmer et al., 1986).

Poliovirus, therefore, has three major N-Ags, of which one (N-AgI) appears to be sequential whereas the other two may be discontinuous. In spite of the insight gained by the crystal structure, the paradox remained as to why the immune system of the BALB/c mouse recognizes mainly N-AgI when confronted with polio 3 (and polio 2; P. D. Minor personal communication), but recognizes N-AgII and N-AgIII when confronted with polio 1 (see a discussion by Wimmer, Jameson & Emini, 1984). The paradox was solved when it was found that the strong preponderance of one N-Ag over others is the property of the immune system of the inbred mouse (Icenogle et al., 1986). That is, N-mcAbs isolated from immunized BALB/c mice show the remarkable preference to N-AgI of polio 2 and polio 3 and to N-AgII and N-AgIII of polio 1, whereas polyclonal sera from outbred animals (such as the horse and monkey) or sera from vaccinated children contain N-Abs that recognize N-AgII, N-AgIII and, to a much lesser extent, N-AgI regardless of the serotype (Icenogle et al., 1986). Considering the close relationship of the poliovirus serotypes in regard to primary and tertiary structure, Icenogle's results came somewhat as a relief since the different humoral responses to poliovirus serotypes in a normal (outbred) mammal simply made no sense.

THE ATTENUATION PHENOTYPE OF POLIOVIRUS

When you and other molecular biologists speak of 'neurovirulence' and 'attenuation' of poliovirus – properties on which I have spent many years of my life to elucidate – I fail to note any appreciation of the tremendous quantitatively graded spectrum of properties that is involved – properties which differ in different species, neurons and other cells.

So wrote Dr Albert Sabin to Dr Olen Kew in February of 1982 in an illuminating letter directed to all of us molecular biologists who had begun to study the molecular basis of pathogenicity of viruses. Five years later a wealth of data has been accumulated on poliovirus structure and function and yet we have no clue why the

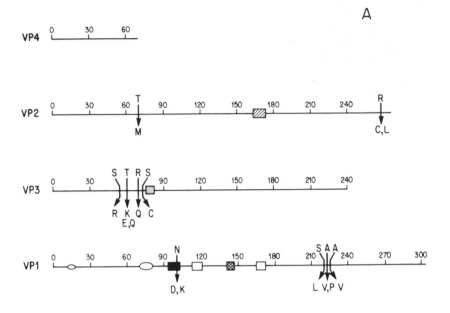

A

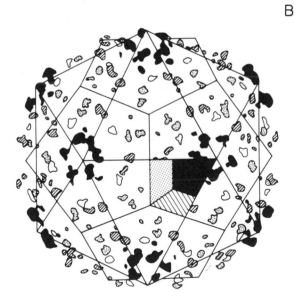

B

Sabin strains grow seemingly well in the gastrointestinal tract but cause no disease of the central nervous system.

Passage of virus in cells related, but not identical, to the virus' original host cells may lead to the selection of viral sub-populations that are adapted to growing efficiently in the new cells. Such adapted sub-populations usually arise as the result of spontaneous mutation in the viral genome. Occasionally, after many cell passages, a new strain emerges from the adapted sub-population which proves to be less virulent for the original host organism. This process is known as attenuation (Fenner et al., 1974).

The attenuated poliovirus strains now in use as vaccines were also progeny virus of adapted sub-populations following numerous passages of virulent strains in Simian extraneural cells and tissues. The type 2 vaccine strain, however, is different in that the parental strain was already intracerebrally avirulent for monkeys (Sabin & Boulger, 1973). The journey from the virulent type 1 (Mahoney) and type 3 (Leon) strains to the attenuated vaccine strains, on the other hand, has been precisely recorded (Sabin & Boulger, 1973). A. Sabin has pointed out, however, that passages per se in non-human tissue culture cells were not sufficient to obtain 'spontaneous mutants' of lowest neurovirulence for monkey spinal neurons; rather, it was crucial at what stage of the infectious cycle the virus was isolated from the monkey kidney epithelial cells (Sabin, 1955). This observation underlines the complexity of the process of selection. At the present time, it is clearly impossible to determine those factors that play a role in the isolation of useful vaccine strains.

Regardless of how the three Sabin vaccine strains were selected,

Fig. 2. Antigenic architecture of poliovirus, type 1 (Mahoney). A. Shown are the four capsid polypeptides of which VP4 is not exposed to the outside of the virion. The solid, hatched, stippled and cross-hatched boxes represent amino acid sequences of which the corresponding synthetic peptides have been found to induce a neutralizing immune response (Wimmer et al., 1986). The solid, hatched and stippled regions belong to N-AgI, N-AgII and N-AgIII, respectively. Also shown are amino acid substitutions found in diverse N-mcAb-resistant variants (Diamond et al., 1985; Blondel et al., 1986). Their relationship to the antigenic sites is as follows: (i) Mutations N > D, K at position 100 of VP1 belong to N-AgI; (ii) mutations at 222–224 of VP1 and at 270 (R > C, L) in VP2 belong to N-AgII; mutations at 58–60 and 71–73 in VP3 belong to N-AgIII. The mutation at 72 (T > M) of VP2 most likely also belongs to N-AgIII (Wimmer et al., 1986). B. Location of the major antigenic sites on the surface of the poliovirion. The small, solid (N-AgI), hatched (N-AgII) and stippled (N-AgIII) regions correspond to those described in A. Several other neutralization escape mutations, not shown in A, contributed to the assignment of the N-Ag (Hogle et al., 1985). The large, solid, hatched and stippled areas are VP1, VP2 and VP3, respectively, which are components of one protomer. (This figure is reproduced from Wimmer et al., 1986, and is based upon computer graphics prepared by A. J. Olsen and J. M. Hogel, Research Institute of Scripps Clinics.)

their properties have now been extensively characterized. The attenuation phenotype has been tested in experimental animals for signs of paralysis, or for the development and spread of lesions in neuronal tissue ('lesion score' and 'spread value') after intracerebral or intraspinal injection. In addition, the Sabin strains differ from their neurovirulent progenitor strains in a large number of other biological characteristics. Some of these biological characteristics are used as *in vitro* markers to analyse the qualities of oral live vaccines (Nakano *et al.*, 1978; WHO Tech. Rep., 1983). These include (i) the sensitivity of viral multiplication to elevated temperature (*rct* marker), (ii) the sensitivity to low concentration of bicarbonate under agar overlay (*d* marker) and (iii) the size of plaques produced in infected monolayers of primate cells.

GENOME STRUCTURE OF THE ATTENUATED VIRUSES

The difference in the biological properties between neurovirulent and attenuated strains must be the result of an altered genomic sequence, i.e. point mutations, deletions or insertions. Indeed, such differences were first observed by fingerprint analyses of RNase T1 or RNase A resistant oligonucleotides (see Fig. 3; an exhaustive treatment of the method of fingerprinting and its application has been published by Kew, Nottay & Obijeski, 1984). This powerful method was applied first to picornaviruses by Frisby *et al.* (1976) and Lee & Wimmer (1976). It was later used to characterize all three types of poliovirus (Lee *et al.*, 1979) and to analyse different isolates of a single serotype (Minor, 1980; Kew *et al.*, 1981). The first identification and mapping of point mutations in the genome of the Sabin 1 vaccine strain was published by Nomoto *et al.* (1979; 1981). It was then estimated that approximately 35 mutations may have accumulated during the process of selecting the Sabin 1 strain (Nomoto *et al.*, 1981). This number correlated well with an estimate by Kew *et al.* (1980) who had compared the four coat proteins of polio 1 (Mahoney) and Sabin 1 by tryptic peptide mapping. The number fell short, however, by some 20 additional point mutations. A precise assessment of the extent of genetic variation between the viral strains was achieved by the elucidation of the complete nucleotide sequence of polio 1 (Mahoney) and Sabin 1 genomes (Kitamura *et al.*, 1981; Racaniello & Baltimore, 1981*a*; Nomoto *et al.*, 1982). These studies revealed 55 nucleotide substitutions (Nomoto *et al.*, 1982; Toyoda *et al.*, 1984), within a total of 7441 heteropolymeric

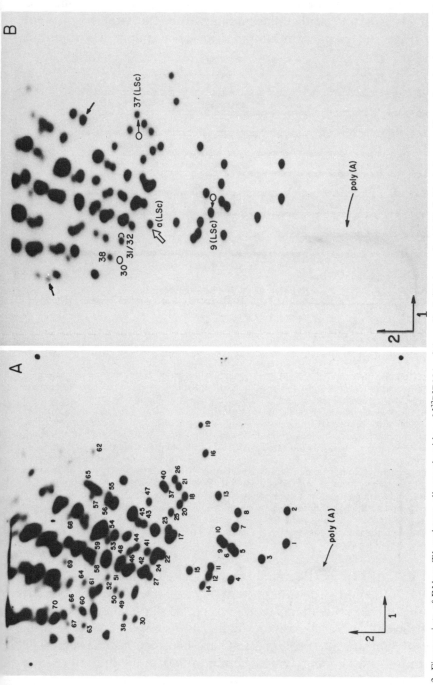

Fig. 3. Fingerprints of RNase T1-resistant oligonucleotides of A. Polio 1 (Mahoney) and B. Sabin 1, by two-dimensional polyacrylamide gel electrophoresis. Open circles: spots missing in B as compared to A; open circle with arrow: shift of a spot due to a pyrimidine transition; open arrow: new spot in B that is absent in A. Spot 31/32 is a mixture of two spots (Lee *et al.*, 1979), one of which is missing in Sabin 1 RNA. The two solid arrows indicate new spots not further characterized. Reproduced from Nomoto *et al.* (1981) with permission.

bases, that are scattered over the entire length of the genome and result in 21 amino acid replacements within the viral polyprotein (Fig. 4). A cluster of seven amino acid replacements was found in

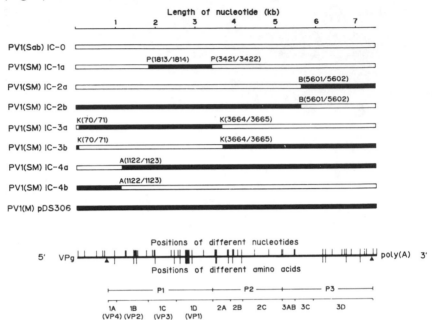

Fig. 4. Genome structure of recombinant type 1 polioviruses and locations of nucleotide and amino acid differences between the Mahoney and Sabin 1 strains. The expected genome structures of the recombinant viruses are shown by the combination of Sabin 1 (□) and Mahoney (■) sequences. K, A, P, and B represent cleavage sites of the restriction enzymes KpnI, AatII, PstI, and BglII, respectively. Numbers in parentheses following the restriction sites indicate nucleotide positions from the 5' end of the viral genome. Length of the entire genome of poliovirus type 1 is indicated at the top of the figure in kilobases (kb) from the 5' terminus. Genomic RNA and its gene organization are shown at the bottom of the figure by using a recently adopted nomenclature (Rueckert & Wimmer, 1984). VPg is a small protein covalently attached to the 5' end of the genome; poly(A) is 3'-terminal. The positions of initiation and termination(▲) of viral polyprotein synthesis are indicated. The locations of nucleotide and amino acid differences between the Mahoney and Sabin 1 strains are indicated by lines over and under the genome RNA, respectively. A total of 55 nucleotide differences exist between genomes of the Mahoney and Sabin 1 strains. Reproduced from Omata et al. (1986) with permission.

VP1; VP2, VP3 and VP4 contain two, two and one mutations, respectively. Of the non-structural proteins the YG-specific proteinase 2A contains three, 2B contains two, and the RNA polymerase 3D four mutations. Amino acid replacements in 2C, 3AB and the glutamine–glycine-specific proteinase 3C are conspicuously absent. The 3'-terminal non-coding region contains two point mutations, the 5'-terminal non-coding region five point mutations. Insertions or deletions have not been observed. Considering the compacting of genetic information and the peculiar mechanism of poliovirus gene expression, this is not surprising (Toyoda *et al.*, 1986).

A similar sequence study was performed by J. W. Almond and his colleagues who determined the primary genome structure of the neurovirulent poliovirus type 3 (Leon) and its attenuated Sabin 3 derivative (Stanway *et al.*, 1983, 1984). At the same time Toyoda *et al.* (1984) had sequenced the genomes of all three Sabin vaccine strains and published a comparison of the genome structures of all three poliovirus serotypes.

The startling result of the studies on polio 3 was that the Sabin 3 strain had accumulated only ten point mutations of which a mere three cause amino acid replacements (Fig. 5). Two point mutations

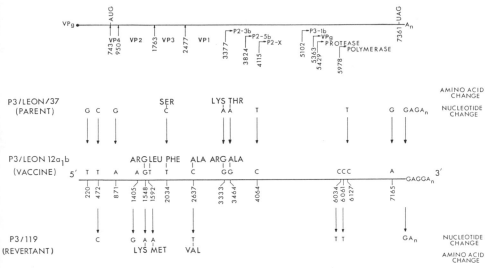

Fig. 5. Nucleotide and predicted amino acid sequence differences between poliovirus type-3 strains, P3/Leon/37 (parent), P3/Leon 12a,b (vaccine), and P3/119 (revertant). Reproduced with permission from Almond *et al.* (1985); Cann *et al.* (1984).

are in each of the non-coding regions; one amino acid replacement each is found in polypeptides VP3, VP1 and proteinase 2A.

A very small but significant incidence of poliomyelitis (four cases in 1984) occurs every year in the United States in spite of extensive vaccination and the virtual elimination of wild-type strains from circulation (Centers of Disease Control (USA), 1986*b*). Most of these cases are classified as vaccine-associated, using both epidemiological and laboratory procedures. However, the involvement of vaccine virus in the development of poliomyelitis is only beyond question if vaccine virus has been isolated from the CNS of the patient. This condition of isolation has not always been met. Nevertheless, the evidence for this conclusion is compelling if one accepts that virus isolated from the patient is the cause of the disease. The genomic RNA of these isolates has been fingerprinted (Kew *et al.*, 1984) and, although changes in the fingerprints compared to the authentic vaccine strain are usually seen, the basic pattern *which is highly diagnostic* remains the same. For example, the patterns shown in Figs 3A and 3B are highly diagnostic for polio 1 (Mahoney) and Sabin 1 RNA. That is, the 55 mutations in the Sabin 1 genome (0.6% difference compared to the polio 1 Mahoney RNA) produce at least five easily discernible changes in the fingerprint. Should the fingerprint of virus from a patient with poliomyelitis show the Sabin 1 pattern, one must conclude that the most likely cause of the post-vaccination infection is a variant (or are variants) of the Sabin 1 vaccine strain. In reality, only one case associated with Sabin 1 has been reported; instead most of these rare cases are related to immunization with the Sabin 3 (and to a lesser extent with the Sabin 2) vaccine. This observation may now have a molecular explanation: The Sabin 3 strain has accumulated only ten mutations, and thus the likelihood is much higher that direct reversions or second site mutations, which may accumulate during viral reproduction in the vaccine recipient, produce a 'less attenuated' or 'more neurovirulent' population of poliovirus than does the Sabin 1 strain. As we shall see, there exists direct evidence for this hypothesis.

ATTEMPTS TO CORRELATE STRUCTURE WITH PATHOGENICITY

The RNA sequence analyses, together with studies on the surface charge of whole virions (Emini *et al.*, 1984) and analyses of tryptic peptides of capsid proteins (Kew *et al.*, 1980), clearly showed that the shell of the attenuated virion was altered when compared to that of the neurovirulent progenitor virus. The neurovirulent and attenuated strains, however, adsorb to primate brain and spinal cord grey matter tissues at similar rates (Harter & Choppin, 1965). The differences in surface structure, therefore, do not appear to interfere with the entry of either virus strain into neuronal cells, at least not at the level of adsorption. It follows that the Sabin strain-specific capsid mutations probably influence the phenotype of the virus at a stage of the replicative cycle that follows adsorption. The discovery of multiple base substitutions scattered over the entire length of the genome, on the other hand, left open the possibility that mutations other than those in the capsid protein strongly influence the attenuation phenotype.

Three strategies have been followed in attempts to map 'attenuation-specific mutations'. First, intra- or intertypic recombinants between neurovirulent and attenuated strains have been isolated either from intracellular recombination events or by genetic engineering, and the phenotype of the recombinants has been determined. Secondly, sequence analyses have been performed on a highly neurovirulent strain (P3/119) that was isolated from a patient who died of poliomyelitis following vaccination. If there are mutations particularly relevant to the attenuated phenotype, it was hoped that in the neurovirulent variant P3/119, specific reversions to bases of the neurovirulent progenitor virus (polio 3, Leon) would be found. This strategy could only be followed with type 3 viruses because (i) there have been no highly neurovirulent type 1 variants isolated from vaccine recipients, and (ii) the neurovirulent progenitor strain to the Sabin 2 vaccine strain is not known. Since we lack a parental neurovirulent virus for the Sabin 2 strain, the first strategy outlined above (recombination) cannot be applied to this system either. V. R. Racaniello and his colleagues have therefore embarked upon a study of the neurovirulence of the specifically adapted polio 2 (Lansing) in mice (LaMonica et al., 1986). It is anticipated that information obtained with the polio/mouse system is relevant to the studies of the biological activities of the poliovirus Sabin strains. Third, serological and sequence studies have been performed on poliovirus strains isolated from vaccine recipients at specific intervals after vaccination to assess the rate of genetic variation occurring during passage in humans and to correlate the genetic variation with virulence.

STUDIES ON POLIOVIRUS TYPE 1

The RNA genomes of picornaviruses are known to undergo intermolecular recombination in vivo as has been documented for poliovirus (Hirst, 1962; Cooper, 1968; Romanova et al., 1980; Tolskaya et al., 1983; Emini et al., 1984) as well as foot-and-mouth disease virus (King et al., 1982). Both inter- and intratypic recombination can be observed with high frequency in infected tissue culture cells, and intertypic recombinants have also been isolated from poliovirus vaccine recipients, as was shown by Kew & Nottay (1984).

V. I. Agol and his colleagues have made use of inter- and intratypic recombinants isolated from infected cells to study the attenuation phenotype of poliovirus (Agol et al., 1984, 1985a, 1985b). The region of the crossover of recombinants between neurovirulent polio 3 and attenuated type 1, or between neurovirulent polio 1 and attenuated polio 1 was mapped by oligonucleotide fingerprinting to the central portion of the genome. Careful neurovirulence tests in monkeys clearly showed that the 5' half of the viral genome carried those mutations that conferred the strongest influence on neurovirulence or attenuation of the viral strain (Agol et al., 1984, 1985b). The other important point is that, although the 3'-terminal portion of the Sabin 1 genome is very weak in influencing neurovirulence, it nevertheless may weaken to some extent the virulence of a recombinant with a wild-type 5'-terminal half (Agol et al., 1985a).

The elegant studies using in vivo recombinants are somewhat restricted in that the crossover is usually found to have occurred inside the P2 region (see Fig. 1) probably due to the lack of suitable selectable markers. Thus, a more refined analysis of individual gene segments is difficult, if not impossible, to achieve. This problem has been overcome by the in vitro recombination of the infectious cDNA clones of polio 1 (Racaniello & Baltimore, 1981b; Semler et al., 1984) and Sabin 1 (Omata et al., 1984). Accordingly, we have replaced a large number of alleles by genetic engineering, constructed infectious cDNA recombinant clones, and produced viable recombinant virus by transfection of suitable mammalian cells (Kohara et al., 1985; Omata et al., 1985, 1986; Nomoto et al., 1986a; see Fig. 4). These recombinants were then tested for neurovirulence by injection into both the right and left thalamuses of seronegative cynomolgus monkeys (the same route of infection was selected by Agol and his colleagues) as well as for the in vitro markers (rct, d and plaque size). The data obtained are shown in Table 1.

The most important result of these studies is that numerous mutations, and not simply one or several, contribute to the phenotypeof attenuation of the Sabin 1 strain (Omata et al., 1986; Nomoto et al., 1986a). None of the recombinant viruses are either entirely neurovirulent or fully attenuated; rather we see a spectrum of phenotypic properties.

Some interesting specific points should be made. First, recombinants 3a and 3b (Fig. 4) reflect most closely the recombinants studied by Agol and his colleagues. In agreement with Agol's results, 3a is more virulent and 3b more attenuated. Recombinant 2b, which has only four amino acid replacements in the polymerase 3D gene

Table 1. *Monkey neurovirulence tests and in vitro markers of recombinant viruses*

Virus	Lesion score (mean ± SE)	Spread value (mean ± SE)	Incidence of paralysis (number paralysed/ number injected)	rct^a	d^b	Plaque size (mm in diameter)^c	Age^e
PV1(Sab)IC-0	0.07 ± 0.01	2.0 ± 0.0	0/4	>6.64	3.96	7.0	S
PV1(SM)IC-1a	0.15 ± 0.05	5.0 ± 1.9	0/4	5.70	1.88	8.5	M
PV1(SM)IC-2a	0.05 ± 0.01	1.5 ± 0.3	0/4	4.09	4.26	8.5	S
PV1(SM)IC-2b	2.03 ± 0.38	37.3 ± 0.6	0/4	2.77	0.66	14.5	M
PV1(SM)IC-3a	0.42 ± 0.17	11.1 ± 4.7	1/8	2.36	0.27	13.5	M
PV1(SM)IC-3b	0.11 ± 0.04	4.3 ± 1.3	0/4	4.73	4.63	6.5	S
PV1(SM)IC-4a	0.72 ± 0.27	18.4 ± 5.7	1/8	3.55	1.06	14.0	M
PV1(SM)IC-4b	0.80 ± 0.27	20.1 ± 6.1	0/8	5.09	3.18	4.5	S
PV1(M)pDS306	2.48 ± 0.34	37.0 ± 1.0	3/4	0.56	0.07	16.0	M
PV1(SM)IC-1b^d	0.07		0/3	>7.64			S

^{a} *rct* marker values shown here are the logarithmic differences of virus titres obtained at 36 and 39.5 °C (except for PV1(M)pDS306, for which the *rct* marker values of titres obtained at 36 and 40 °C are given).

^{b} *d* marker values shown here are the logarithmic differences of virus titres obtained at two different sodium bicarbonate concentrations.

^{c} Plaque size displayed on day 5 of growth.

^{d} Experiments with this recombinant, the reciprocal construct of IC-1a, were carried out in a different test series (Nomoto *et al.*, 1986*a*).

^{e} Antigenic properties tested with Sabin-specific or Mahoney-specific N-mcAbs.

product (and a total of 12 point mutations in the 3'-terminal ~1500 nucleotides), is 'neurovirulent' in all parameters shown, with the exception of paralysis, which developed in none of the experimental animals. Thus, these few 3'-terminal mutations must contribute to the attenuation phenotype, however weakly.

Of particular interest is the recombinant pair 1a and 1b (Omata et al., 1986; Nomoto et al., 1986a). In these, the nine amino acid mutations within VP3 and VP1 have been exchanged (Fig. 4, Table 1); accordingly, strain-specific N-mcAbs recognize 1a as type 1 (Mahoney) and 1b as Sabin 1 (Emini et al., 1983; Crainic et al., 1983; Diamond et al., 1985). To our surprise both recombinants scored as 'attenuated' in terms of paralysis, lesion score and rct marker. We have observed previously that recombinant virus 1a is genetically unstable upon passage in HeLa cells (Kohara et al., 1985). It is possible, therefore, that replacement from one part of a viral coat protein by elements of the coat protein region of other strains, which include a high degree of amino acid variation, might cause alterations in the efficiency of capsid assembly or the correct folding required for the faithful processing of the precursor proteins. Thus, recombinant viruses 1a and 1b may have reduced efficiencies in certain viral replication steps. How this should influence the attenuation phenotype is impossible to assess at this time.

An interesting neutralization escape mutant was found by Diamond et al. (1985; see also Blondel et al., 1986) which contained a single point mutation in capsid protein VP3. This mutant, selected for with a polio 1-specific N-mcAb could be neutralized with Sabin 1-specific N-mcAbs. Indeed, it was found that the mutation in VP3 (T > K) was one of the mutations identified by Nomoto et al. (1982) in the Sabin 1 genome. This single mutation flipped the antigenicity from Mahoney to Sabin, of two phenotypes, which had previously been found to be mutually exclusive (Crainic et al., 1983). Although the mutant carries Sabin 1-specific epitope(s), it was found to be fully neurovirulent (Diamond et al., 1985). Similarly, variants of Sabin 1 selected by in vitro passages at supra-optimal temperatures showed no correlation between neurovirulence and antigenic structure (Crainic et al., 1985).

Of great interest are also the recombinants 4a and 4b (Fig. 4) in which only the 5'-terminal non-coding region, plus a single amino acid replacement in VP4, are exchanged. As can be seen in Table 1, the Sabin 1-specific segment in 4a significantly influences the phenotype of the virus. More recently a recombinant was constructed in which only a segment from nucleotide 71 to 909 of the genomic

RNAs was exchanged (a region upstream of the VP4 mutation). This mutant showed the same properties as recombinant 4b, an observation suggesting that the mutations in the non-translated region (particularly the mutation in nucleotide 480) contribute to the attenuation phenotype (Nomoto *et al.*, 1986*a*). The mechanism for this phenomenon is unknown but could be a modulation in one or all of three steps in viral proliferation; namely, initiation of protein synthesis, RNA replication or morphogenesis. Indeed, the genomes of the Mahoney and the Sabin 1 strains differ in their efficiencies in *in vitro* translation (Svitkin, Maslova & Agol, 1985). As we will see below, a single base change in the 5' non-coding region of type 3 poliovirus has been observed to correlate with the attenuation phenotype of this strain (Evans *et al.*, 1985).

The data of the biological tests on recombinant viruses between the Mahoney and Sabin 1 strains are summarized in Fig. 6. The

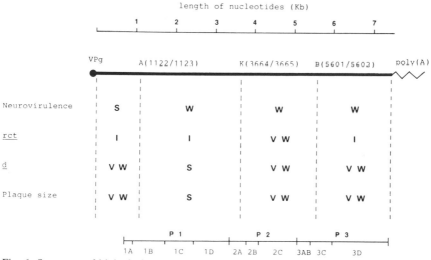

Fig. 6. Summary of biological tests on recombinant viruses between the Mahoney and the Sabin 1 strains. Length of nucleotides from the 5' end of the genome is shown at the top of the figure in kilobases (kb). Gene organization is indicated at the bottom of the figure. A, K, and B represent cleavage sites of the restriction enzymes AatII, Kpn I, and Bgl II, respectively. Numbers in parentheses following the restriction sites indicate nucleotide positions from the 5' end of the viral genome. The extent of influence of the genome region to each biological characteristic is roughly indicated by S (strong), I (intermediate), W (weak), or VW (very weak) in the corresponding genome region. Reproduced with permission from Nomoto *et al.* (1986*a*).

rct marker tests performed on many recombinant virus strains showed that the determinants of temperature sensitivity, like those of neurovirulence, were located across the entire polio-virus genome and expressed fairly strong but not perfect correlation with attenuation (Table 1). Recombinant viruses with Sabin 1-derived capsid proteins showed a small-plaque phenotype, and their plaque-forming ability was strongly dependent on bicarbonate concentration, suggesting that these determinants map to the genome region encoding the viral capsid proteins. Thus, the d marker and plaque size were found to be poor indicators of attenuation (Table 1).

Our results suggest that the extent of viral multiplication in the central nervous system of monkeys might be one of the most important indicators of neurovirulence of the poliovirus. Moreover, we conclude that the expression of the attenuated phenotype of the Sabin 1 strain of poliovirus is the result of several different biological characteristics and that none of the *in vitro* phenotypic markers alone can serve as a good indicator of neurovirulence or attenuation.

STUDIES ON POLIOVIRUS TYPE 3

Given the small number of nucleotide differences between polio 3 (Leon) and Sabin 3, it seemed much easier at first to correlate neurovirulence or attenuation with genome structure. Since certain virulent virus strains were isolated from vaccine recipients who were inflicted with poliomyelitis, it seemed possible that sequence analysis of such virus could prove these to be virulent variants of the Sabin 3 virus which had reverted some or all of the ten mutations to the polio 3 (Leon) genotype. Accordingly, J. W. Almond and his colleagues (Cann *et al.*, 1984; Almond *et al.*, 1985) sequenced a neurovirulent isolate P3/119. Surprisingly, only one of the ten Sabin 3-specific mutations reverted to the polio 3 (Leon) genotype (N472; see below). All other mutations in P3/119 are second-site mutations with respect to Sabin 3, an observation underscoring the difficulty one encounters in correlating genotype and phenotype. The data, however, showed beyond doubt that P3/119 is a derivative of Sabin 3. Almond and his colleagues (Almond *et al.*, 1985; Westrop *et al.*, 1986) then constructed recombinant clones from infectious cDNA clones of polio 3 (Leon) and Sabin 3 and tested recombinant viruses for neurovirulence. It was concluded that only the mutations at N472 and N2034 (S > F), significantly influence the neurovirulent phenotype, but that the mutation in the non-coding region (N472) may be the most significant. This is indeed an intriguing result and may reflect a crucial function of nucleotide 472 either in the formation of a functionally significant secondary structure or by directly interacting with structural or non-structural proteins. The enormous rate by which this mutation reverts in the human gut will be discussed below (Evans *et al.*, 1985).

Another interesting observation was that only the mutation within capsid protein VP3 (N2034, causing a S > F) appears to confer temperature sensitivity to the Sabin 3 strain (Westrop *et al.*, 1986).

STUDIES ON POLIOVIRUS TYPE 2

Polio 2 (Lansing), when inoculated intracerebrally into mice, causes fatal paralytic disease which clinically and pathologically resembles human poliomyelitis (Jubelt *et al.*, 1980). Racaniello and his colleagues (La Monica *et al.*, 1986) have begun to study the molecular basis of the neurovirulence of this polio strain, making use of the observation that most other polio virus strains cannot replicate in the mouse system.

An infectious cDNA clone of polio 2 (Lansing) RNA (Racaniello, 1984) was prepared and allele replacements performed with the infectious cDNA clone of the mouse-avirulent polio 1 (Mahoney). The viral isolates of the recombinant cDNA clones were then tested in 21-day-old Swiss-Webster mice. It was found that virulence mapped to the capsid region of the Lansing strain, an observation suggesting that virulence may be a matter of virus entry via adaptation to specific mouse cell receptors, or require specific steps in uncoating. Based on this observation, the relationship between protein structure and neurovirulence can be analysed because there are only 32 amino acid differences between the capsid polypeptides of the mouse-adapted polio 2 (Lansing) and the mouse-avirulent Sabin 2. Neutralization escape mutants of polio 2 (Lansing), selected in HeLa cells with type 2 N-mcAbs, show reduced or even abolished neurovirulence in mice (V. R. Racaniello, personal communication). It will be of great interest to know whether these neutralization escape mutants have lost their ability to bind to cells of the mouse central nervous system.

GENETIC VARIATION OF THE ORAL POLIOVIRUS VACCINE STRAIN IN MAN

The genomes of RNA viruses, in general, undergo mutation at high rate due to misincorporation of nucleotides during RNA replication and the absence of proof reading and editing functions. Thus RNA viruses have been termed 'quasi-species' in which the wild type is merely the predominant phenotype within a wide spectrum of variants that fits best into a given environment (reviewed by Holland et al., 1982). In view of these considerations, it is not surprising that a significant change in monkey neurovirulence and in vitro phenotype markers was observed with poliovirus isolates from vaccinated children (Melnick, 1961; Benyesh–Melnick et al., 1967). Moreover, a serological analysis of poliovirus recovered from normal vaccine recipients revealed 'antigenic drift' a few days after vaccination (Nakano et al., 1963). These first important studies of genetic variation of vaccine virus in humans were confirmed later with the aid of specific N-mcAbs (Humphrey et al., 1982; Crainic et al., 1983; Minor et al., 1986) and by sequence analysis of genomic RNA of Sabin 1 derived isolates (Kew & Nottay, 1984).

O. Kew and his colleagues (personal communication) have recently found a surprisingly large number of recombinant strains that were isolated from specimens of healthy vaccine recipients. These genetic variants arose by cross-over between genomes of all three types of the Sabin strains. Cross-over had usually occurred within the P2, but occasionally also within the P3 region of the viral genome. Thus, in addition to point mutations, the vaccine strains undergo extensive variation by genetic recombination during viral proliferation in vaccinated individuals.

A particularly intriguing study was reported on the rate of reversion of N472 of the 5' non-coding region from a U in the Sabin 3 strain to a C, the base in the progenitor virulent polio 3 (Leon) strain (Evans et al., 1985; Minor et al., 1986). Remarkably, virus isolated from an infant only 47 h after vaccination had this position completely reverted to wild-type, an observation suggesting strong selection for the wild-type genotype during passage somewhere in the human gut. As pointed out before, a virus isolate with this reversion is significantly more neurovirulent than the original Sabin 3 vaccine (Evans et al., 1985).

The molecular basis for the selection of the N472 revertants is unknown. Clearly, this event must be governed by specific properties of the natural host cells of poliovirus since it is not observed in tissue culture cells such as HeLa, or monkey kidney cells, not even at supra-optimal temperatures.

Finally, if the attenuation phenotype of the Sabin 3 strain is indeed governed by only two mutations one would expect that vaccine recipients excrete a population of fully neurovirulent poliovirus type 3 shortly after vaccination. This is because (i) there should be strong selection against the mutation at N2034, the only mutation that confers temperature sensitivity to the Sabin 3 strain, and because (ii) the N472 mutation is known to revert to the wild type genotype in less than three days. Minor, on the other hand, has found that a vaccine recipient excreted for at least 30 days a population of poliovirus type 3 which was fully temperature sensitive (personal communication). This observation cannot be explained at the present time. It is possible, however, that the ts phenotype of the Sabin 3 strain is a characteristic of *in vitro* assays in tissue culture and may not be expressed in 'normal' host cells in the human gut. This is not an unreasonable consideration since host range mutations, even for genetic traits like temperature sensitivity, are well known among animal viruses (see, for example, Szilágyi, Pringle & McPherson, 1977).

CONCLUSIONS

Poliovirus is probably the best-known virus in regard to primary structure ($C_{327010} H_{492734} O_{130777} N_{97942} P_{7441} S_{2340}$; not considering the poly(A) or cations), crystal structure, antigenic structure, gene organization and gene expression. And yet each step in its replicative cycle still produces a deluge of unanswered questions. For example, the interaction with the unknown entity called cellular receptor, the events leading to penetration and uncoating and the mechanism by which antibodies neutralize the virus are still an enigma.

The study of the antigenic properties of poliovirus, and of the genotypes of attenuated poliovirus strains briefly described here, have yielded a very detailed yet incomplete description of specific physical traits of virions and their genomes. These studies have, however, told us little, if anything, of the mode by which poliovirus causes poliomyelitis in man. Where does the virus replicate prior to its entry into the CNS? How does the virus reach the CNS? Is the replication of poliovirus tuned slightly differently in different tissue cells, and do the accumulated effects of this fine tuning ultimately decide whether disease develops? It may be that the mutations in polypeptide 2A of Sabin 1 have an effect in cell A but not in cell B, whereas the mutations in the non-translated region play a role in cells A *and* B. And might it be that mutations in the capsid proteins influence viral replication only in cells C?

In his illuminating message to molecular biologists of February 19, 1982,* Albert Sabin left no doubt about the predicament and

---

* The above-mentioned letter to Dr O. Kew, copies of which were sent to the authors of this paper.

the Herculean task of a study of pathogenicity of poliovirus, as he wrote:

In my judgment, molecular biologists are limited in their knowledge and understanding of the large number of biological properties of polioviruses which have been measured quantitatively and that each property is not an all-or-none manifestation but varies quantitatively not only in mixed populations of viral particles but also in the progeny of single virus particles. And please, 'virulence' and 'neurovirulence' are not single biological properties. The limitation is not in what can be tested, but in the laborious nature of some of the important tests for which the excellent molecular biologists are neither trained nor inclined to become involved in. Accordingly, they seek easy ways out that are based on untested and therefore doubtful assumptions.

The 'easy ways', cloning, sequencing, genetic engineering, isolation of monoclonal antibodies, etc. may not have given us, as yet, the answers for the awesome question of the molecular basis of pathogenesis. They have provided us, however, with the tools necessary for this study. Moreover, very important by-products emerge. For example, the Sabin-specific genomes can be stored away in the form of stable infectious cDNA clones and thus be an unlimited source for the oral polio live vaccine (Kohara et al., 1986). More significantly, based upon our knowledge of antigenicity and genotypes of the Sabin strain, a new type 3 live vaccine can be engineered in vitro. Allele replacements within the P1 region (coding for the capsid proteins) between Sabin 1 and Sabin 3 strains have led to unstable (Kohara et al., 1985) or non-viable (Stanway et al., 1986) in vitro recombinant viruses. However, the exchange of the complete Sabin 1-specific P1 region with the Sabin 3-specific P1 region yielded a stable, inter-typic Sabin 1/Sabin 3 recombinant virus that may be a candidate to replace the presently used oral Sabin 3 vaccine (Nomoto et al., 1986b).

Although new and somewhat bewildering for most of us molecular biologists, research on the molecular basis of viral disease is a welcome challenge to conquer new land. We are grateful to Albert Sabin for having taught us an invaluable lesson in advance.

ACKNOWLEDGEMENT

We are deeply indebted to our colleagues who have contributed to these studies and who have enlightened us in many discussions.

We thank J. W. Almond for providing us with Fig. 5; J. L. Melnick, H. Toyoda, M. J. H. Nicklin and S. E. Pincus for their criticism of the text, and Lynn Zawacki for the preparation of the manuscript. This work was supported in part by a grant

from the Ministry and Education, Science and Culture of Japan to Akio Nomoto, and by US Public Health Service grants AI 15122 and CA 28146 for the National Institutes of Health to Eckard Wimmer.
This paper has been dedicated to Dr Albert B. Sabin on the occasion of his 80th birthday.

REFERENCES

AGOL, V. I., DROZDOV, S G., FROLOVA, M. P., GRACHEV, V. P., KOLESNIKOVA, M. S., KOZLOV, V. G., RALPH, N. M., ROMANOVA, L. I., TOLSKAYA, E. A. & VIKTOROVA, E. Q. (1985a). Neurovirulence of the intertypic poliovirus recombinant v^3/al-25: characterization of strains isolated from the spinal cord of diseased monkeys and evaluation of the contribution of the 3' half of the genome. *Journal of General Virology*, **66**, 309–16.

AGOL, V. I., DROZDOV S. G., GRACHEV, V. P., KOLESNIKOVA, M. S., KOZLOV, V. G., RALPH, N. M., ROMANOVA, L. I., TOLSKAYA, E. A., TYUFANOV, A. V. & VIKTOROVA, E. G. (1985b). Recombination between attenuated and virulent strains of poliovirus type 1: derivation and characterization of recombinants with centrally located crossover points. *Virology*, **143**, 467–77.

AGOL, V. I., GRACHEV, V. P., DROZDOV, S. G., KOLESNIKOVA, M. S., KOZLOV, V. G., RALPH, N. M., ROMANOVA, L. I., TOLSKAYA, E. A., TYUFANOV, A. V. & VIKTOROVA, E. G. (1984). Construction and properties of intertypic poliovirus recombinants: first approximation mapping of the major determinants of neurovirulence. *Virology*, **136**, 41–55.

ALMOND, J. W., WESTROP, G. D., CANN, A. J., STANWAY, G., EVANS, D. M. A., MINOR, P. D. & SCHILD, G. C. (1985). Attenuation and reversion to neurovirulence of the Sabin poliovirus type-3 vaccine. In *Vaccines 85: Molecular and Chemical Basis of Resistance to Parasitic, Bacterial and Viral Diseases*, eds. R. A. Lerner, R. M. Chanock, F. Brown, Cold Spring Harbor, N.Y.: Cold Spring Harbor Laboratory, pp. 271–7.

BENYESH-MELNICK, M., MELNICK, J. L., RAWLS, W. E., WIMBERLY, I., BARRERA ORO, J., & RENNICK, V. (1967). *American Journal of Epidemiology* **86**, 112–36.

BLONDEL, B., AKACEM, O., CRAINIC, R., COUILLIN, P. & HORODNICEANU, F. (1983). Dectection by monoclonal antibodies of an antigenic determinant critical for poliovirus neutralization present on VP1 and on heat-inactivated virions. *Virology*, **126**, 707–10.

BLONDEL, B., CRAINIC, R., FICHOT, O. D., DUFRAISSE, G., CANDREA, A., DIAMOND, D., GIRARD, M. & HORAUD, F. (1986). Mutations conferring resistance to neutralization with monoclonal antibodies in type 1 poliovirus can be located outside or inside the antibody-binding site. *Journal of Virology*, **57**, 81–90.

CANN, A. Y., STANWAY, G., HUGHES, P. J., MINOR, P. D., EVANS, D. M. A., SCHILD, G. C. & ALMOND, J. W. (1984). Reversion to neurovirulence of the live-attenuated Sabin type 3 oral poliovirus vaccine. *Nucleic Acids Research*, **12** no. 20, 7787–92.

CENTERS OF DISEASE CONTROL (USA): (1986a). *Morbidity and Mortality Weekly Report*. **35** no. 6, 82–6.

CENTERS OF DISEASE CONTROL (USA): (1986b). *Morbidity and Mortality Weekly Report*. **35** no. 11, 180–2.

COOPER, P. D. (1986). A genetic map of poliovirus temperature-sensitive mutants. *Virology*, **35**, 584–96.

CRAINIC, R., BLONDEL, B., CANDREA, A., DUFRAISSE, G. & HORAUD, F. (1985). Antigenic modification of attenuated Sabin 1 poliovirus by *in vitro* passage at suboptimal temperatures. *Developmental Biological Standards*, **60**, 343–7.

130   A. NOMOTO AND E. WIMMER

CRAINIC, R., COUILLIN, P., BLONDEL, B., CABUN, N., BONE, A. & HORODNICEANU,
F. (1983). Natural variation of poliovirus neutralization epitopes. *Infection and Immunology*, **41**, 1217–25.
CROWELL, R. L., REAGAN, K. J., SCHULTZ, M., MAPOLES, J. E., GRUN, J. B.
& LANDAU, B. J. (1985). Cellular receptors as determinants of viral tropism. In *Genetically Altered Viruses and the Environment*, ed. B. Fields, M. A. Martin & D. Kamely, pp. 147–64. Cold Spring Harbor Banbury Report, 22.
DIAMOND, D. C., JAMESON, B. A., BONIN, J., KOHARA, M., ABE, S., ITOH, H.,
KOMATSU, T., ARITA, M., KUGE, S., OSTERHAUS, A. D. M. E., CRAINIC, R., NOMOTO, A. & WIMMER, E. (1985). Antigenic variation and resistance to neutralization in poliovirus type 1. *Science*, **229**, 1090–3.
DORNER, A. J., DORNER, L. F., LARSEN, G. R., WIMMER, E. & ANDERSON, C.
W. (1982). Identification of the initiation site of poliovirus polyprotein synthesis. *Journal of Virology*, **42**, 1017–28.
EMINI, E. A., ELZINGA, M. & WIMMER, E. (1982). Carboxy-terminal analysis of
poliovirus proteins: termination of poliovirus RNA translation and location of unique poliovirus polyprotein cleavage sites. *Journal of Virology*, **42**, 194–9.
EMINI, E. A., JAMESON, B. A. & WIMMER, E. (1985). Antigenic structure of poliovir-
us. In *Immunochemistry of Viruses – The Basis of Serodiagnosis and Vaccines*, ed. A. R. Neurath & M. H. V. van Regenmortel, pp. 281–94. Holland: Elsevier Biomedical Press.
EMINI, E. A., KAO, S.-Y., LEWIS, A. J., CRAINIC, R. & WIMMER, E. (1983). The
functional basis of poliovirus neutralization determined with monospecific neutralizing antibodies. *Journal of Virology*, **46**, 466–74.
EMINI, E. A., LEIBOWITZ, J., DIAMOND, D. C., BONIN, J. & WIMMER, E. (1984).
Recombinants of Mahoney & Sabin strain poliovirus type 1: analysis of *in vitro* phenotypic markers and evidence that resistance to guanidine maps in the non-structural proteins. *Virology*, **137**, 74–85.
EVANS, D. M. A., DUNN, G., MINOR, P. D., SCHILD, G. C., CANN, A. J., STANWAY,
G., ALMOND, J. W., CURREY, K. & MAIZEL, J. V., JR, (1985). Increased neurovirulence associated with a single nucleotide change in a noncoding region of the Sabin type 3 poliovaccine genome. *Nature*, **314**, 548–50.
FENNER, F., MCAUSLAN, B. R., MIMS, C. A., SAMBROOK, J. & WHITE, D. O.
(1974). *The Biology of Animal Viruses*. 2nd ed., pp. 317–18. New York: Academic Press.
FRISBY, D. P., NEWTON, C., CAREY, N. H., FELLNER, P., NEWMAN, J. F. E., HARRIS,
T. J. R. & BROWN, F. (1976). Oligonucleotide mapping of picornavirus RNAs by two-dimensional electrophoresis. *Virology*, **71**, 379–88.
HIRST, G. K. (1962). Genetic recombination with Newcastle disease virus, poliovirus
and influenza. In *Cold Spring Harbor Symposium on Quantitative Biology* **27**, pp. 303.
HARTER, D. H. & CHOPPIN, P. W. (1965). Adsorption of attenuated and neuroviru-
lent poliovirus strains to central cervous system tissues of primates. *Journal of Immunology*, **95**, 730–6.
HOGLE, J. M., CHOW, M. & FILMAN, D. J. (1985). Three-dimensional structure
of poliovirus at 2.9 angstrom resolution. *Science*, **229**, 1358–63.
HOLLAND, J., SPINDLER, K., HORODYSK, F., GRABAU, E., NICHOL, S. & VAN DE
POL, S. (1982). Rapid evolution of RNA genomes. *Science*, **215**, 1577–85.
HORSTMANN, D. M., QUINN, T. C. & ROBBINS, F. C. (1984). Reviews of infectious
diseases. In *International Symposium on Poliomyelitis Control*, **6**, supplement 2.
HUMPHREY, D. D., KEW, O. M. & FEORINO, P. M. (1982). Monoclonal antibodies
of four different specificities for neutralization of type 1 poliovirus. *Infection*

and Immunology, **36**, 841–3.

ICENOGLE, J. P., MINOR, P. D., FURGUSON, M. & HOGLE, J. M. (1986). Modulation of the humoral response to a 12 amino acid site on the poliovirus. *Journal of Virology*, **60**, 297–301.

JUBELT, B., GALLEZ-HAWKIS, B., NARAYAN, O. & JOHNSON, R. T. (1980). Pathogenesis of human poliovirus infection in mice. I. clinical and pathological studies. *Journal of Neuropathology and Experimental Neurology*, **39**, 138–48.

KEW, O. H., NOTTAY, B. K., HATCH, M. H., NAKANO, Y. H. & OBIJESKI, Y. F. (1981). Multiple genetic changes can occur in the oral poliovaccines upon replication in humans. *Journal of General Virology*, **56**, 307–17.

KEW, O. M. & NOTTAY, B. K. (1984). Evolution of the oral polio vaccine strains in humans occur by both mutation and intramolecular recombination. In *Modern Approaches to Vaccines*, ed. R. Chanock & R. Lerner, pp. 357–67. Cold Spring Harbor, N.Y.: Cold Spring Harbor Press.

KEW, O. M., NOTTAY, B. K. & OBIJESKI, K. F. (1984). Applications of oligonucleotide fingerprinting to the identification of viruses. *Methods in Virology*, **8**, 41–84.

KEW, O. M., PALLANSCH, M. A., OMILIANOWOSKI, D. R. & RUECKERT, R. R. (1980). Changes in three of the four coat proteins of oral polio vaccine strain derived from type 1 poliovirus. *Journal of Virology*, **33**, 256–63.

KING, A. M. Q., McCAHON, D., SLADE, W. R. & NEWMAN, J. W. I. (1982). Recombination in RNA. *Cell*, **29**, 921–8.

KITAMURA, N. & WIMMER, E. (1980). Sequence of 1060 3'-terminal nucleotides of poliovirus RNA as determined by a modification of the dideoxynucleotide method. *Proceedings of the National Academy of Sciences USA*, **77**, 3196–3200.

KITAMURA, N., SEMLER, B. L., ROTHBERG, P. G., LARSEN, G. R., ADLER, C. J., DORNER, A. J., EMINI, E. A., HANECAK, R., LEE, J. J., VAN DER WERF, S., ANDERSON, C. W. & WIMMER, E. (1981). Primary structure, gene organization, and polypeptide expression of poliovirus RNA. *Nature*, **291**, 547–53.

KOHARA, M., ABE, S., KUGE, S., SEMLER, B. L., KOMATSU, T., ARITA, M., ITOH, H. & NOMOTO, A. (1986). An infectious cDNA clone of the poliovirus Sabin strain could be used as a stable repository and inoculum for the oral polio live vaccine. *Virology*, **150**, 21–30.

KOHARA, M., OMATA, T., KAMEDA, A., SEMLER, B. L., ITOH, H., WIMMER, E. & NOMOTO, A. (1985). *In vitro* phenotypic markers of a poliovirus recombinant constructed from infectious cDNA clones of the neurovirulent Mahoney strain and the attenuated Sabin 1 strain. *Journal of Virology*, **53**, 786–92.

KÖHLER, G. & MILSTEIN, C. (1975). Continuous cultures of fused cells secreting antibody of predefined specificity. *Nature*, **256**, 495–7.

KUHN, R. J. & WIMMER, E. (1986). The replication of picornaviruses. In *The Molecular Biology or Positive Strand RNA Viruses*, eds. D. J. Rowlands, B. W. J. Mahy and M. Mayo, Academic Press, London. In press.

LA MONICA, N., MERIAN, C. & RACANIELLO, V. R. (1986). Mapping of sequences required for mouse neurovirulence of poliovirus type 2 Lansing. *Journal of Virology* **57**, 515–25.

LARSEN, G. R., ANDERSON, C. W., DORNER, A. J., SEMLER, B. L. & WIMMER, E. (1982). Cleavage sites within the poliovirus capsid protein precursors. *Journal of Virology*, **41**, 340–4.

LEE, Y. F., KITAMURA, N., NOMOTO, A. & WIMMER, E. (1979). Sequence studies of poliovirus RNA. IV. Nucleotide sequence complexities of poliovirus type 1, type 2 and two type 1 defective interfering particles RNAs, and fingerprint of the poliovirus type 3 genome. *Journal of General Virology*, **44**, 311–22.

LEE, Y. F. & WIMMER, E. (1976). 'Fingerprinting' high molecular weight RNA by two-dimensional gel electrophoresis: application to poliovirus RNA. *Nucleic*

Acids Research, **3**, 1647–58.

MELNICK, J. L. (1961). Attenuated poliovirus vaccine: virus stability. In *Poliomyelitis*, International Poliomyelitis Congress, pp. 384–402. J. B. Lippincott Co.

MELNICK, J. L. (1984). Live attenuated oral poliovirus vaccine. *Review of Infectious Diseases*, **6**, Suppl. 2, S323–7.

MENDELSOHN, C., JOHNSON, B., LIONETTI, K. A., NOBIS, P., WIMMER, E. & RACANIELLO, V. R. (1986). Transformation of a human poliovirus receptor gene into mouse cells. *Proceedings of the National Academy of Sciences USA*. In press.

MINOR, P. D. (1980). Comparative Biochemical Studies of type 3 poliovirus. *Journal of Virology*, **34**, 73–84.

MINOR, P. D., EVANS, D. M. A., FERGUSON, M., SCHILD, G. C., WESTROP, G. & ALMOND, J. W. (1985). Principal and subsidiary antigenic sites of VP1 involved in the neutralization of poliovirus type 3. *Journal of General Virology*, **65**, 1159–65.

MINOR, P. D., SCHILD, G. C., BOOTMAN, J., EVANS, D. M. A., FERGUSON, M., REEVE, P., SPITZ, M., STANWAY, G., CANN, A. J., HAUPTMANN, R., CLARKE, L. D., MOUNTFORD, R. C. & ALMOND, J. W. (1983). Location and primary structure of the antigenic site for poliovirus neutralization. *Nature*, **301**, 674–9.

MINOR, P. D., JOHN, A., FERGUSON, M. & ICENOGLE, J. P. (1986). Antigenic and molecular evolution of the vaccine strain of type 3 poliovirus during the period of excretion by a primary vaccine. *Journal of General Virology*, **67**, 693–706.

NAKANO, J. H., GELFAND, H. M., COLE, J. T. (1963). The use of a modified Wecker technique for the sero-differentiation of type 1 polioviruses related and unrelated to Sabin's vaccine strain. II. Antigenic segregation of isolates from specimens collected in field studies. *American Journal of Hygiene*, **78**, 214–26.

NAKANO, J. H., HATCH, M. H., THIEME, M. L. & NOTTAY, B. (1978). Parameters for differentiating vaccine-derived and wild poliovirus strains. *Progress in Medical Virology*, **24**, 178–206.

NICKLIN, J. H., TOYODA, H., MURRAY, M. G. & WIMMER, E. (1986). Proteolytic processing in the replication of polio and related viruses. *Biotechnology*, **4**, 33–42.

NOMOTO, A., KAJIGAYA, S., SUZUKI, K. & IMURA, N. (1979) Possible point mutation sites in LSc, 2ab poliovirus RNA and a protein covalently linked to the 5′ terminus. *Journal of General Virology*, **45**, 107–17.

NOMOTO, A., KOHARA, M., ABE, S., KUGE, S., SEMLER, B. L., KOMOTSU, T., ARITA, M. & ITOH, H. (1986*b*). Abstract, V International Conference on Comparative Virology, Chateau Lake Louise, Alberta, Canada, May 4–9, 1986, p. 23.

NOMOTO, A., KOHARA, M., KUGE, S., KAWAMURA, N., ARITA, M., KOMATSU, T., ABE, S., SEMLER, B. L., WIMMER, E. & ITOH, H. (1986*a*). Study on virulence of poliovirus type 1 using *in vitro* modified viruses. In *Positive Strand RNA Viruses*, eds. M. B. Brinton and R. R. Rueckert, UCLA Symposia on Molecular and Cellular Biology. In press.

NOMOTO, A., KITAMURA, N., LEE, J. J., ROTHBERG, P. G., IMURA, N. & WIMMER, E. (1981). Identification of point mutations in the genome of poliovirus Sabin vaccine LSc 2ab, and catalogue of RNase T1- and RNase A-resistant oligonucleotides of poliovirus type 1 (Mahoney) RNA. *Virology*, **112**, 217–27.

NOMOTO, A., OMATA, T., TOYODA, H., KUGE, S., HORIE, H., KATAOKA, Y., GENBA, Y., NAKANO, Y. & IMURA, N. (1982). Complete nucleotide sequence of the attenuated poliovirus Sabin 1 strain genome. *Proceedings of the National Academy of Sciences USA*, **79**, 5793–7.

OMATA, T., KOHARA, M., ABE, S., ITOH, H., KOMATSU, T., ARITA, M., SEMLER, B. L., WIMMER, E., KUGE, S., KAMEDA, A. & NOMOTO, A. (1985). Construction of recombinant viruses between Mahoney and Sabin strains of type 1 poliovirus and their biological characteristics. In *Vaccines 85: Molecular and Chemical Basis* of Resistance to Parasitic, Bacterial and Viral Diseases, eds. R. A. Lerner, R. M. Chanock, and F. Brown, p. 279–83. Cold Spring Harbor, N.Y.: Cold Spring Harbor Laboratory.

OMATA, T., KOHARA, M., KUGE, S., KOMATSU, T., ABE, S., SEMLER, B. L., KAMEDA, A., ITOH, H., ARITA, M., WIMMER, E. & NOMOTO, A. (1986). Genetic analysis of the attenuation phenotype of poliovirus type1. *Journal of Virology*, **58**, 348–58.

OMATA, T., KOHARA, M., SAKAI, Y., KAMEDA, A., IMURA, N. & NOMOTO, A. (1984). Cloned infectious complementary DNA of the poliovirus Sabin 1 genome: Biochemical and biological properties of the recovered virus. *Gene*, **32**, 1–10.

RACANIELLO, V. R. & BALTIMORE, D. (1981*a*). Molecular cloning of poliovirus cDNA and the determination of the complete nucleotide sequence of the viral genome. *Proceedings of the National Academy of Sciences USA*, **78**, 4887–91.

RACANIELLO, V. R. & BALTIMORE, D. (1981*b*). Cloned poliovirus complementary DNA is infectious in mammalian cells. *Science*, **214**, 916–19.

ROMANOVA, L. I., TOLSKAYA, E. A., KOLESNIKOVA, M. S. & AGOL, V. I. (1980). Biochemical evidence for intertypic recombination of poliovirus. *FEBS Letters* **118**, 109–12.

ROSSMAN, M. G., ARNOLD, E., ERICKSON, J. W., FRANKENBERGER, E. A., GRIFFITH, J. P., HECHT, H.-J., JOHNSON, J., KAMER, G., LUO, M., MOSSER, A. G., RUECKERT, R. R., SHERRY, B. & VRIEND, G. (1985). Structure of a human common cold virus and functional relationship to other picornaviruses. *Nature*, **317**, 145–53.

RUECKERT, R. R. (1985). Picornaviruses and their replication. In *Virology*, eds. B. N. Fields *et al.*, New York: Raven Press, Chapter 32, pp. 705–38.

RUECKERT, R. R. & WIMMER, E. (1984). Systematic nomenclature of picornavirus proteins. *Journal of Virology*, **50**, 957–9.

SABIN, A. B. (1955). Characteristics and genetic potentialities of experimentally produced and naturally occurring variants of poliomyelitis virus. *Annals of the New York Academy of Sciences*, **61**, 924–38.

SABIN, A. B. & BOULGER, C. R. (1973). History of Sabin attenuated poliovirus oral live vaccine strains. *Journal of Biological Standards*, **1**, 115–18.

SALK, J. E. (1960). Persistence of immunity after administration of formalin-treated poliovirus vaccine. *Lancet*, **2**, 715–23.

SEMLER, B. L., ANDERSON, C. W., KITAMURA, N., ROTHBERG, P. G., WISHERT, W. L. & WIMMER, E. (1981*a*). Poliovirus replication proteins: RNA sequence encoding 1b and the sites of proteolytic processing. *Proceedings of the National Academy of Sciences USA*, **78**, 3464–8.

SEMLER, B. L., DORNER, A. J. & WIMMER, E, (1984). Production of infectious poliovirus from cloned cDNA is dramatically increased by SV40 transcription and replication signals. *Nucleic Acids Research*, **12**, 5123–41.

SEMLER, B. L., HANECAK, R., ANDERSON, C. W. & WIMMER, E. (1981*b*). Cleavage sites in the polypeptide precursors of poliovirus protein P2-X. *Virology*, **114**, 589–94.

STANWAY, G., CANN, A. J., HAUPTMANN, R., MOUNTFORD, R. C., CLARKE, L. D., REEVE, P., MINOR, P. D., SCHILD, G. C. & ALMOND, J. W. (1983). Nucleic acid sequence of the region of the genome encoding capsid protein VP1 of neurovirulent and attenuated type 3 polioviruses. *European Journal of Biochemistry*, **135**, 529–33.

STANWAY, G., HUGHES, P. J., MOUNTFORD, R. C., REEVE, P., MINOR, P. D., SCHILD, G. C. & ALMOND, J. W. (1984). Comparison of the complete nucleotide sequences of the genomes of the neurovirulent poliovirus P3/Leon/37 and its attenuated Sabin vaccine derivative P3/Leon 12ab. *Proceedings of the National Academy of Sciences USA*, **81**, 1539–43.

STANWAY, G., HUGHES, P. J., WESTROP, G. D., EVANS, D. M. A., DUNN, G., MINOR, P. D., SCHILD, G. C. & ALMOND, J. W. (1986). Construction of poliovirus intertypic recombinants by use of cDNA. *Journal of Virology*, **57**, 1187–90.

SVITKIN, Y. V., MASLOVA, S. V. & AGOL, V. I. (1985). The genomes of attenuated and virulent poliovirus strains differ in their *in vitro* translation efficiencies. *Virology*, **147**, 243–52.

SZILÁGYI, F. J., PRINGLE, C. R. & McPHERSON, T. M. (1977). Temperature-dependent host range mutation in vesicular stomatitis virus affecting polypeptide L. *Journal of Virology*, **22**, 381–8.

TOLSKAYA, E. A., ROMANOVA, L. I., KOLESNIKOVA, M. S. & AGOL, V. I. (1983). Intertypic recombination in poliovirus: Genetic and biochemical studies. *Virology*, **124**, 121–32.

TOMASSINI, J. E. & COLONNO, R. J. (1986). Isolation of a receptor protein involved in attachment of human rhinoviruses. *Journal of Virology*, **58**, 290–5.

TOYODA, H., KOHARA, M., KATAOKA, Y., SUGANUMA, T., OMATA, T., IMURA, N. & NOMOTO, A. (1984). The complete nucleotide sequence of all three poliovirus serotype genomes: Implication for the genetic relationship, gene function and antigenic determinants. *Journal of Molecular Biology*, **174**, 561–85.

TOYODA, H., NICKLIN, M. J. H., MURRAY, M. G., ANDERSON, C. W., DUNN, J. J., STUDIER, F. W. & WIMMER, E. (1986). A second virus-encoded proteinase involved in proteolytic processing of poliovirus polyprotein. *Cell*, **45**, 761–70.

VAN DER WERF, S., BRADLEY, J., WIMMER, E., STUDIER, F. W. & DUNN, J. J. (1986). Synthesis of infectious poliovirus RNA by purified T7 RNA polymerase. *Proceedings of the National Academy of Sciences USA*, **83**, 2330–4.

VAN DER WERF, S., WYCHOWSKI, C., BRUNEAU, P., BLONDEL, B., CRAINIC, R., HORODNICEANU, F. & GIRARD, M. (1983). Localization of a poliovirus type 1 neutralization epitope in viral capsid polypeptide VP1. *Proceedings of the National Academy of Sciences USA*, **80**, 5080–4.

WHO Tech, Rep. Ser. (1983). **687**, 134.

WESTROP, G. D., EVANS, D. M. A., MINOR, P. D., MAGRATH, D., SCHILD, G. C. & ALMOND, J. W. (1986). In *The Molecular Biology of Positive Strand Viruses*, eds. D. J. Rowlands, B. W. J. Mahy and M. Mayo, Academic Press, London.

WIMMER, E., EMINI, E. A. & DIAMOND, D. C. (1986). Mapping neutralization domains of viruses. In: *Concepts in Clinical Pathogenesis II*, ed. A. L. Notkins and M. B. A. Oldstone, pp. 159–73. New York: Springer Verlag.

WIMMER, E., JAMESON, B. A. & EMINI, E. A. (1984). Poliovirus antigenic sites and vaccines. *Nature*, **308**, 19.

MOLECULAR MECHANISMS IN ARBOVIRUS DISEASE

DAVID H. L. BISHOP*, ROBERT E. SHOPE† AND BARRY J. BEATY‡

NERC Institute of Virology, Oxford, UK
† Yale Arbovirus Research Unit, Yale University Department of Epidemiology & Public Health, New Haven, Connecticut, USA
‡ Department of Microbiology, Colorado State University, Fort Collins, Colorado, USA

INTRODUCTION

The scope of this chapter is limited to a discussion of the molecular mechanisms of disease and infection by representatives of one group of arthropod transmitted viruses, the bunyaviruses (Bunyaviridae family, *Bunyavirus* genus) (Bishop *et al.*, 1980). This group of viruses has been selected because of its diversity (some 150 different viruses have been described: Karabatsos, 1985) and because it is possible to employ both genetic and molecular tools to study the infection processes in permissive vertebrate and invertebrate hosts. Because arthropod-borne viruses (arboviruses) are inoculated into vertebrates by hematophagous insects, they are very different in their routes of transmission from viruses that are passed directly or incidentally between two hosts (e.g., influenza viruses that are spread by aerosols, or polio virus that spreads by contamination involving the faecal–oral route). Arboviruses replicate in both the permissive arthropod species as well as in the vertebrate host. In this regard they differ from viruses that are mechanically transmitted by insects (e.g., carnation latent virus that is transmitted between plant species by aphids: Matthews, 1982). Arboviruses do not cause overt damage to the arthropod vector and so are unlike, for example, the DNA genome baculoviruses which characteristically are pathogenic for the larval forms of particular arthropod species (e.g., caterpillars). Since arthropod-borne viruses are injected into a vertebrate when an insect takes a blood meal, or are taken up from an infected vertebrate when an insect feeds, the viruses do not have to survive outside a living organism. By contrast, baculoviruses rely on their physical characteristics for survival in the environment while they are between host species. Discussions of the properties of the insect specific baculoviruses, and the plant viruses that are mechanically transmitted by arthropods, are to be found in the chapters by Kelly and Harrison in this book.

Although arboviruses replicate in representatives of more than one phylum of species, various strategies are employed. For example, there are arboviruses which have a positive-stranded RNA genome (e.g., flaviviruses, alphaviruses), or a negative-stranded RNA genome (e.g., bunyaviruses, rhabdoviruses), or a double-stranded RNA genome (e.g., orbiviruses) (Matthews, 1982). To cover the molecular biology of the infections and diseases that characterize representatives of each of these virus families is beyond the scope of this review. Instead, the molecular mechanisms involved in the infection processes by particular members of one group, the bunyaviruses, will be discussed primarily in relation to what is known about the infection processes in the arthropod species but also with regard to what is understood about the processes of infection in the vertebrate host. For representative arboviruses belonging to other virus families the molecular mechanisms of replication have been studied in great detail in cell culture and the genome sequences obtained for several of the viruses (see Fields, 1985). However, for many of these viruses the processes involved in transmission, maintenance and replication in the arthropod have not been well documented at either the genetic or the molecular level. Some information on these subjects is available for bunyaviruses as described below.

THE BUNYAVIRIDAE

The hierarchy of relationships that have been proposed for members of the Bunyaviridae is illustrated in Table 1 (Bishop, 1985). Viruses are classified in subfamilies, genera and serogroups. Four genera of viruses have been recognized (*Bunyavirus*, *Phlebovirus*, *Nairovirus* and *Uukuvirus*), a fifth, *Hantavirus*, has been proposed (McCormick *et al.*, 1982; White *et al.*, 1982; Schmaljohn & Dalrymple, 1983). The diversity that exists among the members of the Bunyaviridae includes differences in both host and vector preferences, involving many ecological niches (Karabatsos, 1985). Viruses of this family have been obtained from every continent of the world, except Antarctica. Most of the viruses have only been isolated from a limited

Table 1. *The hierarchy of virus relationships*

Family	Bunyaviridae	(No viruses)
Subfamily	Bunyavirinae	
Genus	Bunyavirus	
Serogroup	Anopheles A Group	(12)
	Anopheles B Group	(2)
	Bunyamwera Group	(26)
	Bwamba Group	(2)
	C Group	(16)
	California Group	(15)
	Capim Group	(10)
	Gamboa Group	(7)
	Guama Group	(12)
	Koongol Group	(2)
	Minatitlan Group	(2)
	Olifantsvlei Group	(3)
	Patois Group	(6)
	Simbu Group	(25)
	Tete Group	(5)
	Turlock Group	(5)
Subfamily	Phlebovirinae	
Genus	Phlebovirus	
Serogroup	Sandfly fever Naples Group	(4)
	Bujaru Group	(3)
	Candiru Group	(8)
	Chilibre Group	(2)
	Frijoles Group	(2)
	Rift valley fever Group	(4)
	Salehabad Group	(2)
	Sandfly fever Sicilian Group	(2)
Unassigned		(11)
Genus	Uukuvirus	
Serogroup	Uukuniemi Group	(7)
Subfamily	Nairovirinae	
Genus	Nairovirus	
Serogroup	Crimean–Congo hemorrhagic fever Group	(3)
	Dera Ghazi Khan Group	(6)
	Hughes Group	(8)
	Nairobi sheep disease Group	(3)
	Qalyub Group	(3)
	Sakhalin Group	(6)
Other members not assigned to a subfamily, or genus		
Serogroup	Bakau Group	(2)
	Kaiodi Group	(3)
	Hantaan Group	(4)
	Maputta Group	(4)
	Yogue Group	(2)
Unassigned		(12)

number of arthropod species (e.g., certain *Aedes* species). Only in some cases have these arthropods been shown to be permissive for virus transmission. Of course, in addition to such vectors, other insects will acquire virus when they take an infected blood meal. Some may not be permissive for virus replication at all. The identification of viruses in such insects is incidental to the normal cycle of virus maintenance. In other arthropods there may only be a low level of virus replication in the requisite tissues so that their ability to be an effective vector is limited (i.e., these insects may be considered to be semi-permissive). Where virus replication is restricted still further (e.g., just to insect midgut tissues) there may be no opportunity for virus transmission to a vertebrate host. The molecular basis of the permissiveness of particular cells in the arthropod host for virus replication is unknown. It is reasonable to postulate, though, that for many of the arbovirus members of the family there are insect species that represent every possible stage between permissiveness and non-permissiveness for virus replication and transmission.

For many members of the Bunyaviridae, certain warm-blooded vertebrates have been shown to be amplifying hosts (e.g., man, domestic animals and/or wildlife: Karabatsos, 1985). Such hosts aid in virus dissemination (e.g., through their migration). Obviously viruses are introduced on occasion into vertebrates that are not the usual amplifying hosts. In some of these hosts there may be no virus replication and little or no immunological response. In others, the virus may only replicate to a limited extent. Clearly, the infection of a vertebrate will depend on the host preferences of the insect species and the ability of the virus to replicate in cells of that vertebrate. The extent of these attributes is the scope to which a virus is able to invade particular hosts. Where an occasional introduction into a vertebrate leads to a transient virus infection, and where that vertebrate is the preferred target of another species of blood-sucking insect, there will be opportunity for the virus to infect a different arthropod species. Without doubt, the existence of such opportunities has contributed to the evolution and diversity seen among the Bunyaviridae both in terms of virus isolation and the identified preferred vector and host species.

Some members of the Bunyaviridae family (e.g., the etiologic agent of Korean hemorrhagic fever, Hantaan virus) do not have an arthropod vector. For other members of the family no vertebrate host has been identified by virus isolation, serology, or other surveys

(Karabatsos, 1985). However, this latter circumstance probably simply reflects the lack of identification of a vertebrate host.

As discussed later, horizontal transmission of virus has been observed for representative members of the Bunyaviridae. Such transmission includes the acquisition of infected blood by an arthropod, or an infected arthropod inoculating a vertebrate host, or an infected male insect inseminating and infecting a female insect. Vertical transmission of virus in arthropods has been demonstrated for some members of the family. The *in ovo* transmission of virus is one way in which a virus may survive over winter and persist in the environment.

Although the arthropods that have been implicated in vectoring viruses belonging to the Bunyaviridae only include hematophagous insects (e.g., mosquitoes, gnats, tabanids, ticks, phlebotomines, etc.), this observation really only represents a conclusion based on the results of the procedures and surveys that have been employed so far to recover and identify viruses. As new procedures are adopted and older procedures further refined, it is likely that bunyaviruses that infect other arthropods, or different organisms (e.g., plants) will be identified.

STRUCTURAL CHARACTERISTICS OF MEMBERS OF THE BUNYAVIRIDAE

Concerning their structural and genetic properties, members of the Bunyaviridae are enveloped, spherical and approximately 90–100 nm in diameter (Bishop & Shope, 1979; Fig. 1). In view of the presence of an envelope, virus infectivity is readily destroyed by treatment with lipid solvents. Virus particles contain three internal nucleocapsids, each consisting of viral nucleoprotein (N), a unique single-stranded species of RNA and transcriptase enzyme. The N protein has group-specific antigenic determinants. Viruses have an outer surface layer of glycoproteins (e.g., bunyaviruses: G1, G2) on which are located the type-specific antigens. Viral morphogenesis involves budding into the Golgi saccules and vesicles of infected cells (Murphy *et al.*, 1968; Smith & Pifat, 1982). Usually during infection there is a virus-induced proliferation of the cisternae of the Golgi apparatus.

The molecular events involved in virus replication have only been defined in broad outline for a few members of the family (Bishop,

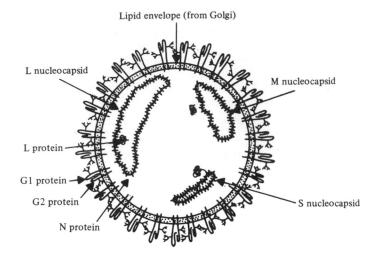

Schematic phlebovirus particle

Fig. 1. Schematic phlebovirus particle. Phlebovirus particles are spherical (approx. 100 nm in diameter), enveloped in lipid (stippled) derived from the Golgi membranes of infected cells, with an outer fringe of glycoprotein (G1, G2) and inner components including three nucleocapsids consisting of RNA (L,M,S), a major structural protein (N) and minor quantities of a large protein (L), believed to be a transcriptase. Although apparently circular, the RNA species in the nucleocapsids have complementary 3' and 5' end sequences that may be responsible for holding the structures in the circular configuration.

1985; notably certain bunyaviruses and phleboviruses). Bunyavirus replication has been reported to occur in the cytoplasm of infected cells (Goldman, Presser & Sreevalsan, 1977). After adsorption and penetration involving viral phagocytosis, the nucleocapsids are released into the cell cytoplasm and viral mRNA synthesis is initiated (Vezza *et al.*, 1979; Eshita *et al.*, 1985). Short (12–17 nucleotides long) non-viral sequences have been identified at the 5' termini of the phlebovirus and bunyavirus mRNA species (Bishop, Gay & Matsuoko, 1983; Eshita *et al.*, 1985; Ihara, Matsuura & Bishop, 1985*a*). Similar data have been obtained for uukuviruses (R. Pettersson, personal communication). Presumably the non-viral sequences are acquired by the viral transcriptase from RNA species of the host cell and used to prime the synthesis of viral mRNA (Patterson, Holloway & Kolakofsky, 1984). Overall, however, the mRNA species are shorter than their corresponding viral RNAs due to the fact that mRNA transcription terminates before the 3' end of the template RNA is reached (Eshita *et al.*, 1985; Ihara, Matsuura & Bishop, 1985*a*). It is not known what signals transcription termination. The

Bunyavirus S RNA: coding, transcription, and replication strategies

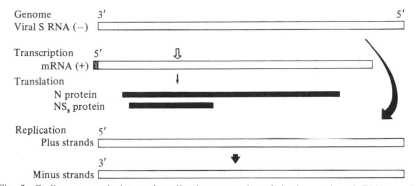

Fig. 2. Coding, transcription and replication strategies of the bunyavirus S RNA species. Transcription of the single S coded mRNA from the negative sense (−) viral RNA species involves the use of primers (stippled) to initiate mRNA synthesis. The S mRNA codes for N and a non-structural protein (NS$_S$) in overlapping reading frames. Replication of the viral S RNA involves the production of a complete viral-complementary intermediate to the synthesis of viral S RNA.

mRNA species of hantaviruses and nairoviruses have yet to be analyzed.

It appears that bunyaviruses have a simple negative-stranded coding arrangement for their L, M and S RNA species with proteins translated from viral-complementary mRNA sequences (Clerx-van Haaster & Bishop, 1980; Bishop, Fuller & Akashi, 1983; Eshita & Bishop, 1984). The N and a non-structural protein (NS$_S$) of bunyaviruses are translated from overlapping reading frames, apparently involving only a single S mRNA species (Bishop, Fuller & Akashi, 1983 and unpublished data; Fig. 2). No splicing of mRNA has been demonstrated by sequence analysis of the bunyavirus S mRNA species (Bishop et al., 1982; Bishop, Fuller & Akashi, 1983). An unanswered question is: what regulates the reading of the two gene products from the single bunyavirus S mRNA species?

Unlike bunyaviruses, the phlebovirus S RNA species (and possibly that of uukuviruses) exhibits an ambisense coding arrangement with proteins coded in viral-complementary (N protein) and viral-sense (NS$_S$ protein) mRNA sequences (Ihara, Akashi & Bishop, 1984; Fig. 3). Both types of the phlebovirus S coded mRNA species are subgenomic in size and possess short non-viral 5′ termini (Ihara, Matsuura & Bishop, 1985). The unique coding arrangement of the phlebovirus S RNA indicates that the viral-sense mRNA and encoded NS$_S$ protein cannot be made until after viral RNA replication has commenced (Ihara, Akashi & Bishop, 1984). This is in

Coding and replication strategies of phlebovirus S RNA

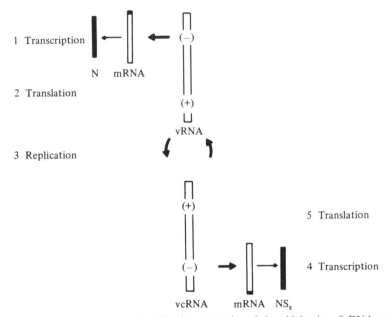

Fig. 3. Coding, transcription and replication strategies of the phlebovirus S RNA species. Transcription involves the use of primers (stippled) to initiate the synthesis of the subgenomic, viral-complementary N mRNA and subgenomic, viral-sense NS_S mRNA species. The templates of the two S coded mRNA species are the viral RNA and the complete, viral-complementary S RNA species, respectively.

contrast to the bunyavirus situation where the NS_S protein can be made from the beginning of the infection. Whether the NS_S proteins of bunyaviruses and phleboviruses serve the same function is not known. Although the coding arrangements for the S RNA species of nairoviruses or hantaviruses have yet to be reported, data obtained by C. Schmaljohn & J. Dalrymple (personal communication) indicate that the hantavirus S RNA species has a simple negative-stranded coding arrangement.

Concerning the M RNA gene products, the bunyavirus glycoproteins and a second non-structural protein (NS_M) are coded in a single viral-complementary mRNA sequence transcribed from the viral M RNA (Eshita & Bishop, 1984). The function of the NS_M protein or its location in an infected cell is unknown. The M coded mRNA species has a 5′ terminus consisting of a small non-viral sequence and a 3′ end that is some 100 nucleotides shorter than the viral

RNA (Eshita *et al.*, 1985). Similar data have been recorded for phlebovirus M mRNA species (Ihara *et al.*, 1985). The molecular properties of the bunyavirus (or phlebovirus) L mRNA species have not been reported.

It is assumed that after mRNA synthesis from the infecting viral RNA species (primary transcription), the mRNA species are translated by the cellular machinery. With the availability of new gene products viral RNA replication then commences (see Figs 2,3). Secondary transcription, dependant on RNA replication, results in the synthesis of larger amounts of the viral mRNA species and proteins (Vezza *et al.*, 1979). This is followed by the onset of viral morphogenesis which, as noted above, involves the budding of virus particles into the cisternae of the cell's Golgi apparatus. Presumably the reason why morphogenesis is restricted to the Golgi system of a cell is because the viral glycoproteins are not transported to the outer plasma membrane of the cell. Viral egress from infected cells is assumed to involve fusion of the vesicles containing virus particles with the surface membrane of the cell followed by release of the progeny virions (Smith & Pifat, 1982).

BUNYAVIRUS GENUS

Serological tests have been used to define the antigenic relationships of members of the *Bunyavirus* genus (Bishop & Shope, 1979; Karabatsos, 1985). They have established that viruses in a genus can be grouped together. Since serology has been the basis for such groupings, the term serogroup is applied (Table 2). For the most part cross-neutralization of infectivity, cross-hemagglutination inhibition, competition radioimmune precipitation, cross-complement fixation and immunodiffusion tests have been used to establish the serogroup relationships presented in Table 2. In general, a member of a *Bunyavirus* serogroup is neutralized by homologous antisera and (with different specificities) by antisera raised to other members of that serogroup (Bishop & Shope, 1979). A virus in a specific serogroup is not neutralized by antisera raised to members of other bunyavirus serogroups. Hemagglutination-inhibition studies have given similar results. The relationships of viruses included in the *Bunyavirus* groups, listed in Table 2, reflect a consensus of results obtained from a variety of reciprocal serological tests. The relative closeness of such relationships is indicated by the degrees of indentation (Table 2; representing virus complexes, serotypes, subtypes

Table 2. *Proposed serological classification of viruses of family bunyaviridae, genus Bunyavirus[a]*

Anopheles A Group	*Bwamba Group*	*Gamboa Group*	*Simbu Group*
Anopheles A	Bwamba	Gamboa	Simbu
CoAr 3624[b]	Pongola	Pueblo Viejo (75-2621)[b]	Akabane
ColAn 57389[b]		Alajuela[b]	Yaba-7[b]
Las Maloyas	*C Group*	San Juan (78V2441,[b] 75V-	Manzanilla
Lukuni	Caraparu	2374)[b]	Ingwavuma
Trombetas[b]	Caraparu (BeH5546,[b]		Inini
Tacaiuma	Trinidad)[b]	*Guama Group*	Mermet
H-32580[b]	Ossa	Guama	Buttonwillow
SPAr 2317[b] (Virgin River)	Apeu	Ananindeua	Nola
CoAr 1071[b] (CoAr 3627)[b]	Vinces	Moju	Oropouche
	Bruconha[b]	Mahogany Hammock	Facey's Paddock[b]
Anopheles B Group	Madrid	Bertioga	Utinga
Anopheles B	Marituba	Cananeia	Utive[b]
Boraceia	Murutucu	Guaratuba	Sabo
	Restan	Itimirim	Tinaroo
Bunyamwera Group	Nepuyo (63U11)[b]	Mirim	Sathuperi (Douglas)
Bunyamwera	Gumbo Limbo	Bimiti	Shamonda
Batai (Calovo)	Oriboca	Catu	Sango
Birao	Itaqui	Timboteua	Peaton
Cache Valley (Tlacotalpan)			Shuni
Maguari (CbaAr 426)[b]	*California Group*	*Koongol Group*	Aino (Kaikalur, Samford)[b]
Playas	California encephalitis	Koongol	Thimiri
Xingu[b]	Inkoo	Wongol	

Germiston
Ilesha
Lokern
Northway
Santa Rosa
Shokwe[b]
Tensaw
Kairi
Main Drain
Wyeomyia
Anhembi (BeAr 314206,[b]
  BeAr 328208)[b]
Macaua[b]
Sororoca
Taiassui[b]

La Crosse (snowshoe hare)
San Angelo
Tahyna (Lumbo)[b]
Melao
Keystone
Jamestown Canyon (South
  River,[b] Jerry Slough)
Serra do Navio
trivittatus
Guaroa
Capim Group
Capim
Acara
Moriche
Benevides
BushBush
  Benfica
  GU71U344[b]
Juan Diaz
Guajara (GU71U350)[b]

Minatitlan Group
Minatitlan
Palestina
Olifantsvlei Group
Olifantsvlei (Bobia)
Botambi
Patois Group
Patois
Abras
Babahoyo
Shark River
Zegla
Pahayokee

Tete Group
Tete
Bahig
Matruh
Tsuruse
Batama
Turlock Group
Turlock
Lednice
Umbre
M'Poko
Yaba-1[b]

[a] Viruses are classified in three steps indicated by degrees of indentation – complex, virus, and subtype; viruses in parentheses are varieties.
[b] These viruses are not in the published or working *International Catalogue of Arboviruses* (Berge, 1975; Karabatsos, 1978).

RNA species Bunyaviridae

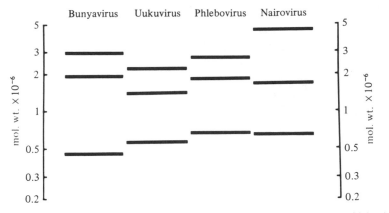

Fig. 4. Consensus sizes of the viral RNA species of bunyaviruses, uukuviruses, phleboviruses and nairoviruses.

and varieties). In addition to confirming such relationships, cross-complement fixation studies, competitive radioimmune precipitation and immunodiffusion analyses have revealed relationships involving antigens of members representing different serogroups (see Klimas *et al.*, 1981*b*). For this reason the various serogroups have themselves been grouped together into the *Bunyavirus* genus.

Similar observations have been made for members of the other genera of the Bunyaviridae (Bishop & Shope, 1979). By comparison to the bunyavirus results there are differences in the usefulness and applicability of the various serological tests employed to characterize the relationships of the members of these other genera.

Members of a Bunyaviridae genus do not share antigenic epitopes with members representing another genus. In addition, members of a virus genus, although they have similarly-sized components (RNA and protein species as illustrated in Figs 4 and 5), are for the most part structurally distinct when compared to members of another genus (Bishop, 1985). So far, the limited RNA sequence analyses that have been reported support the assignments and separation into genera shown in Table 1.

BUNYAVIRUS EVOLUTION BY GENETIC DRIFT

Evidence has been presented that for La Crosse (LAC) bunyavirus (California serogroup, Table 2) no two virus isolates recovered from

Virion polypeptides Bunyaviridae

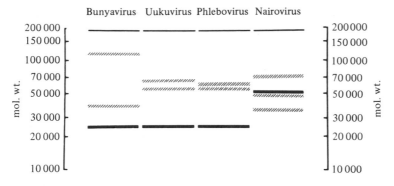

Fig. 5. Consensus sizes of the viral nucleocapsid (solid bars) and glycoproteins (hatched bars) of bunyaviruses, uukuviruses, phleboviruses and nairoviruses.

nature have identical genome sequences as evidenced by RNA oligo-nucleotide fingerprinting, or by RNA sequencing (El Said *et al.*, 1979; Klimas *et al.*, 1981*a*; Clerx-van Haaster *et al.*, 1982). This observation applies to viruses isolated from the same place but at different times, or at the same time but different places. However, by such procedures most of the LAC virus isolates have been shown to be closely related to each other (albeit they are also distinguishable). No doubt LAC viruses which have identical sequences could be isolated; however, the relationships that have been seen are taken as evidence for the continuous evolution of the virus by the accumulation of point mutations.

BUNYAVIRUS EVOLUTION BY GENETIC RECOMBINATION (REASSORTMENT)

In view of the observation that bunyaviruses have a tripartite RNA genome, with segments coding for different gene products, the possibility that recombinant viruses may be generated in dual virus infections by RNA segment reassortment (or by other mechanisms) has been investigated. Genetic studies using temperature-sensitive (*ts*) mutants of snowshoe hare (SSH) bunyavirus (California serogroup, Table 2) have confirmed that intratypic wild-type recombinant viruses can be formed from *ts* mutants representing different RNA species (Gentsch & Bishop, 1976; Gentsch *et al.*, 1977; Gentsch,

Robeson & Bishop, 1979). Similar results have been reported for LAC, Tahyna, trivittatus, Lumbo and other California group bunyaviruses (Bishop & Shope, 1979). For example, on coinfection of cells with LAC *ts* I-16 (MRNA mutant) and LAC *ts* II-5 (L RNA), recombinant wild-type viruses were recovered that, unlike the parent *ts* viruses which only gave plaques at 33 °C, produced plaques at both 39.8 °C (40 °C) and 33 °C.

Intertypic genetic recombination has been demonstrated using *ts* mutants representing different California serogroup viruses (Gentsch *et al.*, 1977, 1980; Gentsch, Robeson & Bishop, 1979; Gentsch & Bishop, 1978, 1979; Rozhon *et al.*, 1981; Shope, Rozhon & Bishop, 1981). Using SSH and LAC *ts* mutants, all the possible genotypes of reassortment viruses (2^3, i.e., 8) have been isolated from dual virus infections. Reassortant viruses have been obtained among LAC, SSH, California encephalitis, trivittatus, Lumbo and Tahyna viruses (all members of the California group). Similar data have been reported for bunyamwera serogroup members (Iroegbu & Pringle, 1981*a,b*; Pringle & Iroegbu, 1982) and Group C bunyaviruses (Bishop, Fuller & Akashi, 1983). However, viruses representing different bunyavirus serogroups do not appear to be capable of generating recombinant viruses, also not all viruses assigned to a single bunyavirus serogroup are genetically interactive (see Pringle *et al.*, 1984). No reassortants have been detected from coinfections involving the above California serogroup members and Guaroa virus (a serologically distant California group virus), Oriboca, or Caraparu (Group C viruses), or with several members of the Bunyamwera serogroup. Whether this conclusion extends to all bunyavirus serogroups, or to members of serogroups of other genera, remains to be determined.

A derivatory question from these *in vitro* experimental observations is whether bunyaviruses in their natural environment evolve through genetic recombination (reassortment). This question has to be considered in relation to the preferred vertebrate and invertebrate hosts, the viral determinants of permissive infections, and the opportunities afforded for dual virus infection (see below). Direct evidence for naturally occurring reassortant viruses has been obtained. RNA genome fingerprint analyses of field isolates of LAC virus have provided evidence for intratypic recombinant (reassortant) LAC viruses (Klimas *et al.*, 1981*a*). Reassortant viruses have also been identified among field isolates of members of the Patois serogroup of bunyaviruses (Ushijima, Clerx-van Haaster & Bishop, 1981).

BUNYAVIRUS INFECTIONS OF MOSQUITO SPECIES

The natural relationship between LAC virus and *Ae. triseriatus* mosquitoes has been extensively studied. This mosquito species has been demonstrated to be an efficient oral (Watts *et al.*, 1972), transovarial (Pantuwatana *et al.*, 1974; Watts *et al.*, 1973*a,b*), and venereal transmitter of LAC virus (Thompson & Beaty, 1977). In temperate regions of the United States of America, LAC virus overwinters in diapaused *Ae. triseriatus* eggs (Watts *et al.*, 1974; Beaty & Thompson, 1975).

Immunofluorescence techniques have been used to determine the virogenesis of LAC in *Ae. triseriatus* and to derive anatomical explanations of the vector-virus interactions (transovarial and venereal transmission) observed in this system. Subsequent to oral infection, virus antigen was first detected in the pyloric portion of the midgut (six days post-infection). By ten days the virus had disseminated from the midgut and antigen was detected in most secondary organ systems, including ovaries and salivary glands. It was observed that LAC virus infection was virtually pantropic in the arthropod with most organ systems containing large quantities of virus antigen. Detection of virus antigen in ovarian follicles and in accessory sex gland fluid provided anatomical explanations for the observed transovarial and venereal transmission, respectively (Beaty & Thompson, 1976, 1977).

Although serologically closely related, each of the California serogroup bunyaviruses has a distinct epizootiology, often involving particular but not exclusive vector and vertebrate hosts (Turell & Le Duc, 1983). For example, in the United States of America, trivittatus virus is closely associated with an *Ae. trivittatus*-cottontail rabbit feeding cycle. Keystone virus is associated with an *Ae. atlanticus*-squirrel cycle. SSH virus (which is serologically almost indistinguishable from LAC virus), is associated with an *Ae. canadensis* and *Ae. communis* group-snowshoe hare cycle, while LAC virus in the midwest of the USA (and elsewhere) is associated with an *Ae. triseriatus*-chipmunk-tree squirrel cycle (Sudia *et al.*, 1971; Pantuwatana *et al.*, 1972; Le Duc, 1979). California serogroup viruses have been isolated on occasion from alternate arthropods. An example is LAC virus which has been isolated from *Ae. canadensis* and *Ae. communis* group mosquitoes, as well as from tabanids (Karabatsos, 1985). SSH virus has been isolated infrequently from *Ae. triseriatus* mosquitoes. In addition, at least six of the California group viruses have been isolated from *Ae. vexans*. The possibility that the viruses may evolve in nature into species that exploit new ecological niches cannot, therefore, be ignored.

VIRAL DETERMINANTS OF PERMISSIVE BUNYAVIRUS INFECTIONS OF MOSQUITOES

Since all the possible genotype combinations of LAC and SSH virus reassortants are available (Gentsch *et al.*, 1977; Gentsch, Robeson & Bishop, 1979; Rozhon *et al.*, 1981), and because *Ae. triseriatus* mosquitoes are not the normal vectors of SSH virus, the question of the viral determinants for a permissive replication of LAC and SSH virus in that arthropod species has been investigated by employing either LAC, or SSH virus, or LAC-SSH reassortant viruses to infect *Ae. triseriatus* mosquitoes. The results obtained from several studies are summarized in Table 3 (Beaty *et al.*, 1981a, 1982). The data have been interpreted to indicate that the LAC viral M RNA gene products (the glycoprotein species) are the principal determinants of the efficiency of LAC virus, both to establish a disseminated infection and to be transmitted by *Ae. triseriatus* mosquitoes. By contrast, viruses with a SSH M RNA were inefficiently transmitted. Although attenuating mutations in other LAC RNA species may affect the LAC M gene property, as determined with an attenuated LAC/LAC/SSH reassortant virus (L/M/S RNA species, Table 3 and Rozhon *et al.*, 1981), the major viral determinants of efficient vector transmission appear to be the LAC viral glycoproteins (Beaty *et al.*, 1981a, 1982).

BUNYAVIRUS RECOMBINATION IN MOSQUITOES AFTER INTRATHORACIC INFECTION

From the bunyavirus isolation data it can be concluded that even though the viruses are distinct epizootiologically, many are sympatric throughout much of their respective ranges. In theory this would allow ample opportunity for dual virus infections to occur in nature. In order to investigate experimentally the question of whether dual California group virus infections of mosquitoes would result in intertypic recombinant virus formation, dual infections of laboratory

Table 3. *The role of the M RNA in infection, dissemination and transmission of LAC and SSH parent and SSH-LAC reassortant viruses by* Aedes triseriatus *mosquitoes*

Virus L/M/S/ genotype	% Disseminated infection[a]	% Transmission[b]
LAC/LAC/LAC	100	100
SSH/LAC/LAC	97	96
SSH/LAC/SSH	97	90
LAC/LAC/SSH	12	64
SSH/SSH/SSH	17	33
LAC/SSH/LAC	42	42
LAC/SSH/SSH	8	36
SSH/SSH/LAC	29	31

[a] Mosquitoes were allowed to engorge on blood-virus mixtures containing either wild-type LAC, or SSH, or LAC-SSH reassortants of the indicated L/M/S RNA genotypes. After 14 days of extrinsic incubation, the mosquitoes were then analysed. Viral antigen in mosquito tissues was identified by immunofluorescence and the % disseminated infections was scored on the basis of the numbers of mosquitoes in which antigen was observed in all tissues, divided by those for which antigen was only located in midgut cells. The LAC/LAC/SSH data used in the dissemination analyses are probably atypical since they involved a reassortant which subsequent analyses (Rozhon *et al.*, 1981) demonstrated carried a silent attenuating L defect. In each of the analyses, representing the data from several experiments, between 13 and 45 individual mosquitoes were used (for details see Beaty *et al.*, 1982).

[b] The % transmission was scored on the basis of the numbers of individual mosquitoes that transmitted virus and induced disease in the suckling mice (moribund or dead mice) following intrathoracic inoculation of 10^3 plaque-forming units of virus and after 10 days (average) of extrinsic incubation, divided by the number of feeding mosquitoes that exhibited disseminated infections when the engorged mosquitoes were sacrificed and analysed by immunofluorescence. Viruses were inoculated intrathoracically to bypass the mesenteron and thereby preclude variables associated with midgut passage. In each of the analyses, representing the data from several experiments, between 14 and 60 mosquitoes were used (for details see Beaty *et al.*, 1981*a*).

stocks of colonized *Ae. triseriatus* mosquitoes were undertaken using intrathoracic inoculation of *ts* mutants of LAC and SSH viruses. The study yielded evidence for wild-type SSH-LAC reassortant virus formation and for the transmission of the recombinant viruses to a vertebrate host (Table 4; Beaty *et al.*, 1981*b*). In parenthesis, one combination of mutants (SSH II-21 × LAC I-20) did not yield the expected reassortant progeny. Similar results were obtained in tissue culture (Gentsch *et al.*, 1977). It was subsequently shown that in tissue culture the expected reassortants are only infrequently obtained (Rozhon *et al.*, 1981). In summary, it has been demonstrated that at least certain intertypic reassortant bunyaviruses can be generated in the arthropod host.

Table 4. *Viruses recovered from intrathoracic dually infected* Aedes triseriatus *mosquitoes and from mice on which the infected mosquitoes were allowed to feed*[a]

Virus cross			Mosquitoes		Mice	
			% ts	% wt	% ts	% wt
SSH I-1	×	SSH II-22	95	5	90	10
LAC I-20	×	LAC II-4	45	55	20	80
SSH I-1	×	LAC II-5[b]	35	65	55	45
SSH II-21	×	LAC I-20[c]	100	0	100	0

[a] Viruses recovered from dually infected mosquitoes that had been inoculated with *ts* viruses of LAC or SSH virus representing different RNA segments (mutants SSH I-1, SSH II-22, etc.), or viruses obtained from derived moribund and dead mice, were plated on BHK-21 cells at 33 °C, virus plaques picked and reassayed at both 33 °C and 39.8 °C to score for temperature-sensitive (*ts*) and wild-type (*wt*) viruses. The results for each cross represent the averages of analyses of several mosquitoes recovered after 7, 14 and 21 days post-inoculation, or of one or more mice obtained after the 7, 14 and 21 day mosquito feedings (for details see Beaty *et al.*, 1983).
[b] For different progeny *wt* virus clones, both virus induced intracellular polypeptides, and RNA oligonucleotide fingerprint analyses, indicated that the expected SSH/LAC/SSH reassortants were present (Gentsch *et al.*, 1977; Beaty *et al.*, 1981b).
[c] The *wt* progeny that would be expected from this cross are rarely obtained even in tissue culture, presumably due to inefficient gene product interactions of the heterologous viral RNA species (Gentsch *et al.*, 1979; Rozhon *et al.*, 1981).

BUNYAVIRUS RECOMBINATION IN MOSQUITOES AFTER ORAL INFECTION

Studies were conducted to determine if *Ae. triseriatus* mosquitoes would yield recombinant viruses if they were allowed to orally acquire the appropriate viruses either simultaneous or following interrupted feeding (Beaty *et al.*, 1985). Mosquitoes were allowed to partially engorge on blood-virus mixtures containing either LAC *ts* I-16, or LAC *ts* II-5, or wild-type LAC virus (to serve as three control experiments), or mixtures of LAC *ts* I-16 and wild-type LAC virus, or LAC *ts* I-16 and LAC *ts* II-5 viruses. As shown in Table 5, after 14 days of incubation, wild-type viruses were only obtained from the dually infected mosquitoes that fed on blood containing the two *ts* mutants, or from those that received the mutant and wild-type viruses. As expected, only mutant viruses were obtained from the controls that ingested a single *ts* mutant; likewise wild-type virus was recovered from the mosquitoes that only received the wild-type virus inoculum (data not shown). The results of the simultaneous *ts* mutant infections indicated therefore that recombination

Table 5. *Dual infection of* Aedes triseriatus *mosquitoes and generation of recombinant viruses*[a]

Infection protocol	Infection rates	
	33 °C assay	40 °C assay
Simultaneous		
LAC *ts* I-16 + LAC *wt*	15/15 (100%)	15/15 (100%)
LAC *ts* I-16 + LAC *ts* II-5	8/8 (100%)	2/8 (25%)
Interrupted feeding		
LAC *ts* I-16 then LAC *wt*	19/19 (100%)	18/19 (95%)
LAC *ts* I-16 then LAC *ts* II-5	20/20 (100%)	4/20 (20%)

[a] Infection rates are expressed as the number of mosquitoes that were found to contain virus (≥ 10 pfu) as detected by plaque assay at 33 °C or 40 °C, divided by the number tested. Virus-blood meals on which the mosquitoes were initially allowed to feed (partially for the interrupted feeding protocol) contained 6.5–7.3 logs of each of the indicated viruses/ml. At two hours post-ingestion the mosquitoes in the interrupted feeding experiment were allowed to engorge to completion on blood-virus mixtures containing 6.5–7.3 logs/ml of the second virus (LAC *ts* II-5 or LAC wild-type, *wt*, virus). All mosquitoes were held for 14 days, triturated, then assayed for virus. For further details see Beaty and associates (1985).

had occurred in the dually infected mosquitoes.

For the interrupted feeding studies, mosquitoes were allowed to engorge partially on blood meals containing LAC *ts* I-16 and two hours later permitted to engorge to completion on meals containing LAC *ts* II-5 (Table 5). Wild-type viruses were detected in the mosquitoes after 14 days of incubation. As expected, mosquitoes that received wild-type LAC virus through the interrupted feeding protocol also yielded wild-type virus (Table 5). Thus analyses of the mosquitoes which were superinfected through interrupted feeding protocol showed that recombinant viruses were produced by this method.

Since bunyavirus RNA segment reassortment would only be epidemiologically significant if the recombinant viruses were transmitted to a vertebrate host, mosquitoes from the above experiment were permitted to feed 14 days post-infection on groups of five to seven baby mice. Brains were extracted from the resulting moribund or dead mice and assayed at the permissive (33 °C) and non-permissive (40 °C) temperatures for virus replication. Viruses producing plaques at 40 °C (i.e., with a wild-type phenotype) were isolated from the mice on which mosquitoes were allowed to feed that had previously ingested LAC *ts* I-16 followed two hours later by wild-type LAC virus, indicating that prior infection by the mutant did not preclude transmission of the superinfecting virus. Also, wild-type virus was recovered from the mice which were substrates for feeding by mosquitoes initially infected with LAC *ts* I-16 then super-infected with LAC *ts* II-5. These results indicated, therefore, that recombinant viruses could be recovered from a vertebrate host on which a dually infected mosquito had fed.

The preceding experiments indicated a low level of recombinant virus formation (Table 5) in both the simultaneous and interrupted feeding protocols. The reason for the low level of recombinants may be trivial (e.g., the particular experimental conditions that were employed), or may be related to the relative numbers of the infecting viruses. The effect of these and other factors, such as the incubation time, the number of gonadotrophic cycles, or the use of alternative vectors, upon the production of recombinant bunyaviruses has yet to be determined.

BUNYAVIRUS INTERFERENCE IN MOSQUITO SPECIES

In nature, the opportunities for simultaneous infection of vector species involving the ingestion of two or more viruses in a single blood meal are probably severely limited by the acute (short term) character of bunyavirus infections in a vertebrate host. In most instances, critical viraemia threshold titres for infection of mosquitoes are present for only a few days duration before the virus is cleared. Infection and production of maximum viraemia in a particular vertebrate host by two viruses simultaneously is probably relatively rare, although it may occur. It can be argued that the opportunities for sequential infection of a vector species are more probable if several blood meals are taken by the female mosquito. It is known that *Ae. triseriatus* females may ingest several blood meals during their lifetime, thereby allowing the possibility for dual virus infection to occur by that route (DeFoliart, 1983). Other means of dual virus infections are possible, for instance when an infected male insect inseminates and infects a previously infected female, or when an infected mosquito that has acquired virus transovarially from its mother ingests a blood meal containing an alternate virus.

A series of experiments was therefore conducted to analyse the potential for bunyavirus superinfection of *Ae. triseriatus* and to determine possible temporal and phylogenetic constraints on these phenomena. In initial experiments, mosquitoes were inoculated intrathoracically with a *ts* mutant of LAC virus (*ts* II-5, an L RNA mutant) and three, seven, or 14 days later superinfected by the same route with SSH *ts* I-1 (an M RNA mutant). After further extrinsic incubation the mosquitoes were triturated in tissue culture medium and the homogenates assayed at 33 °C and 40 °C to quantitate the numbers of *ts* and reassortant wild-type viruses (respectively). Despite the presence of 10^{3-5} plaque forming units of *ts* viruses, no wild-type viruses were detected (Beaty *et al.*, 1983). The reverse virus inoculation schedule (i.e., SSH *ts* I-1 followed by LAC *ts* II-5) gave similar results, suggesting that genetic interaction between the two viruses had been inhibited.

Reassortment can be an event that occurs at a low frequency; therefore, in order to examine the interference phenomenon with a more sensitive procedure, mosquitoes were inoculated intrathoracically with a *ts* mutant of LAC virus and subsequently challenged seven days later by intrathoracic inoculation of either an homologous or heterologous wild-type virus. Since wild-type virus progeny can easily be quantitated by *in vitro* plaque assays (at 40 °C), the ability of the superinfecting virus to replicate in the mosquito could be determined. The results of the analyses are shown in Table 6. They indicated that the mosquitoes were resistant to superinfection with related California group viruses but not to viruses of other families (Table 6), or to viruses representing other bunyavirus gene pools (Beaty *et al.*, 1983). The data can be interpreted in terms of an interference phenomenon that is specific to viruses that are members of a gene pool.

If mosquitoes became resistant to superinfection by natural routes of infection, the opportunities for dual infection of vectors and, for bunyaviruses, virus evolution through RNA segment reassortment, would be limited. In nature, many mosquito species exhibit a behaviour pattern called interrupted feeding. If the defensive reaction of a host causes the mosquito to interrupt its feeding, the vector may complete engorgement at a later time on an alternate host. Thus mosquitoes could ingest blood meals from two different vertebrate hosts that are viremic with two different viruses in a period of time brief enough to preclude interference. In the light of these considerations, an experiment was conducted to determine when interference to oral superinfection occurs in mosquitoes (Beaty *et al.*, 1985).

In order to determine if mosquitoes could be superinfected by the oral route, *Ae. triseriatus* mosquitoes were permitted to ingest

Table 6. *Parenteral superinfection of* Aedes triseriatus *mosquitoes previously inoculated with a La Crosse (LAC) ts mutant virus*[a]

Virus inoculum		Geometric mean titer (log pfu/mosquito)		
Day 0	Day 7	33 °C assay	40 °C assay	Log difference
LAC I-16	None	4.5 ± 0.3	≤1.0	
None	LAC wt	4.5 ± 0.4	4.2 ± 0.2	
LAC I-16	LAC wt	4.4 ± 0.1	≤1.0	≥3.2
None	SSH wt	3.3 ± 0.5	3.0 ± 0.5	
LAC I-16	SSH wt	4.1 ± 0.4	≤1.0	≥2.0
None	TAH wt	4.1 ± 0.6	3.3 ± 0.5	
LAC I-16	TAH wt	4.1 ± 0.4	≤1.0	≥2.3
None	TVT wt	4.6 ± 0.3	3.7 ± 0.2	
LAC I-16	TVT wt	3.8 ± 0.2	≤1.0	≥2.7
None	WN wt	5.1 ± 0.3	4.4 ± 0.2	
LAC I-16	WN wt	5.2 ± 0.3	4.4 ± 0.4	0
None	VSV wt	4.9 ± 0.3	4.0 ± 0.4	
LAC I-16	VSV wt	4.5 ± 0.2	3.4 ± 0.7	0.6

[a] Groups of (minimally) 4 mosquitoes were inoculated intrathoracically with 1.4 log plaque forming units (pfu) of a LAC *ts* mutant representing the M RNA segment (LAC I-16) and superinfected 7 days later by inoculation with 2–4 log pfu of wild-type viruses representing LAC, or other California group bunyaviruses (snowshoe hare, SSH, Tahyna, TAH, trivitattus, TVT), or a flavivirus (West Nile, WN), or a rhabdovirus (vesicular stomatitis virus, VSV). After a further 7 days of extrinsic incubation, the presence of wild-type and *ts* viruses in the mosquitoes was determined by plaque assays. For further details see Beaty and associates (1983).

a partial or complete blood meal containing a LAC *ts* mutant virus (LAC I-16). At predetermined times post-feeding, the mosquitoes were permitted to engorge to repletion on a blood meal containing wild-type LAC virus. One cohort of mosquitoes ingested a meal containing both LAC *ts* I-16 and wild-type virus. As shown in Table 7, mosquitoes that received the wild-type virus challenge in the first 24 hours replicated the superinfecting virus. Mosquitoes that ingested wild-type virus after 48 hours were resistant to superinfection (Beaty *et al.*, 1985). Control groups of mosquitoes that received only *ts* virus yielded only *ts* progeny viruses (i.e., the virus was phenotypically stable; Beaty *et al.*, 1985). It was concluded from these studies and the earlier experiment (Table 5), demonstrating that reassortment could occur in mosquitoes superinfected two hours after the initial virus infection, that with the greater elapsed time

Table 7. *Interference to LAC virus oral superinfection of* Aedes triseriatus *mosquitoes*[a]

Time until ingestion of challenge virus	Infection rates	
	33 °C assay	40 °C assay
Simultaneous	15/15 (100%)	15/15 (100%)
30 min	8/8 (100%)	8/8 (100%)
2 h	19/19 (100%)	18/19 (95%)
4 h	7/7 (100%)	7/7 (100%)
1 day	18/18 (100%)	11/18 (60%)
2 day	11/11 (100%)	3/11 (27%)
7 day	6/6 (100%)	0/6 (0%)
14 day	5/5 (100%)	0/5 (0%)
21 day	3/3 (100%)	0/3 (0%)
28 day	4/4 (100%)	0/4 (0%)

[a] Infection rates are expressed as the number of mosquitoes that were found to contain virus (≥ 10 pfu) as detected by plaque assay at 33 °C or 40 °C, divided by the number tested. Virus blood meals on which the mosquitoes were initially allowed to feed (partially) contained 6.5–7.8 logs of LAC *ts* mutant I-16 per ml. At the indicated times post-ingestion the mosquitoes were allowed to engorge to completion on blood-virus mixtures containing 7–7.8 logs per ml of the challenge wild-type virus. All mosquitoes were held for 14 days after the second meal, triturated, than assayed for virus. For further details see Beaty and associates (1985).

between the two blood meals the mosquitoes became refractory to homologous virus superinfection.

The molecular basis for the interference phenomenon remains to be determined. Possible mechanisms of interference requiring further investigation include the removal of cellular receptors, used for virus infection of a cell, or the development of defective interfering viruses which prevent the replication of a second, genetically interactive, virus. Using the available genetic tools, further experimentation is required to explore these possibilities and to determine whether interference is operative in transovarially-infected mosquitoes.

Whatever the mechanism, the experimental observation of virus interference between genetically permissive viruses, if indeed it has a counterpart in nature, may restrict the ability of a bunyavirus to evolve by RNA segment reassortment. That reassortment occurs has been shown by analyses of natural virus isolates (Klimas *et al.*, 1981*a*; Ushijima, Clerx-van Haaster & Bishop, 1981). However, of some 25 LAC virus isolates analysed, only one intratypic recombinant virus was identified, which may be interpreted to indicate that recombination is a rare event. It is not clear at this time whether interference is a viral or host mediated phenomenon.

VIRULENCE CHARACTERISTICS OF REPRESENTATIVE BUNYAVIRUSES

Bunyaviruses exhibit a wide range of virulence patterns in vertebrate species. LAC virus is known to cause a severe and sometimes fatal encephalitis in children (Fauvel *et al.*, 1980; Thompson, Kalfayan & Anslow, 1965). Other members of the California group (Table 2) such as SSH, Jamestown Canyon, California encephalitis, Inkoo and Tahyna (TAH) viruses, as well as members of certain other bunyavirus serogroups, have also been implicated on occasion in the etiology of human encephalitis in various regions of the world. Among the Group C, Guama, Bwamba, Tataguine and Bunyamwera viruses are agents that are responsible for minor illnesses in humans which almost always involve uncomplicated fevers, or symptoms of fever and rash (Shope, 1985). Such viruses are primarily found in the tropics and do not usually cause large outbreaks of human disease. The geographically limited nature of these diseases is probably due to the habitat limits of the normal vector and vertebrate hosts of the individual viruses and the abilities of these hosts to support virus replication. Oropouche virus (Simbu serogroup, Table 2) has caused significant outbreaks of human disease in Brazil (Pinheiro *et al.*, 1962) involving several thousands of cases both in 1962 and subsequently (Shope, 1985).

Among the other members of the family there are viruses that cause hemorrhagic fever (Crimean–Congo hemorrhagic fever virus), hemorrhagic fever with renal syndrome (Hantaan viruses), sandfly fever (phlebotomus fever viruses) and Rift valley fever which, although primarily of significance to the welfare of domestic animals such as sheep and cattle, is also responsible for severe and sometimes fatal human disease (Iman, Karamany & Darwish, 1978; Meegan, 1979). A review of what is known about the pathogenesis, pathology, clinical features and diagnosis of these viruses has been provided recently by Shope (1985).

GENETIC ASPECTS OF THE VIRULENCE POTENTIAL OF CALIFORNIA GROUP BUNYAVIRUSES

In model animal systems the virulence characteristics of LAC and TAH viruses differ. Intraperitoneal inoculation of outbred Swiss

Table 8. *Intracerebral (ic) and intraperitoneal (ip) virulence in mice of viruses containing LAC, SSH, or TAH M RNA*

Virus	log pfu/ml	log LD50/ml[a]		Index of LD50/pfu[b]
		ic	ip	
Viruses with LAC M RNA				
LAC/LAC/LAC	9.0	7.7	4.4	1.0
SSH/LAC/SSH	8.4	7.4	4.5	5.0
SSH/LAC/LAC	9.0	8.3	4.0	0.4
TAH/LAC/TAH	8.9	7.7	4.8	3.2
TAH/LAC/LAC	8.1	7.7	3.0	0.3
LAC/LAC/SSH[c]	8.2	8.3	2.5	0.08
Viruses with SSH M RNA				
SSH/SSH/SSH	8.1	8.3	4.3	6.3
SSH/SSH/LAC	7.7	7.2	3.5	2.5
LAC/SSH/SSH	7.4	7.7	2.8	1.0
Viruses with TAH M RNA				
TAH/TAH/TAH	9.1	9.0	0[d]	0.0013
LAC/TAH/LAC	7.3	7.1	0	0.063
SSH/TAH/SSH	7.3	6.5	0	0.063
LAC/TAH/TAH	8.9	8.5	0	0.002
SSH/TAH/TAH	8.2	7.7	0	0.01

[a] 5 4-week-old mice were inoculated per dilution.
[b] ip 50% lethal dose (LD50) per plaque forming unit (pfu) of La Crosse virus/ip LD50 per pfu of virus X.
[c] LAC/LAC/SSH was derived from a cross involving LAC/LAC/LAC I-20 × SSH/LAC/SSH II-13 *ts* mutants and was found to contain an attenuating L RNA mutation (Rozhon *et al.*, 1981).
[d] 0 equals less than 1.6 LD50/ml.

mice by LAC virus elicits a fatal disease in 4-week-old animals at virus concentrations which do not cause death when TAH virus is employed (Shope *et al.*, 1981). SSH virus resembles LAC virus in this regard. By using the available LAC, SSH and TAH reassortant viruses, the genetic basis for these properties has been shown to reside with the bunyavirus M RNA gene products, presumably the viral glycoproteins (Table 8). Parental and reassortant viruses that contain a LAC or SSH M RNA species were found to be more virulent by the intraperitoneal route of inoculation than those with a TAH M RNA species (Shope *et al.*, 1981, 1982). One virus that initially appeared to behave aberrantly was a LAC/LAC/SSH reassortant. However, backcross analyses with a distinguishable wild-type LAC virus showed that this particular reassortant contained

Table 9. *Survival of mice inoculated intracerebrally with viruses having LAC, or TAH, M RNA*

A. Small dose (1.0–1.9 log LD50)[a]	Days ± SEM		Days ± SEM	Difference[b]
Viruses with LAC M RNA		*Viruses with TAH M RNA*		
LAC/LAC/LAC	5.4 ± 0.6	TAH/TAH/TAH	2.6 ± 0.2	
TAH/LAC/LAC	6.8 ± 0.4	LAC/TAH/TAH	3.6 ± 0.4	
TAH/LAC/TAH	6.0 ± 1.0	LAC/TAH/LAC	2.2 ± 0.4	
SSH/LAC/SSH	6.8 ± 0.8	SSH/TAH/SSH	3.0 ± 0.4	
SSH/LAC/LAC	3.6 ± 0.2	SSH/TAH/TAH	4.6 ± 0.4	
LAC/LAC/SSH[c]	6.6 ± 0.5			
Average	5.8 ± 0.3	Average	3.2 ± 0.2	2.6

B. Large dose (3.0–3.9 log LD50)[a]	Days ± SEM		Days ± SEM	Difference[b]
Viruses with LAC M RNA		*Viruses with TAH M RNA*		
LAC/LAC/LAC	3.4 ± 0.2	TAH/TAH/TAH	1.8 ± 0.2	
TAH/LAC/LAC	4.0 ± 0.4	LAC/TAH/TAH	2.8 ± 0.4	
TAH/LAC/TAH	3.9 ± 0.3	LAC/TAH/LAC	1.6 ± 0.2	
SSH/LAC/SSH	4.8 ± 0.4	SSH/TAH/SSH	2.2 ± 0.4	
SSH/LAC/LAC	2.8 ± 0.2	SSH/TAH/TAH	3.8 ± 0.4	
LAC/LAC/SSH[c]	4.4 ± 0.4			
Average	3.9 ± 0.2	Average	2.4 ± 0.2	1.5

[a] 5 4-week-old mice per dilution. Data are mean ± SEM.
[b] Average difference in days between survival of mice receiving viruses having LAC M RNA and mice receiving viruses having TAH M RNA.
[c] LAC/LAC/SSH was derived from a cross involving LAC/LAC/LAC I-20 × SSH/LAC/SSH II-13 *ts* mutants and was found to have an attenuating L RNA mutation (Rozhon *et al.*, 1981).

an attenuating L RNA species. Although it could have been acquired spontaneously, it is likely that the attenuating mutation was induced when mutagen was used to obtain the LAC *ts* virus used to derive the reassortant (Rozhon *et al.*, 1981). Whatever the origins, the observation demonstrates that in addition to the M RNA species the gene products specified by the other bunyavirus RNA species can affect the viral phenotype.

It has been observed that mice which receive a lethal dose of TAH virus by the intracerebral route of inoculation do not live as long as those receiving an equivalent dose of LAC or SSH virus. This is illustrated in Table 9, together with data from dose-related studies with reassortant LAC, SSH and TAH viruses. Such data again show that the bunyavirus M RNA gene is responsible for this virulence phenotype (Shope *et al.*, 1981).

From the above studies it is clear that the M RNA species of certain California group bunyaviruses, presumably through their gene products, govern the invasion potentials and other virulence characteristics of the viruses. Similar results have been obtained with Group C viruses (Bishop, Fuller & Akashi, 1983). Whether these observations extend to other bunyaviruses, or to other members of the Bunyaviridae, remains to be determined. Apart from such overall characteristics the results say nothing about the tissue tropisms of the individual viruses, nor why LAC but not TAH virus is lethal following intraperitoneal inoculation, nor why TAH virus kills mice faster than LAC virus after intracerebral inoculation.

In a study designed to follow the infection course of California group bunyaviruses following subcutaneous inoculation in mice, Tignor and associates (1983) have reported that there is an antigenemia (viremia) which extends from 18 to 40 hours post-inoculation. Viral antigen was observed in the lumena of capillaries and in phagocytic cells adjacent to these vessels. For some viruses, antigen was also observed in chondrocytes as well as fibroblasts and basement membranes adjoining connective tissues. LAC virus antigen has been observed in muscle fibre cells and at sites that appeared to represent neuromuscular junctions, raising the possibility that such sites may have a role to play in the entrance of the virus into the central nervous system. Such sites did not appear to be infected by TAH or LAC/TAH/LAC reassortant viruses (Tignor *et al.*, 1983).

Distinct histopathologic lesions in the brains of mice inoculated intracerebrally with various California serogroup viruses have been identified for LAC, TAH and trivittatus viruses (Shope *et al.*, 1982). LAC virus induced a lytic, mildly inflammatory encephalitis with marked necrosis in the olfactory cortex and in the pyramidal cells of the hippocampus. TAH virus infection was also characterized by lesions in these tissues, but in addition there was a striking focal necrosis in the granular layer of the cerebellum. In most mice inoculated intracerebrally with TAH virus, the amount of perivascular inflammation by the time of death was not as marked as that observed for the LAC virus infected brains. No lesions were identified in the brains of mice which received TAH virus by the intraperitoneal route of inoculation. Mice infected intracerebrally with trivittatus virus survived longer than those infected with LAC or TAH virus. It was observed that for trivittatus-infected brains there was a loss of neurons, as well as gliosis and demyelinating signs in the olfactory region. In addition there was focal necrosis in the granular layer of the cerebellum and a marked inflammatory response throughout the brain. One lesion that appeared to be specific to trivittatus virus infections was neurolysis in the dendate gyrus.

CONCLUSIONS

Analyses of the biological, molecular and genetic attributes of California group members have contributed significantly to what is known about the infection potentials of arboviruses both in the vertebrate host and in potential arthropod vectors. The two host systems represent considerably different substrates for virus infections. In the vertebrate the virus has to elicit within a short period of time a sufficient viremia for transmission to a blood-sucking arthropod to occur. The incidence of diseases such as encephalitis may be incidental to the production of a viremia, or it may predispose the infected animal to becoming an easier target for the arthropod vector. In the arthropod host the virus has to replicate in various tissues in order to be transmitted horizontally (to a vertebrate) or vertically (to an offspring). The availability of genetic and molecular tools to address the questions pertaining to each of these situations has allowed considerable progress to be made in understanding virus–vector–host relationships.

REFERENCES

BEATY, B. J., BISHOP, D. H. L., GAY, M. & FULLER, F. (1983). Interference between bunyaviruses in *Aedes triseriatus* mosquitoes. *Virology*, **127**, 83–90.

BEATY, B. J., HOLTERMAN, M., TABACHNICK, W., SHOPE, R. E., ROZHON, E. J. & BISHOP, D. H. L. (1981*a*). Molecular basis of bunyavirus transmission by mosquitoes: Role of the middle-sized RNA segment. *Science*, **211**, 1433–5.

BEATY, B. J., MILLER, B. R., SHOPE, R. E., ROZHON, E. J. & BISHOP, D. H. L. (1982). Molecular basis of bunyavirus *per os* infection of mosquitoes: The role of the M RNA segment. *Proceedings of the National Academy of Sciences, USA*, **79**, 1295–7.

BEATY, B. J., ROZHON, E. J., GENSEMER, P. & BISHOP, D. H. L. (1981*b*). Formation of reassortant bunyaviruses in dually-infected mosquitoes. *Virology*, **111**, 662–5.

BEATY, B. J., SUNDIN, D. R., CHANDLER, L. & BISHOP, D. H. L. (1985). Evolution of bunyaviruses via genome segment reassortment in dually-infected (per os) *Aedes triseriatus* mosquitoes. *Science*, **230**, 548–50.

BEATY, B. J. & THOMPSON, W. H. (1975). Emergence of La Crosse virus from endemic foci: Fluorescent antibody studies of overwintered *Aedes triseriatus*. *American Journal of Tropical Medicine and Hygiene*, **24**, 685–91.

BEATY, B. J. & THOMPSON, W. H. (1976). Delineation of La Crosse virus in developmental stages of transovarially infected *Aedes triseriatus. American Journal of Tropical Medicine and Hygiene*, **25**, 505–12.

BEATY, B. J. & THOMPSON, W. H. (1977). Tropisms of La Crosse virus in *Aedes triseriatus* following infective blood meals. *Journal of Medical Entomology Honolulu*, **14**, 499–503.

BERGE, T. O. (1975) International catalogue of arboviruses including certain other viruses of vertebrates. *US Department of Health, Education and Welfare, DHEW Publicn. No. (COC) 75–8301.*

BISHOP, D. H. L. (1985). The genetic basis for describing viruses as species. *Intervirology*, **24**, 79–93.

BISHOP, D. H. L., CALISHER, C., CASALS, J., CHUMAKOV, M. P., GAIDAMOVICH, S. Y. A., HANNOUN, C., LVOV, D. K., MARSHALL, I. D., OKER-BLOM, N., PETTERSSON, R. F., PORTERFIELD, J. S., RUSSELL, P. K., SHOPE, R. E. & WESTAWAY, E. G. (1980). Bunyaviridae. *Intervirology*, **14**, 125–43.

BISHOP, D. H. L., FULLER, F. J. & AKASHI, H. (1983). Coding assignments of the RNA genome segments of California serogroup viruses. In *Progress in Clinical and Biological Research*, **123**, ed. C. H. Calisher & W. H. Thompson, pp. 107–17. New York: A. Liss.

BISHOP, D. H. L., GAY, M. E. & MATSUOKO, Y. (1983). Nonviral heterogeneous sequences are present at the 5' ends of one species of snowshoe hare bunyavirus S complementary RNA. *Nucleic Acids Research*, **11**, 6409–18.

BISHOP, D. H. L., GOULD, K. G., AKASHI, H. & CLERX-VAN HAASTER, C. M. (1982). The complete sequence and coding content of snowshoe hare bunyavirus small (S) viral RNA species. *Nucleic Acids Research*, **10**, 3703–13.

BISHOP, D. H. L. & SHOPE, R. E. (1979). Bunyaviridae. In *Comprehensive Virology*, Vol. *14*, ed. H. Fraenkel-Conrat & R. R. Wagner, pp. 1–156. New York: Plenum Press.

CLERX-VAN HAASTER, C. M., AKASHI, H., AUPERIN, D. D. & BISHOP, D. H. L. (1982). Nucleotide sequence analyses and predicted coding of bunyavirus genome RNA species. *Journal of Virology*, **41**, 119–28.

CLERX-VAN HAASTER, C. M. & BISHOP, D. H. L. (1980). Analyses of the 3' terminal sequences of snowshoe hare and La Crosse bunyaviruses. *Virology*, **105**, 564–74.

DEFOLIART, G. R. (1983). *Aedes triseriatus:* Vector biology in relationship to the persistence of La Crosse virus in endemic foci. In *Progress in Clinical and Biological Research*, Vol. **123**, ed. C. H. Calisher & W. H. Thompson, pp. 89–104. New York: A. Liss.

EL SAID, L. H., VORNDAM, V., GENTSCH, J. R., CLEWLEY, J. P., CALISHER, C. H., KLIMAS, R. A., THOMPSON, W. H., GRAYSON, M., TRENT, D. W. & BISHOP, D. H. L. (1979). A comparison of La Crosse virus isolates obtained from different ecological niches and an analysis of the structural components of California encephalitis serogroup viruses and other bunyaviruses. *American Journal of Tropical Medicine and Hygiene*, **28**, 364–86.

ESHITA, Y. & BISHOP, D. H. L. (1984). The complete sequence of the M RNA of snowshoe hare bunyavirus reveals the presence of internal hydrophobic domains in the viral glycoproteins. *Virology*, **137**, 227–40.

ESHITA, Y., ERICSON, B., ROMANOWSKI, V. & BISHOP, D. H. L. (1985). Analyses of the mRNA transcription processes of showshoe hare bunyavirus S and M RNA species. *Journal of Virology*, **55**, 681–9.

FAUVEL, M., ARTSOB, H., CALSIHER, C. H., DAVIGNON, L., CHAGNON, A., SKVORC-RANKO, R. & BELLONCIK, S. (1980). California group virus encephalitis in three children from Quebec: Clinical and serologic findings. *Journal of Canadian Medicine*, **122**, 60–70.

FIELDS, B. N. (ed.) (1985). Virology. Raven Press: New York.

GENTSCH, J. & BISHOP, D. H. L. (1976). Recombination and complementation between temperature-sensitive mutants of the bunyavirus, snowshoe hare virus. *Journal of Virology*, **20**, 351–4.

GENTSCH, J. & BISHOP, D. H. L. (1978). Small viral RNA segment of bunyaviruses codes for viral nucleocapsid protein. *Journal of Virology*, **28**, 417–19.

GENTSCH, J. R. & BISHOP, D. H. L. (1979). M viral RNA segment of bunyaviruses codes for two unique glycoproteins: G1 and G2. *Journal of Virology*, **30**, 767–76

GENTSCH, J. R., ROBESON, G. & BISHOP, D. H. L. (1979). Recombination between snowshoe hare and La Crosse bunyaviruses. *Journal of Virology*, **31**, 707–17.

GENTSCH, J. R., ROZHON, E. J., KLIMAS, R. A., EL SAID, L. H., SHOPE, R. E. & BISHOP, D. H. L. (1980). Evidence from recombinant bunyavirus studies that the M RNA gene products elicit neutralizing antibodies. *Virology*, **102**, 190–204.

GENTSCH, J., WYNNE, L. R., CLEWLEY, J. P., SHOPE, R. E. & BISHOP, D. H. L. (1977). Formation of recombinants between showshoe hare and La Crosse bunyaviruses. *Journal of Virology*, **24**, 893–902.

GOLDMAN, N., PRESSER, L. & SREEVALSAN, T. (1977). California encephalitis virus: some biological and biochemical properties. *Virology*, **76**, 352–64.

IHARA, T., AKASHI, H. & BISHOP, D. H. L. (1984). Novel coding strategy (ambisense genomic RNA) revealed by sequence analyses of Punta Toro phlebovirus S RNA. *Virology*, **136**, 293–306.

IHARA, T., MATSUURA, Y. & BISHOP, D. H. L. (1985). Analyses of the mRNA transcription processes of Punta Toro phlebovirus (Bunyaviridae). *Virology*, **147**, 317–25.

IHARA, T., SMITH, J., DALRYMPLE, J. M. & BISHOP, D. H. L. (1985). Complete sequences of the glycoprotein and M RNA of Punta Toro phlebovirus compared to those of Rift Valley fever virus. *Virology*, **144**, 246–59.

IMAN, I. Z. E., KARAMANY, R. E. & DARWISH, M. A. (1978). Epidemic of Rift Valley fever (RVF) in Egypt: isolation of RVF virus from animals. *Journal of the Egyptian Public Health Association*, **265–9.**

IROEGBU, C. U. & PRINGLE, C. R. (1981*a*). Genetic interactions among viruses of the Bunyamwera complex. *Journal of Virology*, **37**, 383–94.

IROEGBU, C. U. & PRINGLE, C. R. (1981*b*). Genetics of the Bunyamwera complex. In *The Replication of Negative Strand Viruses*, ed. D. H. L. Bishop & R. W. Compans, pp. 159–65. New York: Elsevier.

KARABATSOS, N. (1978). Supplement to International Catalogue of Arboviruses including certain other viruses of vertebrates. *American Journal of Tropical Medicine and Hygiene*, **27**, 372–440.

KARABATSOS, N. (ed.) (1985). *International Catalogue of Arboviruses*, 3rd Edition. San Antonio, Texas: American Society of Tropical Medicine and Hygiene.

KLIMAS, R. A., THOMPSON, W. H., CALISHER, C. H., CLARK, G. G., GRIMSTAD, P. R. & BISHOP, D. H. L. (1981*a*). Genotypic varieties of La Crosse virus isolated from different geographic regions of the continental United States and evidence for a naturally occurring intertypic recombinant La Crosse virus. *American Journal of Epidemiology*, **114**, 112–31.

KLIMAS, R. A., USHIJIMA, H., CLERX-VAN HAASTER, C. M. & BISHOP, D. H. L. (1981*b*). Radioimmune assays and molecular studies that place Anopheles B and Turlock serogroup viruses in the *Bunyavirus* genus (Bunyaviridae). *American Journal of Tropical Medicine and Hygiene*, **30**, 876–87.

LE DUC, J. (1979). The ecology of California group viruses. *Journal of Medical Entomology*, **16**, 1–17.

MATTHEWS, R. E. F. (1982). Classification of nomenclature of viruses. *Fourth Report of the International Committee on Taxonomy of Viruses*. Basel: Karger.

McCormick, J. G., Palmer, E. L., Sasso, D. R. & Kiley, M. P. (1982). Morphological identification of the agent of Korean hemorrhagic fever (Hantaan virus) as a member of Bunyaviridae. *Lancet*, **1**, 765–68.

Meegan, J. M. (1979). The Rift Valley fever epizootic in Egypt 1977–1978. 1. Description of the epizootic and virological studies. *Transactions of the Royal Society of Tropical Medicine and Hygiene*, **73**, 618–723.

Murphy, F. A., Whitfield, S. G., Coleman, P. H., Calisher, C. H., Rabin, E. R., Jenson, A. B., Melnick, J. L., Edwards, M. R. & Whitney, E. (1968). California group arboviruses: Electron microscopic studies. *Experimental Molecular Pathology*, **9**, 44–56.

Pantuwatana, S., Thompson, W. H., Watts, D. M. & Hanson, R. P. (1972). Experimental infection of chipmunks and squirrels with La Crosse and trivittatus viruses and biological transmission of La Crosse by *Aedes triseriatus*. *American Journal of Tropical Medicine and Hygiene*, **21**, 476–81.

Pantuwatana, S., Thompson, W. H., Watts, D. M., Yuill, T. M. & Hanson, R. P. (1974). Isolation of La Crosse virus from field collected *Aedes triseriatus* larvae. *American Journal of Tropical Medicine and Hygiene*, **23**, 246–50.

Patterson, J. L., Holloway, B. & Kolakofsky, D. (1984). La Crosse virions contain a primer-stimulated RNA polymerase and a methylated cap-dependent endonuclease. *Journal of Virology*, **52**, 215–222.

Pinheiro, F. P., Pinheiro, M., Bensabath, G., Causey, O. R. & Shope, R. E. (1962). Epidemia de virus Oropouche im Belem. *Review Series Saude Publica*, **12**, 15–23.

Pringle, C. R., Clark, W., Lees, J. F. & Elliott, R. M. (1984). Restriction of sub-unit reassortment in the bunyaviridae. *Molecular Biology of Negative Strand Viruses*, ed. R. W. Compans & D. H. L. Bishop, pp. 45–50. Orlando: Academic Press.

Pringle, C. R. & Iroegbu, C. U. (1982). A mutant identifying a third recombination group in a bunyavirus. *Journal of Virology*, **42**, 873–79.

Rozhon, E. J., Gensemer, P., Shope, R. E. & Bishop, D. H. L. (1981). Attenuation of virulence of a bunyavirus involving an L RNA defect and isolation of LAC/SSH/LAC and LAC/SSH/SSH reassortants, *Virology*, **111**, 125–38.

Schmaljohn, C. S. & Dalrymple, J. M. (1983). Analysis of Hantaan virus RNA: Evidence for a new genus of Bunyaviridae. *Virology*, **131**, 482–91.

Shope, R. E. (1985). Bunyaviridae. In *Virology*, ed. B. N. Fields, pp. 1055–82. New York: Raven Press.

Shope, R. E., Rozhon, E. J. & Bishop, D. H. L. (1981). Role of the middle-sized bunyavirus RNA segment in mouse virulence. *Virology*, **114**, 273–76.

Shope, R. E., Tignor, G. H., Jacoby, R. O., Watson, H., Rozhon, E. J. & Bishop, D. H. L. (1982). Pathogenicity analyses of reassortant bunyaviruses: coding assignments. In *International Symposium on Tropical Arboviruses and Hemorrhagic Fever*, ed. F. Pinheiro, pp. 135–46. Belem, Brazil: Impresso Nationale Fundamental Science and Technical Developments.

Smith, J. F. & Pifat, D. Y. (1982). Morphogenesis of sandfly viruses (Bunyaviridae family). *Virology*, **121**, 61–81.

Sudia, W. D., Newhouse, V. F., Calisher, C. H. & Chamberlain, R. W. (1971). California group arboviruses: Isolations from mosquitoes in North America. *Mosquito News*, **31**, 576–600.

Thompson, W. & Beaty, B. (1977). Venereal transmission of La Crosse (California encephalitis) arbovirus in *Aedes triseriatus* mosquitoes. *Science*, **196**, 530–1.

Thompson, W. H., Kalfayan, B. & Anslow, R. O. (1965). Isolation of California encephalitis group virus from a fatal human disease. *American Journal of Epidemiology*, **81**, 245–53.

TIGNOR, G. H., BURRAGE, T. G., SMITH, A. L., SHOPE, R. E. & BISHOP, D. H. L. (1983). California serogroup gene structure–function relationships: virulence and tissue tropisms. *Proceedings of the International Symposium on California Serogroup Viruses*, ed. C. H. Calisher & W. H. Thompson, pp. 129–38. New York: A. Liss.

TURELL, M. J. & LEDUC, J. W. (1983). The role of mosquitoes in the natural history of California serogroup viruses. *Progress in Clinical and Biological Research*, vol. 123, ed. C. H. Calisher & W. H. Thompson, pp. 43–55. New York: A. Liss.

USHIJIMA, H. CLERX-VAN HAASTER, C. M. & BISHOP, D. H. L. (1981). Analyses of Patois group bunyaviruses: evidence for naturally occurring recombinant bunyaviruses and existence of viral coded nonstructural proteins induced in bunyavirus infected cells. *Virology*, **110**, 318–32.

VEZZA, A. C., REPIK, P. M., CASH, P. & BISHOP, D. H. L. (1979). *In vivo* transcription and protein synthesis capabilities of bunyaviruses: wild-type snowshoe hare virus and its temperature-sensitive Group I, Group II and Group I/II mutants. *Journal of Virology*, **31**, 426–36.

WATTS, D. M., GRIMSTAD, P. R., DEFOLIART, G. R., YUILL, T. M. & HANSON, R. P. (1973a). Laboratory transmission of La Crosse encephalitis virus by several species of mosquitoes. *Journal of Medical Entomology*, **10**, 583–6.

WATTS, D. M., MORRIS, C. D., WRIGHT, R. E., DEFOLIART, G. R. & HANSON, R. P. (1972). Transmission of La Crosse virus (California encephalitis group) by the mosquito *Aedes triseriatus*. *Journal of Medical Entomology*, **9**, 125–7.

WATTS, D., PANTUWATANA, S., DEFOLIART, G., YUILL, T. M. & THOMPSON, W. H. (1973b). Transovarial transmission of La Crosse virus (California encephalitis group) in the mosquito, *Aedes triseriatus*. *Science*, **182**, 1140–1.

WATTS, D. M., THOMPSON, W. H., YUILL, T. M., DEFOLIART, G. R. & HANSON, R. P. (1974). Over-wintering of La Crosse virus in *Aedes triseriatus*. *American Journal of Tropical Medicine and Hygiene*, **23**, 694–700.

WHITE, J. D., SHIREY, F. G., FRENCH, G. R., HUGGINS, J. W. & BRAND, O. M. (1982). Hantaan virus aetiological agent of Korean haemorrhagic fever, has Bunyaviridae-like morphology. *Lancet*, **1**, 768–71.

MOLECULAR AND CELLULAR ASPECTS OF RETROVIRUS PATHOGENESIS

ROBIN A. WEISS

Institute of Cancer Research, Chester Beatty Laboratories, Fulham Road, London SW3 6JB, UK

INTRODUCTION

Retroviruses infect vertebrate hosts ranging from fish to man and, as listed in Table 1, different strains cause diverse diseases. The first disease attributable to a filtrable agent later classified as a retrovirus was equine infectious anaemia described in 1905. In 1908 Ellerman & Bang reported that erythroid leukaemia in chickens was infectiously transmissible, and from 1910 to 1914 Rous published his series of studies on the sarcoma agent that we can now view as the first experimental retrovirus to transduce a cellular oncogene. But the medical world was not prepared to accept that viruses could cause cancer, and Rous' discovery waited 55 years for the Nobel prize.

Meanwhile, the importance of oncogenic viruses in mammals was indicated by Bittner's observation in 1936 of the transmission of murine mammary carcinoma by a factor in milk. The experiments by Gross in 1951 on the induction of tumours following inoculation of new-born C3H mice with thymoma filtrates revealed latent infections and the pathogenesis both of murine leukaemia virus (MLV) and of polyoma virus. Following Gross' seminal work, a renaissance of tumour virus occurred, largely inspired by Dulbecco and others who applied the new cell culture techniques to viral transformation. Quantitative assays allowing biological cloning of Rous sarcoma (RSV) by Temin & Rubin and MLV by Rowe & Huebner set the stage for the genetic and molecular analysis of retroviruses that has flowered in the last 20 years (Weiss *et al.*, 1985).

The first authentic retroviral pathogen of man was not, however, reported until 1980 (Poiesz *et al.*, 1980). Human T-cell leukaemia virus type 1 (HTLV-1) causes adult T-cell leukaemia-lymphoma (ATLL), first identified as a malignancy endemic in southwest Japan (Uchiyama *et al.*, 1977) and among West Indians (Catovsky *et al.*, 1982). A related virus, HTLV-2, was identified in cells derived from a patient with hairy cell leukaemia (Kalyanaraman *et al.*, 1982),

Table 1. *Diseases caused by retroviruses*

Leukaemia
Lymphoma
Sarcoma
Carcinoma

Aplastic anaemia
Haemolytic anaemia
Myelodysplasia
Autoimmunity
Immunodeficiency

Osteopetrosis
Arthritis

Encephalitis
Slow neuropathy
Paralysis

Pneumonia

and further isolates now indicate that HTLV-2 is causally related to a rare type of hairy cell leukaemia with T-cell markers.

A human retrovirus disease, acquired immune deficiency syndrome (AIDS), has emerged as a new pestilence of alarming proportions (Curran *et al.*, 1985; Pinching & Weiss, 1986). The causative agent, human immunodeficiency virus type 1 (HIV-1) was first isolated by Barré-Sinoussi *et al.* (1983) and its causal association with AIDS was firmly established by Gallo *et al.* (1984). The virus has been variously named LAV, IDAV, HTLV-3 and ARV, but HIV has recently been adopted as the accepted nomenclature (Coffin *et al.*, 1986). A second, related, virus associated with AIDS, named HIV-2 (LAV-2), has recently been identified in West Africans (Clavel *et al.*, 1986). A virus named HTLV-4 from healthy subjects also in West Africa (Kanki *et al.*, 1986) may well be the same agent, as both are closely related to a virus isolated from African green monkeys (Kanki *et al.*, 1985). The distribution and virulence of HIV-2 are as yet unclear, but HIV-1 has spread in just a few years to become one of man's most formidable microbiological foes.

CLASSIFICATION

Retroviruses are classified into three subfamilies (Weiss *et al.*, 1985). The *Spumavirinae* or foamy viruses are the least understood. They occur as latent infections of cats, monkeys and man, and several serotypes have been isolated from the brain and other tissues (Hooks & Gibbs, 1975). In culture, foamy viruses induce the appearance of large vacuolated cells – hence their name. They are difficult to propagate to high titre and attempts to generate molecular clones have only recently been undertaken. They are not known to be pathogenic, although it has been suggested that human foamy virus may be associated with a rare autoimmune disease, De Quervain subacute thyroiditis (Werner & Gelderblom, 1979). Foamy viruses deserve more attention but the lack of information precludes further discussion here.

The *Lentivirinae* are so called because the prototype virus, visna virus (VV), can cause a 'slow' degeneration of the central nervous system in sheep (Haase, 1986). Equine infectious anaemia virus (EIAV) already mentioned, and caprine arthritis encephalitis virus (CAEV) also belong to this group. Because HIV and related simian immunodeficiency viruses (SIV) also resemble lentiviruses, there has been a resurgence of research into this previously rather neglected subfamily.

The *Oncovirinae* have been the most extensively studied retroviruses. These include oncogenic retroviruses, but also some cytopathic strains causing anaemia or immunodeficiency, and others that are not pathogenic at all in their natural hosts. Oncoviruses are subdivided on morphological criteria into B-type, C-type and D-type viruses (A-type particles are incomplete virions).

TRANSMISSION AND LATENCY

Retroviruses are, of course, RNA viruses that replicate via a DNA intermediate, mediated by the viral enzyme reverse transcriptase. The DNA proviral genome of oncoviruses integrates into host chromosomal DNA, and may remain latent in the infected cell and its descendants, or be expressed as messenger and genomic RNA. The production of progeny oncovirus is not usually cytocidal and infected cells will persist throughout the life of the host. Lentiviruses also establish persistent infections at the cellular level in which integrated proviruses are found, but it appears the chromosomal integration is not an obligatory step for lentivirus replication.

In addition to horizontal spread of infection, retroviruses are passed vertically through milk transmission (e.g. mouse mammary tumour virus and probably HTLV-1 and HIV), and congenital transmission (e.g. avian leukosis virus and HIV). For some retroviruses, cell-free transmission is rare and they depend on the spread of infected lymphocytes or macrophages (e.g. HTLV-1 and VV). Others (e.g. HIV) clearly can be transmitted cell free, at least by inoculation, as became tragically apparent with the widespread infection of haemophiliacs through the administration of contaminated batches of clotting factor VIII.

The most remarkable mode of transmission in the case of some oncoviruses is in the form of proviral sequences integrated in the DNA of the gametes (endogenous viruses). These are inherited by the host in a Mendelian way, and may remain wholly latent, partially expressed by the synthesis of viral antigens in certain tissues, or, more rarely, fully expressed as an endogenous viraemia.

In a sense, oncoviruses represent the transposable elements of vertebrates (Temin, 1980), colonising the host chromosomes, including the germ-line, and occasionally hijacking host genes as seen in oncogene transduction. The persistence and transmission of retrovirus infection has been reviewed in a previous SGM Symposium (Weiss, 1982) and will be discussed here only in relation to latency and also the seemingly unlikely role of endogenous genomes in pathogenesis.

Retrovirus latency is not only a feature of the endogenous viruses, but also of infectiously transmitted retroviruses of all three subfamilies. Infection, once acquired, remains in the host for life but may be inapparent as with foamy viruses, and also with some oncoviruses only 1% of individuals infected perinatally with HTLV-1 develop ATLL, (usually after the age of 50 years). Cycles of latent infection and viraemia are a feature of lentiviruses, especially EIAV, in which recurrent fever occurs at intervals when antigenically variant substrains arise. Antigenic and genomic drift has also been recorded during the course of VV infection of sheep but, unlike EIAV, the original strain persists alongside the new antigenic variants. Host genetics also plays a role in determining latency. The German sheep by which visna was introduced into Iceland with such disastrous effects were themselves widely infected with no evidence of disease (Haase, 1986).

STRUCTURE AND FUNCTION OF THE GENOME AND
PROTEINS

The DNA provirus of retroviruses is bounded at either end by the long terminal repeats (LTR), containing promotor and enhancer sequences for viral RNA transcription, as well as sequences required for integration. The LTRs are sensitive to host transcriptional control as well as viral signals, both of which may determine the specificity of virulence and pathogenesis.

Until recently, naturally occurring retroviruses were thought to have only three genes, *gag*, encoding the structural proteins for core antigens, *pol*, the reverse transcriptase and endonuclease and protease, and *env*, the envelope glycoproteins. Oncogene-bearing retroviruses are generally defective, and are not transmitted from one host to another except by experimental inoculation.

The internal core or *gag* proteins are cleaved from a polypeptide precursor translated from a full-length mRNA resembling genomic RNA. A viral protease mediates cleavage mainly during virion assembly while budding from the plasma membrane, so that the major protein found inside infected cells is the precursor.

The *pol* protein is translated from the same mRNA, but much less efficiently, as there is a stop codon between *gag* and *pol*. Read-through or frame-shift translation (according to the retrovirus strain) allows a *gag–pol* precursor to be synthesised and this is then cleaved to the functional proteins. The protease at the 5′ end of *pol* mediates *gag* cleavage and probably *gag–pol* cleavage too (Yoshinaka *et al.*, 1985), so that it is autocatalytic and may serve to allow accumulation of substantial amounts of precursor proteins before cleavage and packaging into virions late in the replication cycle. This may be particularly important in lytic retrovirus replication as with virulent phases of lentivirus expression.

The *env* gene encodes the envelope proteins. A transmembrane protein, sometimes glycosylated, serves to anchor a disulphide-linked external glycoprotein. Both envelope proteins are cleaved from a larger precursor by cellular proteases. The external glycoprotein recognises specific receptors on cells susceptible to infection and a remarkable feature of retroviruses is the great diversity of receptors and hence host range among different species and strains.

The lentivirus external *env* glycoproteins are particularly heavily glycosylated. In natural infection they are also more variable in sequence than some oncovirus glycoproteins and are larger, having extra domains. These may possibly serve to mask functional domains from access to neutralising antibodies until required for receptor interaction when the virus interacts with target cells for infection (Coffin, 1986). Although receptor-mediated endocytosis appears to be the usual mechanism of entry into cells, some retroviruses, e.g. HTLV-1 and HIV-1 frequently generate cell-to-cell fusion of receptor-bearing cells to form giant, multinucleated syncytia (Nagy et al., 1983; Montagnier et al., 1984). If the phenomenon occurs in vivo it may contribute to pathogenesis (Sodroski et al., 1986a).

The genetic variability of retroviruses may be partially explained by the error-prone process of reverse transcription. However, that does not account for the greater polymorphism seen with lentiviruses (e.g. HIV-1) than with oncoviruses (e.g. HTLV-1). Perhaps the obligatory integrative step in viral genome replication of oncoviruses, and the chronic production of progeny genomes from these stably integrated proviruses, limit the high rate of genetic divergence seen with lytically replicating retroviruses.

TRANSACTIVATING PROTEINS

When the HTLV-I genome was first cloned and sequenced, it became apparent that this virus contained sequences located between the env gene and the 3' LTR that were not present in other sequenced oncoviruses (Seiki et al., 1983). This region of the genome, X, contained four open reading frames, of which the fourth and longest codes for a nuclear protein. The X^{IV} protein of HTLV-1 and HTLV-2 behaves as a transactivating factor for viral transcription by binding specifically to an enhancer sequence in the LTR (Sodroski, Rosen & Haseltine, 1984). This was demonstrated by the construction of plasmids containing the bacterial chloramphenicol acetyl transferase (CAT) gene and the LTR of HTLV-I so the CAT expression depended on LTR activation. By transfecting such plasmids into cells expressing different HTLV-1 genes, it was shown that CAT expression was dependent on the pX^{IV} sequence, which could cause up to 1000-fold enhancement of CAT expression. This new, nonstructural control gene has been called tat, standing for transactivation of transcription (Sodroski et al., 1985), and it acts on a specific enhancer sequence in the 5' LTR (Fujisawa et al., 1986).

Viruses related to HTLV-1, such as HTLV-2 and BLV, also carry tat genes. Expression of tat acts to enhance further transcription of the viral genome. In the absence of tat activity, the gag, pol and env genes are barely expressed at all, and infection remains

latent. But, if *tat* is expressed, then substantial viral gene expression
and virus replication can occur. The *tat* gene of HTLV-1 actually
has three exons, with double splice sites within the *gag* and *env*
regions (Wong-Staal & Gallo, 1985). Clearly it must be under some
degree of independent expression from the other genes in order
to act in positive feed-back upon the LTR. The *tat* gene plays an
important role in the fine tuning of viral regulation for latent and
productive infection. As discussed later, it may also transactivate
cellular genes (Greene *et al.*, 1986) and this could be an important
factor in oncogenesis by HTLV and BLV.

With the cloning of HIV-1, further open reading frames (orf)
representing non-structural genes were revealed (Rabson & Martin,
1985). These genes, shown in Fig. 1, had not previously been seen

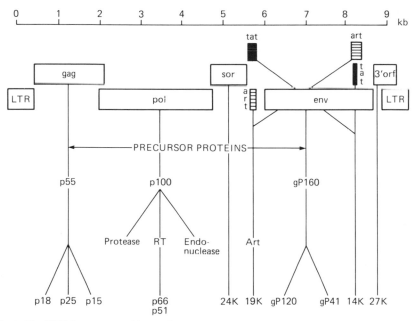

Fig. 1. The HIV-1 genome and its products.

in retroviruses. However, HIV-1 was the first lentivirus to be
sequenced and the VV genome was soon shown to have a similar
structure (Sonigo *et al.*, 1985), though possibly lacking the orf that
overlaps with the 3' LTR.

Transcriptional analysis utilising cDNA clones of HIV-1 has established that there are three functional open reading frames between

the *pol* and *env* genes as well as the 3' orf (Arya & Gallo, 1986; Sodroski *et al.*, 1986*b*). Each has double-spliced mRNA. The functions of *sor* (*orfA*) and 3'*orf* are unknown, but they are expressed in natural infection because infected subjects produce serum antibodies specific to proteins encoded by these genes (Arya & Gallo, 1986). Deletion of the *sor* or 3'*orf* genes causes only a modest reduction in viral replication in T-lymphocytes in culture. It has recently been established that at least two transactivating proteins are encoded by HIV-1 genes name *tat* and *art*, as described below (Fisher *et al.*, 1986; Sodroski *et al.*, 1986*b*). Thus, HIV-1 has at least seven genes (Fig. 1), whereas oncoviruses have only three.

The *tat* gene HIV-1 was at first assumed to act like *tat* of HTLV-1 to stimulate transcription by binding to enhancer sequences in the LTR. Deletion analysis had established the presence of a *cis*-acting regulatory sequence in the LTR, which is activated by the *tat* product. However, using LTR-CAT recombinants, it was noted that *tat* of HIV-1 causes a large increase in CAT protein synthesis without a large effect on the mRNA level. It therefore appears that transactivation by *tat* of HIV-1 is mediated through a post-transcriptional regulation, probably by the *tat* protein binding to sequences at the 5' end of *gag* and *env* mRNAs (Rosen *et al.*, 1986). Deletion of *tat* prevents replication of HIV-1, but it can be complemented in *trans* by cells expressing *tat*.

The second transactivator gene, called *art* for activation of retrovirus translation, also acts to promote protein synthesis from pre-transcribed *gag* and *env* mRNAs (Sodroski *et al.*, 1986*b*). The bipartite coding sequences for *art* overlap *tat* and *env*, in different reading frames (Fig. 1). The *art* gene was discovered by Haseltine and colleagues following observations that certain deletions in the first exon of the *tat* region could not be complemented by a functional *tat* provided in *trans*. The *tat* protein transactivates any gene manipulated to be initiated by the HIV LTR. On the other hand, the *art* protein acts on a sequence in the mRNAs of *gag* and *env* which is 3' to the LTR and is therefore specific to these genes alone.

The *tat* and *art* genes of HIV-1 serve as important post-transcriptional regulators of the synthesis of structural proteins of the virus. These control mechanisms are giving us new insight into the regulation of latency and virulence at the cellular level in lentivirus infection. When *tat* and *art* are expressed, large amounts of viral *gag* and *env* proteins are quickly synthesised and are probably cytocidal. These controls suggest an 'early–late' switch or perhaps more aptly, a 'late–very late' switch in the control of virus expression, by regulating the translation of viral mRNAs already accumulated in the cell. And this may help to explain aspects of lentivirus pathogenesis, in which very few cells actively replicate the virus at any one time.

PATHOGENESIS OF LENTIVIRUSES

This subfamily of viruses is linked with a multiplicity of syndromes (Haase, 1986). Perhaps the most salient feature is degenerative disease of the central nervous system (CNS), characteristic of visna virus (VV), caprine arthritis encephalitis virus (CAEV) and HIV. CNS disease is apparently not caused by EIAV or by progressive pneumonia virus (PPV), an as yet ill-defined retrovirus of sheep and goats, distinct from VV but often present in the same flock. In the case of VV, the CNS disease is characterised by progressive demyelination, one of the target cells for infection being the oligodendrocyte. However, much of the CNS disease in sheep as in CAEV-infected goats results from inflammatory encephalitis as the lesions tend to disappear with immunosuppression.

In AIDS, it is not yet clear whether CNS disease develops independently of immunosuppression. Paralysis is less evident than in visna, but progressive loss of mental, sensory and motor functions occurs, and is linked cytologically with brain tissue atrophy and the appearance of vacuolated and multinucleated cells. There is sometimes a transitory encephalitis or aseptic meningitis 2–4 weeks after primary HIV-1 infection, but the slow, non-inflammatory degenerative disease only becomes apparent much later (Pinching & Weiss, 1986).

Pneumonia is both a primary symptom of lentivirus infection (VV, CAEV, HIV-1) and a secondary one of opportunistic infection by *Pneumocystis carinii* following immunosuppression in human and simian AIDS. Visna virus is also known as Maedi virus, from the Icelandic for breathlessness, and the pneumonic or maedi form of the disease was not at first identified with visna. As in the brain, VV and CAEV infection in the lung leads to inflammatory responses of lymphocytes and macrophages which occlude the pulmonary alveoli. Primary infection of alveolar macrophages by HIV-1 has been reported in AIDS by Gartner *et al.* (1986) who have discerned evidence for preferential tropism of brain and lung HIV-1 isolates for macrophage infection and peripheral blood lymphocytes isolates for T-cells.

The arthritis of CAEV-infected goats is again associated with invasion of the joints by infected and inflammatory cells. It is seen to a lesser extent in visna and transiently in HIV-1 infection.

Equine infectious anaemia is a rather different syndrome, on the one hand, from VV and CAEV infections in sheep, and, on the other, from AIDS. The characteristic anaemia of EIAV infection is a haemolytic one rather than a bone marrow defect. Erythrocytes become coated with viral antigen and antibody and are then subject to phagocytosis and to complement-mediated haemolysis. Circulating immune complexes are correlated with recurrent fever and also glomerulo-nephritis. The glomerular disease and thrombocytopaenic purpura sometimes seen in HIV-1 infections may similarly be linked with immune complexes. The molecular biology of EIAV is less well known than HIV-1 and VV, and its relationship with other lentiviruses, apart from its morphology and cyclical variation, remains to be determined.

A characteristic of several lentivirus infections is a severe wasting disease or cachexia. The wasting in visna-affected sheep may be partly attributable to starvation as the progressive paralysis prevents normal grazing and food intake. In Uganda, AIDS is locally called 'slim' disease, because it typically presents as an enteropathic disease with severe weight loss (Serwadda et al., 1985). The wasting is accompanied by chronic and copious diarrhoea. While some cases of enteropathic AIDS may be attributed to opportunistic enteric infections such as *Isospora* and *Cryptosporidia*, a direct effect of HIV-1 on the intestinal mucosa and its macrophages also appears likely.

HIV-1 and SIV infections are distinguished from other lentivirus infections by the development of severe immunodeficiency, linked to a depletion of T4$^+$ helper T-lymphocytes and malfunction of macrophages and antigen-presenting cells (Pinching & Weiss, 1986). The immunosuppression allows the colonisation of the host by life-threatening viral, fungal, mycobacterial and protozoan infections which would be controlled as asymptomatic or mild infections in immunocompetent individuals. The cancers associated with AIDS also appear to result from immunodeficiency, although Kaposi's sarcoma sometimes precedes overt immunological compromise. Other lentiviruses are not associated with neoplasia, except for the benign pulmonary adenomatosis of PVV, and this further emphasises the link between immunosuppression and cancer in AIDS.

CELL TROPISM AND VIRAL RECEPTORS

Retroviruses use specific cell surface receptors to initiate infection. Although the host genetics of retrovirus receptors has long been studied in chickens, mice and men, the HIV-1 receptor is the first to be defined as a biochemical entity. The restriction of HIV-1 infection *in vitro* to cells bearing the T4 differentiation antigen suggested that this antigen itself might be a component of the receptor, and this was demonstrated by the ability of monoclonal antibodies (mAbs) to some, but not all, epitopes of T4 to block infection and syncytium induction by HIV-1 (Dalgleish *et al.*, 1984; Klatzmann *et al.*, 1984; Sattentau *et al.*, 1986). The binding of HIV-1 virions or purified *env* gp120 to T4+ cells is blocked by anti-T4 mAbs and gp120-T4 conjugates can be specifically immune precipitated from cells that have bound virions (McDougal *et al.*, 1986).

T4 receptor dependence for HIV-1 infection is not confined to T-helper lymphocytes. Some EBV-transformed B-lymphoblast cell lines susceptible to HIV-1 infection also express T4 antigen (Dalgleish *et al.*, 1984) as do many monocytes and antigen-presenting cells. Studies with the U937 monocytic leukaemia cell line show that its susceptibility to HIV-1 infection is mediated by the T4 antigen receptor (Åsjö *et al.* 1986; Clapham *et al.*, 1986). In experiments employing transfection and expression of the gene for T4 into cells that do not naturally express this antigen, it was found that human T-cells, B-cells and HeLa carcinoma cells became susceptible to infection and replication of HIV-1 (Maddon *et al.*, 1986). By contrast, T4-transfected mouse cells permitted only the binding of HIV-1 virions to the cell surface, without further steps in virus or pseudotype penetration.

While T4 antigen accounts for HIV-1 infection of T-lymphocytes and macrophages (and possibly intestinal mucosa cells and endothelial cells), the precise cell targets for brain infection and their receptor for virus binding remain obscure. Infection of macrophages and lymphocytes in the brain can be readily explained, and microglia representing brain macrophages of bone marrow origin are probably one target. However, we have observed inefficient, usually latent, infection of some glioma cell lines, and it seems likely from *in situ* hybridisation studies (Shaw *et al.*, 1985) that glial and possibly neuronal cells may be infected directly *in vivo*. Maddon *et al.* (in press) have shown that the T4 gene is expressed in the brain, but the T4 mRNA transcript is smaller (1.8 Kb in place of 3.2 Kb) than in lymphocytes and macrophages. If a truncated or modified form of T4 antigen is expressed on the surface of brain cells, it will be interesting to see whether it serves as the HIV-1 receptor.

The identification of T4 antigen, as the receptor, goes some way to explain pathogenesis by HIV-1 in terms of the specific tissues and cell types affected and the function of T4 in immune networks (Dalgleish, 1986). The receptor gene transfer experiments (Maddon *et al.*, in press), and also transfection and replication of the complete HIV-1 genome into fibroblasts bypassing the requirement for receptors (Levy *et al.*, 1986), indicate that stages of HIV-1 infection later than virion penetration are not restricted by cell type.

The precise tissue specificity of receptor expression of other lentiviruses is not known. VV is found in macrophages *in vivo*, and grows best in choroid plexus cells *in vitro*. For oncoviruses, although receptors are frequently species-specific, they do not appear to be differentiation-specific (Teich *et al.*, 1977; Weiss *et al.*, 1985), with the possible exception of T-cell receptors discussed later (McGrath & Weissman, 1979).

MOLECULAR MODELS OF HIV-1 CYTOPATHY AND IMMUNOSUPPRESSION

Although HIV-1 is one of the most recently discovered retroviruses, its intensive study has elicited numerous models for the molecular basis for cytopathic effects.

The cytopathic effect of lentiviruses may be largely but not wholly explained by the killing of infected cells. As already explained, the activity of tat and art proteins as transactivating regulators may allow latently infected cells to switch quickly to a late phase in replication producing large quantities of gag and env proteins, of which env in particular may be cytocidal (Sodroski *et al.*, 1986*a*). Pulse production of virions similarly may kill the cell, and there is also evidence of the synthesis of large numbers of integrated proviral genomes. This has previously been associated with cell death in cytopathic oncovirus infections (Keshet & Temin, 1979; Mullins, Chen & Hoover, 1986).

The envelope glycoproteins might exert a cytopathic effect not only in the cells they have infected but also on non-infected cells, especially if they express virus-specific receptors. There are four models for env protein pathogenesis.

1. Like several strains of oncovirus (e.g. HTLV-1), HIV-1 and other lentiviruses induce cell fusion when virus or virus-infected cells are mixed with uninfected, receptor-bearing cells in culture (Dalgleish *et al.*, 1984; Montagnier *et al.*, 1984). The resulting syncytia do not survive long and cannot multiply (Sodroski *et al.*, 1986*a*). Syncytium induction *in vivo* would therefore be cytopathic, and the observation of large multinucleated cells in affected tissues such as the brain suggest that this phenomenon plays some role in pathogenesis.

2. Secreted gp120 env protein binds to T4 antigen receptors of non-infected cells (McDougal *et al.*, 1986) and may thereby prevent T4$^+$ lymphocytes functioning as helper cells, since interaction of T4 antigen with Class II MHC antigens is believed to be important in immune responses involving T-helper cells (Dalgleish, 1986).

3. There may be autoimmune antibodies reacting and blocking the function of T4 antigen as a result of HIV-1 infection (Klatzmann & Montagnier, 1986). These could be elicited either because binding of viral gp120 to T4 modifies the antigenicity of T4 or because anti-idiotypic antibodies to anti-gp120 antibodies might recognise T4 antigen. If that were the case, Dalgleish (1986) has pointed out that the anti-idiotypes might induce immunosuppression, including tolerance to HIV, just as anti-T4 antibodies have recently been shown to induce immunological tolerance in mice (Benjamin & Waldmann, 1986).

4. The transmembrane envelope glycoprotein gp41 may exert an immunosuppressive effect directly. This has been shown for the homologous protein, p15E of murine oncoviruses (Ciancolo *et al.*, 1983). Again, secreted gp41 could affect many cells that have not been infected by HIV-1.

All these models could help to explain why, even in AIDS patients with severe immunosuppression, a very small proportion of T4 cells in the peripheral blood are infected with HIV-1 (Harper *et al.*, 1986). However, circulating T4 cells actually give us little insight into the whole picture of immunosuppression. Infected cells may be efficiently trapped in lymph nodes and other tissues, as substantial amounts of viral antigen have been seen in lymph nodes. Moreover, infection of antigen-presenting cells, monocytes and macrophages may play an important role in immunosuppression in addition to T-cell depletion. It would be unwise at this stage of our knowledge to extrapolate too much from studies of infection of T-cells in culture and from observations on circulating blood cells.

RETROVIRUS ONCOGENESIS

Oncogenesis by retroviruses has been studied more extensively than any other aspect of their pathogenesis, and it has recently been reviewed in the *SGM Symposium on Viruses and Cancer* (Rigby & Wilkie, 1985). Essentially there are five molecular models of retrovirus oncogenesis, the first three illustrated in Fig. 2.

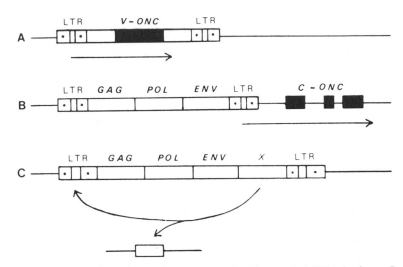

Fig. 2. Molecular models of retrovirus oncogenesis. The proviral DNA is shown flanked by 5′ and 3′ long terminal repeats (LTRs) integrated into host cellular DNA. (A) LTR-initiated transcription of a transduced oncogene (v-*onc*) within a defective viral genome. (B) Integration of a non-defective provirus in the vicinity of a cellular proto-oncogene (c-*onc*) leads to its activation via enhancer or promoter sequences in the adjacent LTR. (C) The HTLV genome carries an *X* gene region whose product, acting in positive feedback, enhances transcription of its own LTR and also other transcription units. The transcription of v-*onc* and c-*onc* genes in A and B is *cis*-controlled, whereas activation of the LTR or cellular genes by the *X* product in C is *trans*. Model B requires site-specific integration of the provirus, while models A and C are independent of the host chromosomal locus.

1. *Viral oncogenes* are transduced from cellular proto-oncogenes, and directly transform the infected cell into a neoplastic state. Viral oncogenes have provided us with the greatest knowledge of molecular oncogenesis generally as they were discovered first and used as probes to identify cellular oncogenes (Bishop, 1984; Rigby & Wilkie, 1985). Oncogenes encode proteins that stimulate cell proliferation by acting on the cell surface, in secondary signal systems, in the cytoplasm, or on DNA synthesis in the nucleus. It is doubtful whether viral oncogenes play a role in oncogenesis by naturally occurring

retroviruses – which lack them – although, in the induction of T-cell leukaemia by feline leukaemia virus, transduction of the c-*myc* gene has been observed (Neil, in Rigby & Wilkie, 1985).

2. *Cellular oncogenes* are activated by integration of retroviruses into DNA in the region of the oncogene. In some cases this is due to promoter insertion so that an RNA transcript initiated in the viral LTR reads through into the cellular oncogene. This is also how cellular oncogenes on rare occasions become transduced as viral oncogenes. More frequently, c-*onc* genes are activated through the enhancer sequences of the viral LTR, which may be inserted upstream or downstream of the c-*onc* transcription unit. In either case, however, the activation is a *cis*-effect, occurring only when an integrated provirus is near the proto-oncogene.

3. *Transactivating* viral gene products may act on regulatory sequences for cellular genes as well as enhancers in the viral LTR. This is an important mechanism for neoplastic transformation by viruses such as HTLV and BLV (Rigby & Wilkie, 1985). There is an overexpression of the interleukin 2 receptor (IL2R) in HTLV-transformed cells, allowing cell proliferation with trace amounts or independence of IL2. The *tat* gene product may act in *trans* on the IL2R gene and to a lesser extent on the IL2 gene (Greene *et al.*, 1986), allowing an autocrine stimulation of T-cells. From these, a clonal IL2R-positive T-cell tumour eventually develops. Experiments on the transfection of cloned HTLV *tat* genes into a variety of cell types indicate that only in certain cells, in which the IL2R gene may already be in a potentially active state, is IL2R overexpressed. *Trans*-acting viral genes in HTLV and BLV would explain why there are no common sites for integration in tumours caused by these retroviruses, in contrast to the *cis*-acting oncoviruses.

4. *The LTR* of the virus itself determines the virulence and specificity of oncovirus pathogenicity. The LTR contains specific enhancer sequences, which make retroviruses subject to host control, and in particular to tissue-specific regulation (Des Groseillers, Rassart & Jolicoeur, 1983; Lenz *et al.*, 1984). This may determine the type of tumour, e.g. thymic or splenic lymphoid leukaemias that develop following infection with different MLV strains (Chatis *et al.*, 1984; Lenz *et al.*, 1984). With avian leukosis viruses (ALV) the enhancer regions of the LTR have also been shown to determine oncogenic properties (Robinson *et al.*, 1982). A non-oncogenic endogenous virus can be converted to high oncogenicity by substituting a short enhancer sequence of the LTR from an oncogenic strain.

5. *Envelope antigens* of certain oncoviruses may act to induce cell transformation in several ways. (a) The transforming gene of the defective spleen focus-forming genome of Friend erythroleukaemia virus is derived from the *env* gene of its helper virus (Machida *et al.*, 1985). (b) McGrath & Weissman (1979) have postulated a kind of autocrine stimulation model of neoplasis by MLV causing thymic lymphomas. They propose that variant MLV genomes encode env gp70 molecules that recognise T-cell immune receptors as viral receptors. The variant virus would thereby gain entry through such a receptor. Expression of its integrated provirus would include gp70 synthesis, stimulating cell proliferation eventually leading to a clonal thymoma. Although there is some immunofluorescence evidence for specific binding of variant MLV to its own thymoma cells, the interaction of gp70 with T-cells receptors has yet to be convincingly demonstrated. Recent studies on a thymic tumour induced by FeLV show that the beta gene of the T-cell receptor has itself been transduced by the virus giving credence to the involvement of T-cell receptors in leukaemogenesis (J. Neil, pers. comm.). (c) Host range variants of gp70 of MLV (and FeLV) arise during the course of infection and are probably associated with the tumours that eventually develop (Hartley *et al.*, 1977; Fischinger *et al.*, 1985; Cloyd & Chattopadhyay, 1986). These are generated by recombination with endogenous viral elements and discussed in the next section.

The models of retrovirus oncogenesis mentioned so far, with the exceptions of viral oncogenes and transactivating viral genes, presuppose a clonal origin of the tumour, in that a particular integration site, oncogene or viral variant induces neoplastic transformation in one progenitor cell. Clonality has been demonstrated by evidence of constant integration sites of proviruses in the tumours, including ATLL induced by *tat*-bearing HTLV-I.

One must bear in mind that the pathogenesis of retroviral cancers is a slow one, indeed, usually slower than the pathogenesis of so-called lentiviruses, except for viruses carrying their own oncogenes. The long period between host infection and tumour development can be explained in two ways. First, neoplastic transformation is a low probability event, and there may be a need for the establishment of widespread infection and viraemia before the appropriate target cell is infected with a virus that happens, by chance in the promoter or enhancer insertion model, to integrate near the particular target proto-oncogene. The generation of oncogenic recombinant viruses may also require prolonged viraemia, and the experimental inoculation of recombinant viruses will accelerate the rate of tumour development (Hartley *et al.*, 1977).

Second, carcinogenesis is a multistep process. The retrovirus may provide only one step, or its recombination to form a variant *env* gene may be one step and proviral integration next to a proto-onco-gene another, both dependent on an active LTR. For example, pre-neoplastic follicles in the bursa of Fabricius in chickens infected with ALV arise long before the overt leukaemia; these transformed folli-cles already have ALV integrated next to the c-myc gene, but require activation from further cellular oncogenes to achieve full malignancy. The tempo and molecular mode of retrovirus oncogenesis is thus complex, involving different interactions with the host cell, and between viral genomes.

Finally, retroviruses can cause cancer by indirect effects not requir-ing infection of the target cell or persistence of the virus in that cell. This appears to be the case with HIV-1, in which the cancer appears to be secondary to the immunosuppression. Kaposi's sar-coma (KS) was one of the first clinical features observed that led to the identification of AIDS as a new disease. Interestingly, KS occurs frequently in AIDS in male homosexuals (46% of US cases), but is virtually absent from haemophiliacs with AIDS and rare in drug-abusers. The KS cells are not typically infected with HIV-1, which is, perhaps, associated with an unknown agent, prevalent in homosexuals but not haemophiliacs. Other tumours arising with a high relative risk in AIDS are EBV-positive B-cell lymphomas, and ano-genital warts and squamous carcinoma, which almost certainly contain human papilloma viruses. The pattern of tumour induction in AIDS thus resembles that in other immunosuppressed conditions, such as renal transplant patients (Weiss, 1984). The immuno-deficiency allows the opportunistic emergence of a selective set of tumours, namely, those arising from cells latently transformed by other oncogenic viruses.

ENDOGENOUS VIRUSES AND PATHOGENESIS

If endogenous proviruses carried as host mendelian elements were highly pathogenic, they would not persist as almost universal 'infec-tions' of their hosts (Weiss, 1982). One would expect endogenous virus to be non-pathogenic. What role then, if any, do they play in retrovirus pathogenesis. The answer is in fact quite complicated, as in some cases the proviral elements may be neutral, in others potentially oncogenic, yet they may also protect the host from the more pathogenic consequences of infection with virulent exogenous strains of related retrovirus.

With one possible exception, lentivirus genomes have not colonised the germ-line of their hosts to become endogenous. On the other hand, numerous strains of endogenous oncovirus genomes exist, falling into B-type, C-type, and D-type classes. In outbred host populations, these viruses remain latent or partially expressed, and viraemia is rare. In some inbred strains of mouse and chicken, however, active viraemic infection occurs. The mouse strains showing active endogenous viruses have frequently been specially selected for having a high incidence of malignancy, e.g. thymic lymphoma in AKR mice which have active MLV, and breast carcinoma in GR mice which have active MMTV. These strains not only have recombinant retrovirus genomes integrated as recently derived endogenous elements, but are homozygous for a number of host controlling genes promoting viral replication, e.g. the $Fv-1$ locus in AKR mice. It is only following replication as full viruses from spontaneously activated endogenous genomes, and the opportunities for recombination that occur with replication, that tumours finally arise.

Infectious replication of endogenous genomes is rare in outbred hosts. Indeed, the majority of endogenous viruses, even if activated by mutagens and carcinogens, are unable to infect their own host species, a phenomenon known as xenotropism (Levy, 1978; Weiss, 1982). Resistance to reinfection by spontaneously activated endogenous viruses may operate at several levels, the chief being a lack of cell surface receptors recognised by the envelope glycoproteins of the endogenous virus. In some cases, however, that does not prevent constitutive virus production, as in New Zealand strain mice with their associated autoimmune disease discussed in the next section.

Envelope recombinants may arise that bypass host resistance mechanisms. Furthermore, opportunities for recombination between exogenous viral genomes and reverse transcripts of endogenous viral mRNA can change the properties of the exogenous virus, including its pathogenesis (Weiss, 1982). For example, infectiously transmitted feline leukaemia virus (FeLV) generally belongs to the envelope subgroup A, representing a particular serotype and receptor recognition type. Subgroup B and C strains of FeLV represent recombinants between exogenous FeLV-A and endogenous *env* genes related to FeLV. And it is these recombinants that are most closely associated with leukaemogenesis (Neil, in Rigby & Wilkie, 1985), anaemia and immunosuppression (Mullins, Chen & Hoover, 1986). In inbred mice, all the leukaemogenic viruses have an endogenous origin, but again, it is recombination between different genomes under conditions of viraemia that gives rise to the more proximate leukaemogenic variants (Hartley *et al.*, 1977). This would seldom occur, or happen very late in the life-span of outbred mice, and hence is of little evolutionary significance in natural selection.

On the other hand, there is growing evidence that endogenous viral genomes might serve in some way to protect the host from the more pathogenic spectrum of disease induced by related exogenous viruses. This has been observed in chickens and related fowl infected with avian leukosis viruses. Endogenous ALV genomes are present in chickens and their wild ancestor, the red junglefowl, but not in other junglefowl of the same genus (Weiss, 1982). When grey junglefowl are infected with ALV, including the activated, endogenous virus of chickens, they suffer a wasting syndrome and thymic atrophy not unlike AIDS. By crossbreeding into these fowl a single endogenous element that expresses an *env* gene, this disease is largely prevented (Weiss & Frisby, 1982). Similar findings are observed in White Leghorn chickens specially bred to segregate specific endogenous viral genetic elements (Crittenden & Fadly, 1985). We can thus see that the widespread inheritance through the germ-line of a relatively non-pathogenic retroviral genome may protect the host from acute pathogenesis by a virulent exogenous strain.

NON-MALIGNANT DISEASE INDUCED BY ONCOVIRUSES

Although neoplasia has been most intensively studied in retrovirus infection, a number of non-malignant diseases have been attributed to oncoviruses, and serve as models for the study of autoimmune disease, immunodeficiency, anaemia and neuropathy.

The wasting disease and thymic atrophy of fowl lacking endogenous viral genes has already been discussed. Similar syndromes develop in cats infected with FeLV, including a spectrum of disease renamed feline AIDS (Mullins, Chen & Hoover, 1986). FeLV substrains and recombinant viruses, especially those with subgroup C envelopes are also associated with aplastic anaemia.

Recent studies indicate that simian D-type viruses cause a severe immune deficiency syndrome known as simian AIDS (SAIDS), though now seen to be distinct from that caused by simian immunodeficiency virus (SIV). The prototype D-type virus, Mason–Pfizer monkey virus (MPMV) was isolated from a biopsy of breast cancer in a rhesus monkey and, together with the taxonomic grouping of D-type viruses with other oncoviruses, it was thought that MPMV may be aetologically related to the tumour. Animal tests, however, indicated that MPMV was not generally pathogenic, except for causing a wasting disease when inoculated into very young monkeys (Fine & Schochetman, 1975).

Two D-type viruses, SRV-1 and SRV-2, have been isolated from macaques with SAIDS in primate centres in the USA (Gardner & Marx, 1985). While showing some degree of nucleotide homology, each is distinct by virus neutralisation from the other and from MPMV (Bryant et al., 1986). A molecular clone of SRV-2 has been recovered in infectious form and induces SAIDS, so there seems to be no doubt as to the causative role of D-type viruses in this type of immune deficiency syndrome.

D-type SAIDS is characterised by panleukopaenia, suppression of T-cell immunity, gingivitis, Pneumocystis carinii pneumonia and retroperitoneal fibromatosis. Malignant tumours are not a feature of D-type SAIDS, although D-type viruses appear to be more closely related to the oncoviruses than to the lentiviruses. Indeed, the env gene of MPMV is related to that of the endogenous cat and baboon C-type viruses, and probably evolved from some recombinational event. The molecular mechanisms of viral replication and latency are not as well understood for D-type viruses as for C-type viruses and HIV-1, but the discovery of D-type SAIDS has aroused new interest in this group of retroviruses.

Autoimmune disease may also be linked with retrovirus activity though a causal association is not established. New Zealand Black and White hybrid mice develop a syndrome resembling systemic lupus erythematosus, with haemolytic anaemia, anti-nuclear antibody-antigen complexes and glomerular nephritis. In these mice an endogenous, xenotropic C-type virus is constitutively expressed with viraemia, even though it cannot re-infect mouse cells (Levy, 1978).

Perhaps the most remarkable disease caused by a C-type virus is the motor neurone syndrome seen in wild mice in California infected with ecotropic and amphotropic strains of MLV (Andrews & Gardner, 1974). These infect motor neurones in the spine causing a polio-like paralysis. Biological cloning shows that these viruses are very closely related to lymphogenic MLV (Gardner *et al.*, 1980), and it will be interesting to learn which particular molecular sequences give these viruses their neurotropism and virulence.

Finally, there is growing evidence that the human T-cell leukaemia virus type-1, or a related virus, is associated with progressive myelopathies resembling multiple sclerosis (MS). Koprowski *et al.*, (1985) reported a serological association of antibodies in MS patients cross-reacting with HTLV-1 p24 antigen, but evidence is lacking for an oncovirus in MS. More intriguing at the time of writing is the association of HTLV-1 with tropical spastic paraparesis (TSP) (Gessain *et al.*, 1985; Rodgers-Johnson *et al.*, 1985; Montgomery, 1986; Hirose *et al.*, 1986). TSP has a higher incidence in the Caribbean and in Japan in those areas where HTLV-1 is endemic. Furthermore, antibodies cross-reacting with HTLV-1 are more prevalent in TSP patients (58%) than in healthy medical staff and blood donors (4%) (Gessain *et al.*, 1985). Further studies of HTLV-like viruses in myelopathies are awaited with great interest.

CONCLUSIONS

Retroviruses are structurally a homogeneous group of viruses that induce a wide variety of disease with different patterns of transmission, latency, virulence and cell tropisms. In addition to revealing oncogenes, the molecular and cell biology of retrovirus infection has thrown light upon gene regulation at transcriptional and post-transcriptional levels, the importance of specific cell receptors for virus infection, and a variety of other factors that play a role in determining pathogenesis. Apart from illuminating basic aspects of cell–virus interactions, a better knowledge of the molecular events in retrovirus infection will be crucial to the development of anti-viral strategies in preventing human diseases such as leukaemia and AIDS.

REFERENCES

ANDREWS, J. M. & GARDNER, M. B.(1974). Lower motor neuron degeneration associated with type C RNA virus infection in mice: neuropathological features. *Journal of Neuropathology and Experimental Neurology*, **33**, 285–307.
ARYA, S. K. & GALLO, R. C. (1986). Three novel genes of human T-lymphotropic virus type III: Immune reactivity of their products with sera from acquired

immune deficiency syndrome patients. *Proceedings of the National Academy of Sciences of the United States of America*, **83,** 2209–13.

ASJO, B., IVHED, I., GIDLUND, M., FUERSTENBERG, S., FENYO, E. M., NILSSON, K. & WIGZELL, H. (1987). Susceptibility to infection by the AIDS virus correlates with T4 expression in monocytoid cell lines. *Virology* (in press).

BARRÉ-SINOUSSI, F., CHERMANN, J. C., REY, F., NUGEYRE, M. T., CHAMARET, S., GRUEST, J., DANGUET, C., AXLER-BLIN, C., VEZINET-BRUN, F., ROUZIOUX, C., ROZENBAUM, W. & MONTAGNIER, L. (1983). Isolation of T-lymphotropic retrovirus from a patient at risk for acquired immune deficiency syndrome (AIDS). *Science*, **220,** 868–70.

BENJAMIN, R. J. & WALDMAN, H. (1986). Induction of tolerance by monoclonal antibody therapy. *Nature*, **320,** 449–51.

BISHOP, J. M. (1984). Exploring carcinogenesis with retroviruses. *In: The Virus*, eds B. W. J. Mahy & J. Pattison, *Symposium of the Society of General Microbiology*, **36,** 121–47.

BRYANT, M. L., MARX, P. A., SHIIGI, S. M., WILSON, B. J., McNULTY, W. P. & GARDNER, M. B. (1986). Distribution of type D retrovirus sequences in tissues of macaques with simian acquired immune deficiency and retroperitoneal fibromatosis. *Virology*, **150,** 149–60.

CATOVSKY, D., GREAVES, M. F., ROSE, M., GALTON, D. A. G., GOOLDEN, A. W. G., McCLUSKEY, D. R., WHITE, J. M., LAMPERT, I., BOURIKAS, G., IRELAND, R., BROWNELL, A. I., BRIDGES, J. M., BLATTNER, W. A. & GALLO, R. C. (1982). Adult T-cell lymphoma-leukemia in Blacks from the West Indies. *Lancet*, **i,** 639–43.

CHATIS, P. A., HOLLAND, C. A., SILVER, J. E., FREDERICKSON, T. N., HOPKINS, N. & HARTLEY, J. W. (1984). A 3′ end fragment encompassing the transcriptional enhancers of nondefective Friend virus confers erythroleukemogenecity on Moloney leukemia virus. *Journal of Virology*, **52,** 248–54.

CIANCOLO, G. J., LOSTROM, M. E., TAM, M. & SNYDERMAN (1983). Murine malignant cells synthesize a 19,000 dalton protein that is physicochemically and antigenically related to the immunosuppressive retroviral protein, p15E. *Journal of Experimental Medicine*, **158,** 885–900.

CLAPHAM, P. R., WEISS, R. A., DALGLEISH, A. G., EXLEY M., WHITBY, D. & HOGG, N. (1987) Human immunodeficiency virus infection of monocytic and T-lymphoid cells: receptor modulation and differentiation induced by Phorbol Esters. *Virology* (in press).

CLAVEL, F., GUETARD, D., BRUN-VEZINET, F., CHAMARET, S., REY, M.-A., SANTOS FERREIRA, M. O., LAURENT, A. G., DAUGUET, C., KATLAHA, C., ROUZIOUX, C., KLATZMANN, D., CHAMPALIMAUD, J. L. & MONTAGNIER, L. (1986). Isolation of a new human retrovirus for West African patients with AIDS. *Science*, **233,** 343–6.

CLOYD, M. W. & CHATTOPADHYAY, S. K. (1986). A new class of retrovirus present in many murine leukemia systems. *Virology*, **151,** 31–40.

COFFIN, J. M. (1986). Genetic variation in AIDS viruses. *Cell*, **46,** 1–4.

COFFIN, J., HAASE, A., LEVY, J. A., MONTAGNIER, L., OROSZLAN, S., TEICH, N., TEMIN, H., TOYOSHIMA, K., VARMUS, H., VOGT, P. & WEISS, R. A. (1986). What to call the AIDS virus. *Nature*, **321,** 10.

CRITTENDEN, L. B. & FADLY, A. M. (1985). Response of chickens lacking or expressing endogenous avian leukosis virus genes to infection with exogenous virus. *Poultry Science*, **64,** 454–63.

CURRAN, J. W., MORGAN, W. M., HARDY, A. W., JAFFE, H. W., DARROW, W. W. & DOWDLE, W. R. (1985). The epidemiology of AIDS: Current status and future prospects. *Science*, **229,** 1352–7.

DALGLEISH, A. G., BEVERLEY, P. C. L., CLAPHAM, P. R., CRAWFORD, D. H., GREAVES, M. F. & WEISS, R. A. (1984). The CD4 (T4) antigen is an essential component of the receptor for the AIDS retrovirus. *Nature*, **312**, 763–7.

DALGLEISH, A. G. (1986). The T4 molecule: Function and structure. *Immunology Today*, **7**, 142–4.

DESGROSEILLERS, L., RASSART, E. & JOLICOEUR, P. (1983). Thymotropism of murine leukemia virus is conferred by its long terminal repeat. *Proceedings of the National Academy of Sciences of the United States of America*, **80**, 4203–7.

FINE, D. L., LANDON, J. C., PIENTA, R. J., KUBICEK, M. T., VALERIO, M. G., LOEB, W. F. & CHOPRA, H. C. (1975). Responses of infant rhesus monkeys to inoculation with Mason–Pfizer monkey virus materials. *Journal of the National Cancer Institute*, **54**, 651 8.

FISCHINGER, P. J., DUNLOP, N. M., ROBEY, W. G. & SCHAFER, W. (1985). Generation of thymotropic envelope gene recombinant virus and induction of lymphoma by ecotropic Moloney murine leukemia virus. *Virology*, **142**, 197–205.

FISHER, A. G., FEINBERG, M. B., JOSEPHS, S. F., HARPER, M. E., MARSELLE, L. M., REYES, G., GONDA, M. A., ALDOVINI, A., DEBOUK, C., GALLO, R. C., & WONG-STAAL, F. (1986). *Nature*, **320**, 367–70.

FUJISAWA, J., SEIKI, M., SATO, M. & YOSHIDA, M. (1986). A transcriptional enhancer sequence of HTLV-I is responsible for trans-activation mediated by p40 of HTLV. *EMBO Journal*, **5**, 713–8.

GALLO, R. C., SALAHUDDIN, S. Z., POPOVIC, M., SHEARER, G. M., KAPLAN, M., HAYNES, B. F., PALKER, T. J., REDFIELD, R., OLESKE, J., SAFAI, B., WHITE, G., FOSTER, P. & MARKHAM, P. D. (1984). Human T-lymphotropic retrovirus, HTLV-III, isolated from AIDS patients and donors at risk for AIDS. *Science*, **224**, 500–3.

GARDNER, M. B., ESTES, J. D., CASAGRANDE, J. & RASHEED, S. (1980). Prevention of paralysis and suppression of lymphoma in wild mice by passive immunization to congenitally transmitted murine leukemia virus. *Journal of the National Cancer Institute*, **65**, 359–64.

GARDNER, M. B. & MARX, P. A. (1985). Simian acquired immunodeficiency syndrome. *In: Advances in Viral Oncology*, ed. G. Klein, Vol. 5, pp. 57–81.

GARTNER, S., MARKOVITS, P., MARKOVITS, D. M., KAPLAN, M. H., GALLO, R. C. & POPOVIC, M. (1986). The role of mononuclear phagocytes in HTLV-III/LAV infection. *Science*, **233**, 215–19.

GESSAIN, A., BARIN, F., VERNANT, J. C., GOUT, O., MAURS, L., CALENDER, A. and DE THÉ, G. (1985). Antibodies to human T-lymphotropic virus type-I in patients with tropical spastic paraparesis. *Lancet*, **ii**, 407–10.

GREENE, W. C., LEONARD, W. J., WANO, Y., SVETLIK, P. B., PEFFER, N. J., SODROSKI, J. G., ROSEN, C. A., GOH, W. C. & HASELTINE, W. A. (1986). Transactivator gene of HTLV-II induces IL-2 receptor and IL-2 cellular gene expression. *Science*, **232**, 877–80.

HAASE, A. T. (1986). Pathogenesis of lentivirus infection. *Nature*, **322**, 130–6.

HARPER, M. E., MARSELLE, L. M., GALLO, R. C. & WONG-STAAL, F. (1986). Detection of lymphocytes expressing human T-lymphotropic virus type III in lymph nodes and peripheral blood from infected individuals by *in situ* hybridization. *Proceedings of the National Academy of Sciences of the United States of America*, **83**, 772–6.

HARTLEY, J. W., WOLFORD, N. K., OLD, L. J. & ROWE, W. P. (1977). A new class of murine leukemia virus associated with the development of spontaneous lymphomas. *Proceedings of the National Academy of Sciences of the United States of America*, **74**, 789–92.

HIROSE, S., UEMURA, Y., FUJISHITA, M., KITIGAWA, T. YAMASHITA, M., IMAMURA,

190   R. A. WEISS

J., OHTSUKI, Y., TAGUCHI, H. and MIYOSHI, I. (1986). Isolation of HTLV-I from cerebrospinal fluid of a patient with myelopathy. *Lancet*, **ii**, 397–8.

HOOKS, J. J. & GIBBS, C. J. (1975). The foamy viruses. *Bacteriological Reviews*, **39**, 169–85.

KALYANARAMAN, V. S., SARNGADHARAN, M. G., ROBERT-GUROFF, M., MIYOSHI, I., BLAYNEY, D., GOLDE, D. & GALLO, R. C. (1982). A new subtype of human T-cell leukemia virus (HTLV-II) associated with a T-cell variant of hairy cell leukemia. *Science*, **218**, 571–3.

KANKI, P. J., MCLANE, M. F., KING, N. W., LETVIN, N. L., HUNT, R. D., SEHGAL, P., DANIEL, M. D., DESROSIERS, R. C. & ESSEX, M. (1985). Serological identification and characterisation of a macaque T-lymphotropic retrovirus closely related to HTLV-III. *Science*, **228**, 1199–201.

KANKI, P. J., BARIN, F., M'BOUP, S., ALLAN, J. S., ROMET-LEMMONE, J. L., MARLINK, R., MCLANE, M. F., LEE, T. H., ARBEILLE, B., DENIS, F. & ESSEX, M. (1986). New human T-lymphotropic retrovirus related to simian T-lymphotropic virus type III (STLV-III$_{AGM}$). *Science*, **232**, 238–43.

KESHET, E. & TEMIN, H. M. (1979). Cell killing by spleen necrosis virus is correlated with a transient accumulation of spleen necrosis virus DNA. *Journal of Virology*, **31**, 376–86.

KLATZMANN, D., CHAMPAGNE, E., CHAMARET, S., GRUEST, J., GUETARD, D., HERCEND, T., GLUCKMAN, J.-C. & MONTAGNIER, L. (1984). T-lymphocyte T4 molecule behaves as the receptor for human retrovirus LAV. *Nature*, **312**, 767–8.

KLATZMANN, D. & MONTAGNIER, L. (1986). Approaches to AIDS therapy. *Nature*, **319**, 10–11.

KOPROWSKI, H., DE FREITAS, E. C., HARPER, M. E., SANDBERG-WOLLHEIM, M., SHEREMATA, W. A., ROBERT-GUROFF, M., SAXINGER, C. W., FEINBERG, M. B., WONG-STAAL, F. and GALLO, R. C. (1985). Multiple sclerosis and human T-cell lymphotropic retroviruses. *Nature*, **318**, 154–60.

LENZ, J., CELANDER, D. L., CROWTHER, R. L., PATARCA, R., PERKINS, D. W. & HASELTINE, W. A. (1984). Determination of the leukemogenicities of a murine retrovirus by sequences within the long terminal repeat. *Nature*, **308**, 467–70.

LEVY, J. A. (1978). Xenotropic type C viruses. *Current Topics in Microbiology and Immunology*, **79**, 111–213.

LEVY, J. A., CHENG-MAYER, C., DINA, D. & LUCIW, P. A. (1986). AIDS retrovirus (ARV-2) clone replicates in transfected human and animal fibroblasts. *Science*, **232**, 998–1001.

MCDOUGAL, J. S., KENNEDY, M. S., SLIGH, J. H., CORT, S. P., MAWLE, A. & NICHOLSON, J. K. A. (1986). Binding of HTLV-III/LAV to T4$^+$ T cells by a complex of the 110K viral protein and the T4 molecule. *Science*, **231**, 382–5.

MCGRATH, M. S. & WEISSMAN, I. L. (1979). AKR leukemogenesis: Identification and biological significance of thymic lymphoma receptors for AKR retroviruses. *Cell*, **17**, 65–75.

MACHIDA, C. A., BESTWICK, R. K., BOSWELL, B. A. & KABAT, D. (1985). Role of a membrane glycoprotein in Friend virus-induced erythroleukemia: Studies of mutant and revertant viruses. *Virology*, **144**, 158–72.

MADDON, P. J., DALGLEISH, A. G., MCDOUGAL, J. S., CLAPHAM, P. R., WEISS, R. A. & AXEL, R. (1986). The T4 gene encodes the AIDS virus receptor and is expressed in the immune system and the brain. *Cell*

MONTAGNIER, L., CHERMANN, J. C., BARRÉ-SINOUSSI, F., CHAMARET, S., GROUEST, J., NUGEYRE, M. T., REY, F., DAUGUET, C., AXLER-BLIN, C., VEZINET-BRUN, F., RAZIOUX, C., SAIMOT, G. A., ROZENBAUM, W., GLUCKMAN, J. C., KLATZMANN, D., VILMER, E., GRISCELLI, C., FOYER, C. & BRUNET, J. B. (1984). A new human T-lymphotropic retrovirus: characterization and possible role in lymphadenopathy and acquired immune deficiency syndrome. *In: Human T-cell leukaemia-lymphoma viruses*, eds R. C. Gallo, M. Essex & L. Gross, pp. 363–79. New York: Cold Spring Harbor Laboratory.

MONTGOMERY, C. D. (1986). HTLV-I, Visna, and tropical spastic paraparesis. *Lancet*, **ii**, 227–8.

MULLINS, J. I., CHEN, C. S. & HOOVER, E. A. (1986). Disease-specific and tissue-specific production of unintegrated feline leukaemia virus variant DNA in feline AIDS. *Nature*, **319**, 333–6.

NAGY, K., CLAPHAM, P., CHEINGSONG-POPOV, R. & WEISS, R. A. (1983). Human T-cell leukemia type 1: Induction of syncytia and inhibition by patients' sera. *International Journal of Cancer*, **32**, 321–8.

PINCHING, A. & WEISS, R. A. (1986). AIDS and the spectrum of HTLV-III/LAV infection. *International Review of Experimental Pathology*, **28**, 1–44.

POIESZ, B. J., RUSCETTI, F. W., GAZDAR, A. F., BUNN, P. A., MINNA, J. D. & GALLO, R. C. (1980). Detection and isolation of type C retrovirus particles from fresh and cultured lymphocytes of a patient with cutaneous T-cell lymphoma. *Proceedings of the National Academy of Sciences of the United States of America*, **77**, 7415–9.

RABSON, A. B. & MARTIN, M. A. (1985). Molecular organization of the AIDS retrovirus. *Cell*, **40**, 477–80.

RIGBY, P. W., R. & WILKIE, N. (eds) (1985). *Viruses and Cancer, Symposium of General Microbiology*, **37**, 1–21. Cambridge: University Press.

ROBINSON, H. L., BLAIS, B. M., TSICHLIS, P. N. & COFFIN, J. M. (1982). At least two regions of the viral genome determine the oncogenic potential of avian leukosis viruses. *Proceedings of the National Academy of Sciences of the United States of America*, **79**, 1225–9.

RODGERS-JOHNSON, P., GAJDUSEK, D. C., MORGAN, O.STC., ZANINOVIC, V., SARIN, P. S. and GRAHAM, D. S. (1985). HTLV-I and HTLV-III antibodies and tropical spastic paraparesis. *Lancet*, **ii** 1248–9.

ROSEN, C. A., SODROSKI, J. G., GOH, W. C., DAYTON, A. I., LIPPKE, J. & HASELTINE, W. A. (1986). Post-transcriptional regulation accounts for the trans-activation of the human T-lymphotropic virus type III. *Nature*, **319**, 555–9.

SATTENTAU, Q. J., DALGLEISH, A. G., WEISS, R. A. & BEVERLEY, P. C. L. (1986). Epitopes of the CD4 antigen and HIV infection. *Science* (in press).

SEIKI, M., HATTORI, S., HIRAYAMA, V. & YOSHIDA, M. (1983). Human adult T-cell leukemia virus: Complete nucleotide sequence of the provirus genome integrated in leukemia cell DNA. *Proceedings of the National Academy of Sciences of the United States of America*, **80**, 3618–22.

SERWADDA, D., MUGERWA, R. D., SEWANKAMBO, N. K., LWEGABA, A., CARSWELL, J. W., KIRYA, G. B., BAYLEY, A. C., DOWNING, R. G., TEDDER, R. S., CLAYDEN, S. A., WEISS, R. A. & DALGLEISH, A. G. (1985). Slim disease: A new disease in Uganda and its association with HTLV-III infection. *Lancet*, **ii**, 849–52.

SHAW, G. M., HARPER, M. E., HAHN, B. E., EPSTEIN, L. G., GAJDUSEK, D. C., PRICE, R. W., NAVIA, B. A., PETITO, C. K., O'HARA, C. H., GROOPMAN, J. E., WONG-STAL, F. & GALLO, R. C. (1985). HTLV-III Infection in brains of children and adults with AIDS encephalopathy. *Science*, **227**, 177–82.

SODROSKI, J. G., ROSEN, C. A. & HASELTINE, W. A. (1984). Trans-acting transcriptional activation of the long terminal repeat of human T lymphotropic viruses in infected cells. *Science*, **225**, 381–5.

SODROSKI, J., ROSEN, C., GOH, W. C. & HASELTINE, W. (1985). A transcriptional activator protein encoded by the x-lor region of the human T-cell leukemia virus. *Science*, **228**, 1430–4.

SODROSKI, J., GOH, W. C., ROSEN, C., CAMPBELL, K. & HASELTINE, W. A. (1986a).

Role of the HTLV-III/LAV envelope in syncytium formation and cytopathicity. *Nature*, **322**, 470–4.

SODROSKI, J., GOH, W. C., ROSEN, C., DAYTON, A., TERWILLIGER, E., & HASELTINE, W. (1986b). A second post-transcriptional *trans*-activator gene required for HTLV-III replication. *Nature*, **321**, 412–7.

SONIGO, P., ALIZON, M., STASKUS, K., KLATZMANN, D., COLE, S., DANOS, O., RETZEL, E., TIOLLAIS, P., HAASE, A. & WAIN-HOBSON, S. (1985). Nucleotide sequence of the visna lentivirus: Relationship to the AIDS virus. *Cell*, **42**, 369–82.

TEICH, N. M., WEISS, R. A., MARTIN, G. R. & LOWY, D. R. (1977). Virus infection of murine teratocarcinoma stem cell lines. *Cell*, **12**, 973–82.

TEMIN, H. M. (1980). Origin of retroviruses from transposable genetic elements. *Cell*, **21**, 599–600.

UCHIYAMA, T., YODOI, J., SAGAWA, K., TAKATSUKI, K. & UCHINO, H. (1977). Adult T-cell leukemia: Clinical and hematoalogical features of 16 cases. *Blood*, **50**, 481–92.

WEISS, R. A. (1982). The persistence of retroviruses. *In: Virus Persistence*, eds B. W. J. Mahy, A. C. Minson & G. K. Darby, *Symposium of the Society of General Microbiology*, **33**, 267–88. Cambridge: University Press.

WEISS, R. A. & FRISBY, D. P. (1982). Are avian endogenous viruses pathogenic. *In: Advances in Comparative Leukemia Research* 1981, ed. D. S. Yohn & J. R. Blakeslee, pp. 303–11, North Holland: Elsevier.

WEISS, R. A. (1984). Viruses and human cancer. *In: The Virus*, eds B. W. J. Mahy & J. Pattison, *Symposium of the Society of General Microbiology*, **36**, 211–40. Cambridge: University Press.

WEISS, R. A., TEICH, N. N., VARMUS, H. E. & COFFIN. J. (1985). *RNA Tumor Viruses*. 2nd edn, 2 vols. New York: Cold Spring Harbor Laboratory.

WERNER, J. & GELDERBLOM, H. (1979). Isolation of foamy virus from patients with de Quervain thyroiditis. *Lancet*, **ii**, 258–9.

WONG-STAAL, F. & GALLO, R. C. (1985). Human T-lymphotropic retroviruses. *Nature*, **317**, 395–403.

YOSHINAKA, Y., KATOH, I., COPELAND, T. D. & OROSZLAN, S. (1985). Murine leukemia virus protease is encoded by the *gag–pol* gene and is synthesized through suppression of an amber termination codon. *Proceedings of the National Academy of Sciences of the United States of America*, **82**, 1618–82.

VIRUS–CELL INTERACTIONS IN SIMIAN VIRUS 40 TRANSFORMATION

PETER W. J. RIGBY

Laboratory of Eukaryotic Molecular Genetics, National Institute for Medical Research, The Ridgeway, Mill Hill, London NW7 1AA

INTRODUCTION

Simian virus 40 (SV40) is a small, icosahedral virus isolated from the rhesus monkey. It is non-oncogenic in its natural host but it will induce tumours in rodents and it will transform cultured cells from a variety of species. SV40 has been the subject of intense study during the last 20 years, both as an oncogenic agent and as a general model for eukaryotic gene expression, and we now have a detailed picture of the structure of its genome, and of the strategies employed during its replication in permissive monkey cells. However, despite considerable efforts, the mechanism(s) by which the virus transforms remains obscure. Transformation involves the covalent integration of viral DNA into the chromosomes of the host cell and continued expression of these integrated sequences is required to maintain the transformed phenotype.

The extensive studies of SV40 have generated a vast literature, a comprehensive discussion of which is beyond the scope of this chapter. The reader is referred to the excellent reviews found in Martin (1981) and Tooze (1981) for background information. I shall cite here only our own work and some more recent relevant papers by others.

The structure of the SV40 genome is shown in Fig. 1. Transfection of restriction fragments of viral DNA shows that only the early region is required for transformation; this encodes two proteins, small t-antigen, of apparent molecular weight 17 kd, and large T-antigen, of apparent molecular weight 94 kd. A variety of genetic analyses have shown that it is large T-antigen which is the transforming protein of SV40. *ts*A mutants, which carry a temperature-sensitive lesion in the region encoding only large T-antigen, cannot transform at the non-permissive temperature. Transformed cell lines can be established at the permissive temperature but, when such lines are shifted up to the non-permissive temperature, then, in the majority of cases, the cells revert to the normal phenotype. Large T-antigen is thus

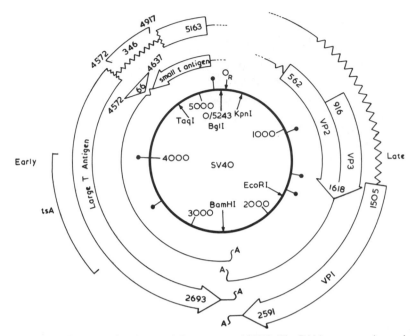

Fig. 1. A simplified functional map of the genome of SV40. The DNA sequence is numbered according to the convention used in Tooze (1981). VP1, VP2 and VP3 are the three capsid proteins of the virus. The symbol (∿) indicates sequences removed from mature mRNAs by splicing. The bracket marked tsA indicates the region of the sequences uniquely encoding large T-antigen in which *tsA* mutations map.

required for both the initiation and the maintenance of transformation. Deletion mutations which remove sequences which uniquely encode small t-antigen but which do not affect the large T-antigen coding sequences, do not, under most assay conditions, affect the ability of the virus to transform cultured cells or to induce tumours *in vivo*. However, it is possible to define culture conditions in which a requirement for small t-antigen is manifest and it is clear that at least one of the features of SV40-transformed cells, namely the disorganization of the actin cables of the cytoskeleton, is due to an activity of small t-antigen. SV40 large T-antigen can transform both established and primary cells.

Much work has been done to elucidate the biochemical properties of large T-antigen (reviewed by Rigby & Lane, 1983) which is predominantly nuclear in location and is a sequence-specific DNA-binding protein. During the productive infection of permissive cells, large T-antigen binds to sequences within the viral origin of DNA replication and this binding is required to initiate each round of viral DNA replication. The promoter for the viral early transcription unit is interdigitated with the origin and consequently when T-antigen binds to the DNA it blocks the promoter of the early transcription unit and the protein thus autoregulates its own synthesis. More recent work has defined another interaction between the transforming protein and the viral regulatory region. Even in the absence of DNA replication, large T-antigen is capable of stimulating transcription from the viral late promoter (Brady & Khoury, 1985; Brady et al., 1984; Hartzell et al., 1984; Keller & Alwine, 1984, 1985). These data indicate that activation of the late promoter requires elements distinct from the presently defined binding sites for large T-antigen. Thus within the productive infection cycle this protein is capable of acting as both a positive activator and a negative repressor of RNA polymerase II transcription.

Large T-antigen is an enzyme; it is an ATPase, and it is generally thought that this ATPase function is connected with the DNA replication activity of the protein. Recent data suggest that the protein is an ATP-dependent helicase (Stahl et al., 1986) and this function is also likely to be involved in DNA replication. Earlier reports that T-antigen is a protein tyrosine kinase have not been confirmed. In both transformed and infected cells large T-antigen is found complexed with a cellular protein, called p53, and it is now known that the gene encoding p53 is an oncogene in its own right (Eliyahu et al., 1984; Jenkins et al., 1984; Parada et al., 1984). It is extremely likely that this complex formation, the effect of which is to dramatically stabilize p53, is important in the mechanism of transformation but there is no direct evidence for this. As well as binding with high affinity to viral sequences, large T-antigen binds with low affinity to cellular DNA and recent work employing an ingenious chromatin immunoprecipitation procedure has shown that specific T-antigen binding sites in the cellular genome can be cloned and characterized (Lane et al., 1985). Large T-antigen can regulate the replication of cellular DNA. If the protein is introduced into quiescent cells in any one of a number of ways, most dramatically by microinjecting pure protein into nuclei, then the cells enter a round of DNA replication. It is widely believed that this ability of T-antigen to stimulate cellular DNA replication is mediated by its ability to bind to such DNA but there is no direct evidence on this point.

Large T-antigen can also regulate the expression of cellular genes. It can activate the transcription of ribosomal RNA genes by RNA polymerase I (Soprano et al., 1983) and it has been shown that this regulation can be observed *in vitro* (Learned *et al.*, 1983). If an RNA polymerase I *in vitro* system derived from Hela cells is programmed with a cloned human ribosomal RNA gene a basal level of transcription is observed which is stimulated when pure large T-antigen is added to the system. Under these circumstances it was not possible to detect any binding of large T-antigen to the template DNA.

At the time that we began our work the evidence for the regulation of polymerase II transcription by large T-antigen was more circumstantial. One of the first effects of large T-antigen on an infected cell is the induction of a number of cellular enzymes concerned with nucleotide metabolism and DNA replication. Postel & Levine (1976) showed that the induction of thymidine kinase required an activity of large T-antigen although there was no direct evidence that this induction is transcriptional. More directly, Williams *et al.* (1977) used mRNA/cDNA hybridization in solution to analyse the cytoplasmic mRNA populations of an SV40-transformed human cell line and its normal parent. They showed that approximately 3% of the mRNAs present in the transformed cell line were absent, or present at a much lower abundance, in the normal cell and that, conversely, approximately 3% of the mRNAs of the normal cell were absent, or present at a much lower abundance, in the transformed cell.

We were interested in exploring these observations in more detail and decided to try to clone some of the cellular genes which are regulated as a result of SV40 transformation. The intellectual basis for these experiments derives from the fact that the transformation of established lines of cultured cells requires only a single oncogene, be it of viral or cellular origin, and is thus mediated by a single protein. It seemed to be extremely unlikely that all of the multitude of biological and biochemical changes which distinguish a transformed cell from its normal parent could be due to direct actions of the transforming protein. Rather, one of the functions of such proteins must be to reprogramme, either directly or indirectly, the pattern of cellular gene expression, such changes then contributing to what we recognize as the transformed phenotype. If one could isolate molecular clones of genes which are regulated by transformation then one could use them in two types of experiment. First, one could artificially activate such genes by placing them under the control of a strong, constitutive promoter and then transfecting them back into normal cells to ask whether the over-expression of a single gene can induce any of the parameters of transformation. Second, one could locate the *cis*-acting DNA sequences which control the expression of the gene and use these in experiments designed to elucidate the mechanisms by which the expression of the gene is altered in transformed cells.

In this chapter I shall briefly review the work done in my laboratory on two of these issues: the mechanism of viral DNA integration and the regulation of cellular gene expression by large T-antigen.

THE INTEGRATION OF SV40 DNA

Southern blotting experiments showed clearly that the integration of SV40 DNA is not specific, i.e. the recombination can occur at a number of places on the viral genome and at many, if not all, sites on the cellular chromosomes. Such experiments could not, however, reveal whether there was nucleotide sequence homology between viral and cellular genomes at the site of integration nor could they discern any structural features of the DNA which might target integration to a particular site. Moreover, the experiments suffered from a logical flaw in that one could only interpret such blotting data on the assumption that the integrated viral genomes are colinear with unintegrated sequences, which subsequent experiments have shown not to be the case.

These early experiments did, however, reveal one general feature, namely that the viral genome is often integrated as partial or complete tandem head-to-tail arrays. A trivial explanation for the existence of such arrays is that one copy of the viral genome integrates non-specifically and then another copy integrates into it by homologous recombination. However, direct tests showed that if the viral genome was offered the opportunity to integrate into a pre-existing copy then it always integrated into a second chromosomal locus. We therefore decided to try to elucidate the mechanism by which these tandem duplications were formed. We had already shown (Rigby & Berg, 1978) that during the productive infection of monkey cells large tandem head-to-tail arrays of the viral genome are produced by a mechanism which probably involves a rolling-circle type of DNA replication. Chia & Rigby (1981) followed the fate of the infecting viral DNA in cultures of non-permissive Balb/c 3T3 cells and could show that in this situation also large tandem head-to-tail oligomers were produced early after infection. By using appropriately marked viruses we could demonstrate that the synthesis of these oligomers could not be accounted for by recombination between parental genomes and thus that it was also likely to involve DNA replication. From these data we derived the model for integration which is shown in Fig. 2A. This model makes a clear prediction. If integrative recombination involves monomeric circles of viral DNA then the cellular sequences which flank the integrated viral genome will be adjacent to one another in normal cells (Fig. 2B). If, however, tandem oligomers are first synthesized and recombination occurs between them and cellular DNA then the recombination must involve at least two cross-overs and cellular sequences will always be deleted at the site of integration (Fig. 2B). In all cases in which this prediction has been tested the latter has been found to be true.

Further analyses of the structure and expression of the integrated viral genomes required that they be isolated by recombinant DNA techniques. Clayton & Rigby (1981) cloned a total of 15 integrated SV40 segments from three lines of transformed Balb/c 3T3 cells. Fig. 3 shows the structures of the segments from SV3T3 C120 which exemplify many of the general principles which we discerned. The early Southern blotting data showed that in most cases the virus–host junctions were located in the late region of the viral genome. This was to be expected as, given that transformation by SV40 requires the continued expression of large T-antigen, one would have thought that integration must occur in such a way as to preserve the structure of the early region which encodes large T-antigen. We were therefore very surprised to find that in only three cases did we recover an intact early region; in all of the others large T-antigen coding sequences were interrupted either by virus–host or virus–virus junctions or by internal rearrangements, deletions or duplications of viral sequences. It was quite possible that all of these aberrant viral early transcription units were merely recombinational debris and that they had no functional significance. However, a variety of experiments described below showed that some of these novel templates are expressed and in many cases that they are competent for transformation.

We (D. Murphy & P. W. J. Rigby, unpublished data) have subsequently sequenced most of the virus–host and virus–virus junctions cloned by Clayton & Rigby (1981). These data show that viral integration does not proceed by homologous recombination. They also confirm a conclusion drawn from the restriction endonuclease mapping experiments of Clayton & Rigby (1981), namely that in a given

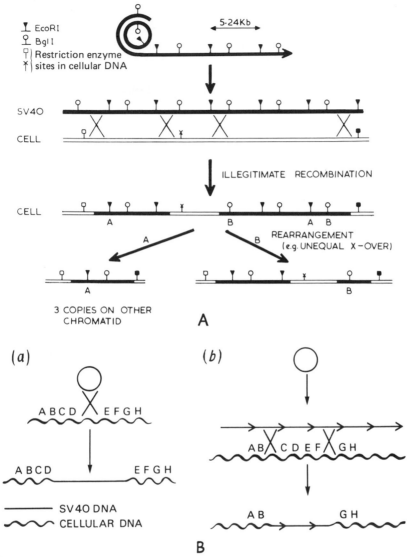

Fig. 2A. A model for the integration of SV40 DNA. In the top panel is indicated the limited replication of viral DNA, probably by a rolling-circle mechanism. The tandem head-to-tail oligomers of viral DNA then recombine with cellular DNA in an illegitimate recombination event which involves at least two cross-overs. The primary integration products are often unstable and rearrange during passage in culture by mechanisms such as unequal crossing over.

B. Comparison of two models for the integration of SV40 DNA. In panel a is shown the scheme for integration by recombination between a monomeric viral DNA molecule and cellular DNA. The sequences (ABCD,EFGH) which flank the integrated viral genome are adjacent in normal cellular DNA. In panel b is shown the scheme for integration by recombination between tandem head-to-tail oligomers of viral DNA and cellular DNA. In this case, the sequences (AB,GH) which flank the integrated viral genome are not adjacent in normal cellular DNA. Taken from Chia and Rigby (1981).

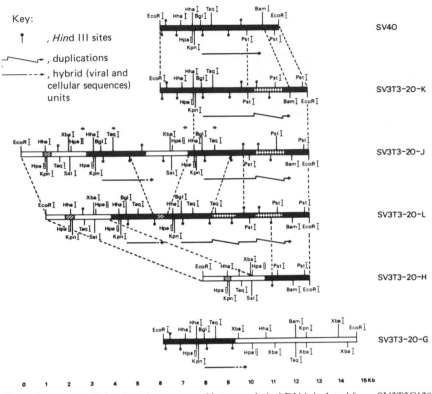

Fig. 3. Structures of the cloned segments of integrated viral DNA isolated from SV3T3C120. At the top is the restriction map of SV40 DNA, linearised at the single EcoRI site. Also shown are the structures of five segments of integrated viral DNA isolated from SV3T3C120 (Rigby et al., 1980). Restriction enzyme sites are indicated by conventional abbreviations, except HindIII sites, which are indicated by the symbol shown in the key. Solid sections are segments colinear with unintegrated SV40 DNA; open sections are mouse DNA. All maps are drawn such that the direction of SV40 early transcription is from left to right. Arrows indicate potential SV40 early transcription units, which may contain duplications (see key) or be hybrid units containing both viral and cellular sequences (see key). Dashed lines indicate similarities between clones. Dotted areas indicate the location of one copy of a tandem duplication of viral sequences. Within each copy of the duplication the restriction sites are colinear with unintegrated viral DNA. Cross-hatched areas define segments that show sequence homology with SV40 DNA by hybridisation but are not of viral origin. Arrowheads above SV3T3-20-J indicate the positions of the inverted repeats in this segment. The structures of SV3T3-20-K and SV3T3-20-G were subsequently refined by nucleotide sequence analysis (Clayton et al., 1982b; Lovett et al., 1982). Reproduced, with permission, from Clayton and Rigby (1981). Copyright MIT Press, 1981.

cell line one can find more than one SV40 insertion next to the same piece of unique cellular DNA with exactly the same virus–host junction. The only explanation for these data is that individual insertions can be amplified after integration has occurred. Moreover, there is a considerable amount of evidence to show that the integration patterns of SV40-transformed cells are not stable. These rearrangements continue as cell lines are passaged in culture and can lead to the disappearance of particular forms of large T-antigen and to the appearance of novel forms (Bender & Brockman, 1981; Hiscott et al., 1980, 1981; Sager et al., 1981; M. Lovett, C. E. Clayton & P. W. J. Rigby, unpublished data). I believe that both amplification and rearrangement are mediated by T-antigen and this point is discussed below.

THE STRUCTURE AND EXPRESSION OF MUTANT LARGE T-ANTIGENS IN SV40-TRANSFORMED CELLS

Probably the most striking observation made from the cloning experiments of Clayton & Rigby (1981) was that most of the integrated viral genomes could, at best, encode mutant forms of large T-antigen. We therefore transfected a number of these mutant templates into Rat-1 cells in order to test their ability to transform and into CV-1 monkey cells in order to test the ability of these T-antigens to initiate viral DNA replication. Our results led to a clear conclusion, namely that these integrated templates encode T-antigens which were competent for transformation but defective for replication, i.e. that the transformation and replication functions of the protein can be separated (Clayton et al., 1982a,b; Lovett et al., 1982; C. E. Clayton, M. Lovett, D. Murphy, P. G. Thomas & P. W. J. Rigby, unpublished data). Similar conclusions have been reached by others (May et al., 1983; Stringer, 1982).

THE REGULATION OF CELLULAR TRANSCRIPTION BY LARGE T-ANTIGEN

We have used differential cDNA cloning techniques to isolate a number of cellular genes which are activated as a result of SV40 transformation (Scott et al., 1983b; Brickell et al., 1983, 1985). These data have been extensively reviewed (Scott et al., 1983a; Skene et al., 1986; Rigby et al., 1984a,b, 1985) and I shall concentrate on a few major points. Most of these cDNA clones have been identified from their nucleotide sequences and these results are summarized in Table 1.

Table 1. *cDNA clones derived by differential*
screening between SV40-transformed and
normal Balb/c 3T3 cells

Set	Prototype	Identity
1	pAG64	H-2D^d Class I MHC Antigen
2	pAG59	Endogenous Retrovirus
3	pAG82	COI
4	pAG88	COII
5	pAG10	COII-related
6	pAG23	Unknown
7	pAG38	B2 repeat, non-Class I

EXPRESSION OF CLASS I MHC GENES IN TRANSFORMED CELLS

The prototype cDNA clone of Set 1,pAG64, has the nucleotide sequence of the class I major histocompatibility complex (MHC) antigen H-2D^d (Brickell *et al.*, 1985). We have sequenced four of the nine clones of Set 1 and all of them have the same sequence, apart from differences due to the alternative splicing of exon VII (Brickell *et al.*, 1983). However, one cannot conclude that it is only the *H-2D^d* gene which is activated in transformed cells. Class I MHC genes form a multi-gene family of which there are 34 members in the Balb/c mouse. When pAG64 is used to probe Northern blots of normal and transformed cell RNA the probe will detect not only transcripts from *H-2D^d* but also transcripts from other, related, class I genes. To decide exactly which class I genes are activated in various transformed cell lines one must use as probes synthetic oligonucleotides which can distinguish between the various genes. We have done this for SV3T3 C138 and find that most, if not all, of the activation can be accounted for by *H-2D^d* (B. I. Skene, A. L. Mellor & P. W. J. Rigby, unpublished data). However, it will be necessary to extend these studies to many other transformed cell lines before one can judge whether it is always the same class I gene which is activated. Nonetheless, the fact that the *H-2D^d* gene is markedly activated in at least one line of SV40-transformed cells has led us to explore the mechanisms by which large T-antigen regulates the expression of this gene. We have constructed chimeric genes in which the 5' flanking DNA of the *H-2D^d* gene is linked to the bacterial gene encoding chloramphenicol acetyl transferase. This construct,

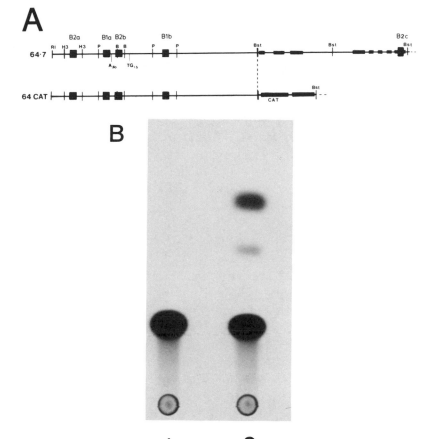

Fig. 4. Regulation of *H-2D^d* gene expression by SV40 large T-antigen.

A. The top line shows the structure of a genomic clone, 64.7, corresponding to the *H-2D^d* gene. ▬▬▬ indicates coding exons as defined by the sequence of Taylor-Sher *et al.* (1985). The upstream DNA sequences contain a number of B1 and B2 repetitive elements indicated by ■. In the chimeric gene, *64cat*, the *H-2D^d* coding sequences have been replaced by the bacterial gene encoding chloramphenicol acetyl transferase. The *Bst*I site used to fuse the two DNA segments lies within the 5′ untranslated leader of each gene.

B. Results of a transient co-transfection experiment into Balb/c 3T3 cells. Cells were transfected with the indicated plasmids and harvested 48 hours later. Extracts were prepared and assayed for CAT activity as previously described. Lane 1,*64cat*: Lane 2,*64cat* plus pTSV3.

called 64-*cat* (see Fig. 4A), has been transfected into a number of different cell types both alone and in the presence of pTSV3, a plasmid encoding large T-antigen. In Fig. 4B are shown the results of a transient expression experiment in which the recipient cells were Balb/c 3T3; 64-*cat* alone is expressed at a low level, consistent with the inefficient expression of the endogenous *H-2D^d* gene in these cells. However, in the presence of pTSV3 there is a marked increase in expression, in this experiment 28-fold. This induction is also seen when the plasmid DNAs are stably integrated into the chromosomes of the recipient cells and we have performed primer extension analyses on mRNA isolated from such stable transformants to show that the *cat* mRNA is properly initiated at the polymerase II promoter of the *H-2D^d* gene (Skene, 1986). Deletion analyses have located at least some of the sequences involved in T-antigen induction to the 0.6 Kb *Hin*dIII fragment indicated in Fig. 3 (Rigby *et al.*, 1985) and we are presently engaged in locating the responsive sequences more precisely and in linking this *Hin*dIII fragment to heterologous promoters to ask if it alone confers T-antigen responsiveness.

REGULATION OF RNA POLYMERASE III TRANSCRIPTION BY LARGE T-ANTIGEN

pAG64 contains a B2 repetitive element within the 3' untranslated region and on Northern blots it hybridizes not only to the 1.6 Kb class I MHC mRNA but also to polydisperse RNAs which migrate as if they are between 0.6 and 0.7 Kb in length. We have shown, by constructing appropriate subclones of pAG64, that the hybridization to these small RNAs is due to the B2 element (Brickell *et al.*, 1983). B2 repeats, in common with many other dispersed repetitive elements in the genomes of higher eukaryotes, contain a consensus promoter for RNA polymerase III (Krayev *et al.*, 1982) and it thus seemed likely that these small RNAs are polymerase III transcripts of B2 elements, a supposition which has since been confirmed (Rigby *et al.*, 1985; Singh *et al.*, 1985a). We have performed *in vitro* transcription experiments which show that pure large T-antigen can stimulate polymerase III transcription from a cloned B2 repeat. DNAase footprinting experiments do not detect binding of large T-antigen to the template DNA. There is specificity to this stimulation as T-antigen does not stimulate transcription from the polymerase III promoter of the VA$_1$ RNA gene of adenovirus type 2 and Singh *et al.* (1985a) showed that in SV40-transformed cell lines which express high levels of the B2 RNAs there is no increase in the level of 5S rRNA, another polymerase III transcript.

ACTIVATION OF ENDOGENOUS RETROVIRUSES IN SV40-TRANSFORMED CELLS

Nucleotide sequence analysis of the prototype clone of Set 2, pAG59, identifies it as being derived from a C-type retrovirus (P. M. Brickell,

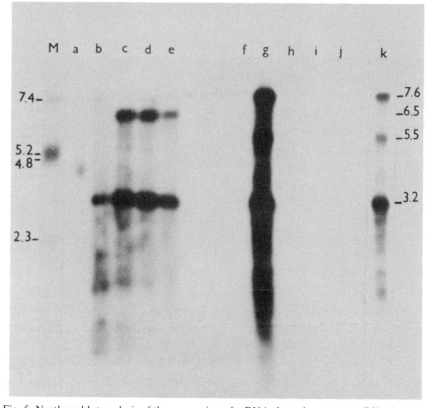

Fig. 5. Northern blot analysis of the expression of mRNAs homologous to pAG59, the proto-type clone of set 2. One microgramme aliquots of polyadenylated, cytoplasmic RNA were fractionated by electrophoresis in a 1.5% agarose gel containing formaldehyde; the RNA was transferred to nitrocellulose filters which were hybridised with ^{32}P-labelled pAG59. The samples were: a, Balb/c 3T3 A31; b, SV3T3 Cl38; c, SV3T3 Cl38.A2; d, SV3T3 Cl38.B4; e, SV3T3 Cl38.B5; f, SV3T3 Cl20; g, SV3T3 Cl26; h, SV3T3 Cl49; i, SV3T3 ClH; j, SV3T3 ClM; k, a shorter exposure of lane g. Transcript sizes are given in Kb. Reproduced, with permission, from Scott *et al.* (1983). Copyright, MIT Press, 1983.

D. S. Latchman & P. W. J. Rigby, unpublished data). One can then interpret the Northern blotting pattern observed with this clone (Fig. 5); the 3.2 Kb transcript is the mRNA which encodes the enve-lope glycoprotein, gp70. In agreement with this, immunoprecipi-tation of extracts of metabolically labelled cells using anti-gp70 monoclonal antibodies shows that cells which contain the 3.2 Kb transcript also contain gp70 (P. M. Timmons & P. W. J. Rigby, unpublished data). In SV40-transformed Balb/c 3T3 cells activation of an endogenous retrovirus occurs in only about 25% of the cases that we have examined although in transformed C3H cells a much higher percentage express the *env* mRNA (L. Gooding & P. W. J. Rigby, unpublished data).

We do not know from which of the endogenous retroviruses in the genome of the Balb/c mouse pAG59 is derived nor do we know whether the expression of retroviral gp70 is of any significance. We can not analyse the mechanisms by which large T-antigen regulates the expression of endogenous retroviruses until we have identified which loci are involved but we have shown that the promoter in the LTR of Moloney murine leukaemia virus is activated by large T-antigen in transient co-transfection experiments (C. M. Gorman, B. I. Skene & P. W. J. Rigby, unpublished data). Singh *et al.*, (1985*b*) have shown that the retrovirus-like VL30 sequences are also activated in some SV40-transformed mouse cell lines.

EXPRESSION OF THE MITOCHONDRIAL GENOME IN SV40-TRANSFORMED CELLS

The cDNA clones of Set 3 have the sequence of cytochrome oxidase subunit I while those of Set 4 have the sequence of cytochrome oxidase subunit II (Scott, 1982; P. M. Brickell, D. S. Latchman & P. W. J. Rigby, unpublished data). We have observed that many transformed mouse cell lines contain not only higher levels of the mature mRNAs but also larger RNAs which we presume to be incompletely processed precursors (Fig. 6). Our observations have acquired renewed interest with the recent report that cells which have been transfected with a plasmid expressing polyoma large T-antigen also express higher levels of the cytochrome oxidase subunit II mRNA (Glaichenhaus *et al.*, 1986). It is striking that two different papovaviral T-antigens appear to be able to regulate the transcription of the mitochondrial genome although one must presume that this is an indirect effect. Because of this and because one cannot transfect mitochondrial DNA back into mitochondria it will be difficult to study the mechanisms involved.

DISCUSSION

While the general principles of SV40 DNA integration are now clear, a number of issues remain to be resolved. Our data (Chia & Rigby, 1981) suggest strongly that the tandem head-to-tail oligomers which accumulate in infected non-permissive cells are products of limited

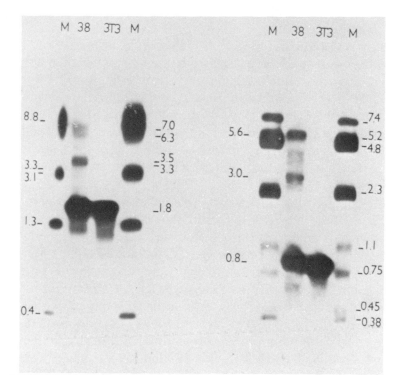

Fig. 6. Northern blot analysis of the expression of mitochondrial genes in normal and SV40-transformed cells. In the left-hand panel the probe was pAG1, a member of Set 3. In the right-hand panel the probe was pAG47, a member of Set 4. The tracks are: M, size markers; 38, polyadenylated, cytoplasmic RNA from SV3T3 Cl38; 3T3, polyadenylated, cytoplasmic RNA from Balb/c 3T3 A31. Transcript sizes are given in Kb. Taken from Scott (1982).

DNA replication but further experiments are required to prove this point. It would also be highly desirable to clone the primary products of integration from infected non-permissive cells in order to be able to analyse in detail the mechanism of recombination. All of the available data are consistent with the notion that integration occurs by non-homologous recombination. However, because of the fact that integration is followed by extensive rearrangements and amplification, it remains a formal possibility that the initial recombination is mediated by limited homology, or some other structural feature of the DNA, and that this is obscured by subsequent events.

Why should there be such a high frequency of mutated viral genomes in transformed cells? I believe that this is related to the observation described above, that there is extensive post-integrational amplification and rearrangement of integrated sequences. Large T-antigen is a DNA replication protein and it is known to be able to stimulate the replication of cellular DNA. Moreover, there is evidence that polyoma large T-antigen is involved in the amplification of integrated viral sequences (Colantuoni *et al.*, 1980). If replication-competent large T-antigen is capable of amplifying and re-arranging cellular sequences then it is likely to be lethal to cells expressing it and thus there will be a strong selective pressure to eliminate the DNA replication function. This idea is supported by the observation that it is extremely difficult to construct cell lines expressing wild-type large T-antigen if the viral genome is introduced in a co-transfection experiment in which there is no selection for transformation (Mulligan & Berg, 1981).

It is clear that quite extensive changes in the pattern of cellular gene expression occur as a result of transformation. However, neither we, nor others who have addressed this question (Schutzbank *et al.*, 1982; Singh *et al.*, 1985a,b) have identified any genes which are clearly on the transformation pathway. Over-expression of the *H-2D^d* gene, for example, induces no detectable alterations in cellular phenotype (Skene, 1986). In our differential cDNA cloning experiments the preparative hybridization procedures that were employed to generate a probe enriched in transformed cell-specific sequences would have led to the elimination of many sequences which correspond to genes which are activated as a result of transformation but which encode mRNAs which are still relatively rare in the transformed cell (Scott *et al.*, 1983). If there are genes which must be activated to induce the transformed phenotype, it is possible that they belong to such a low abundance class. Moreover, we, and others, have only examined genes which are activated as a result of transformation. Perhaps the vital event is the repression of genes which are expressed in the normal cell. The data of Williams *et al.* (1977) show that such genes do exist and one could obviously repeat the differential cDNA cloning in such a way as to isolate them.

The expression of class I MHC genes is regulated in many transformed cell systems. We have shown that elevated levels of class I MHC mRNAs which hybridize to pAG64 are found in a wide variety of transformed mouse cell lines (Brickell *et al.*, 1983); Scott *et al.*, (1983) and Majello *et al.* (1985) have observed high levels of such mRNAs in a variety of polyoma virus-transformed rat cells. In contrast, cells transformed by human adenovirus type 12, but not by adenovirus type 5, express very low levels of class I antigens; this repression of MHC expression is mediated by the viral Ela gene

(Bernards et al., 1983; Schrier et al., 1983). Moreover, adenoviruses also decrease the surface expression of class I antigens by another mechanism in which the 19 kd glycoprotein encoded by the viral E3 transcription unit binds to intracellular class I molecules and prevents their proper processing and transport to the surface (Andersson et al., 1985; Burgert & Kvist, 1985). It has been argued by Bernards et al. (1983) that the ability of adenovirus type 12 to repress class I expression accounts for the oncogenicity of this virus but it is not at all clear what are the consequences, if any, of elevated class I expression in papovavirus-transformed cells. It should be remembered that neither human adenoviruses nor SV40 are oncogenic in their natural hosts and that the evolutionary pressure for these viruses to interact with the MHC is more likely to be related to latency. It is striking that several different viruses have evolved a variety of mechanisms to interfere with the normal expression of class I MHC antigens and one presumes that there must be some advantage to the virus to do this. The precise nature of this advantage will only be elucidated by further experiments.

Our results show that large T-antigen can regulate the expression of the $H\text{-}2D^d$ gene and we are hopeful that by performing further gene transfer experiments, and also by attempting to reconstruct this regulation in in vitro transcription systems, we will be able to determine the precise biochemical mechanisms involved. Our ability to observe the activation of polymerase III transcription of the B2 repeats by large T-antigen in vitro means that we should certainly be able to analyse this regulation in detail. It is interesting to note that in this system and in the activation of polymerase I transcription it has not been possible to demonstrate binding of large T-antigen to the DNA and that in the T-antigen-mediated activation of the viral late promoter, elements other than the T-antigen binding sites appear to be involved. The ability of the protein to repress viral early transcription clearly involves DNA binding (Hansen et al., 1981; Rio & Tjian, 1983) but it seems likely that activation proceeds by a different mechanism.

Large T-antigen is not the only DNA tumour virus-transforming protein which is capable of regulating polymerase III transcription. The Ela products of human adenoviruses can also do this, although in this case there is no evidence for specificity and the mechanism appears to involve an increase in the effective concentration of TFIIIC within the infected cell (Hoeffler & Roeder, 1985; Yoshinaga et al., 1986). It would appear that T-antigen must work by a different mechanism because our data show that pure protein can stimulate transcription in an extract derived from an uninfected cell. We can not comment on the significance for the mechanism of transformation, if any, of this activation of polymerase III transcription because the function of the small B2 RNAs remains unknown.

One can be optimistic that future work will lead to a detailed description of how large T-antigen interacts with the cellular DNA, both to induce DNA replication and to regulate transcription. However, it remains unclear that these activities are essential to the mechanism of transformation. Indeed, despite enormous amounts of work we still can not say with any certainty which of the many biochemical functions of large T-antigen are required for transformation and one can be sure that SV40 will continue to intrigue us for many years to come.

ACKNOWLEDGEMENTS

I am very grateful to all my collaborators who have made such great contributions to this work. Most of the studies reported here were performed while I was a member of the Cancer Research Campaign's Eukaryotic Molecular Genetics Research Group in the Department of Biochemistry at the Imperial College of Science and Technology in London. The Campaign's extremely generous funding, particularly the Career Development Award which supported me for six years, was vital to our efforts and is enormously appreciated.

REFERENCES

ANDERSSON, M., PAABO, S., NILSSON, T. & PETERSON, P. A. (1985). Impaired intracellular transport of class I MHC antigens as a possible means for adenoviruses to evade immune surveillance. *Cell*, **43**, 215–22.

BENDER, M. A. & BROCKMAN, W. W. (1981). Rearrangement of integrated viral DNA sequences in mouse cells transformed by simian virus 40. *Journal of Virology*, **38**, 872–9.

BERNARDS, R., SCHRIER, P. I., HOUWELING, A., BOS, J. L., VAN DER EB, A. J., ZIJLSTRA, M. & MELIEF, C. J. M. (1983). Tumorigenicity of cells transformed by adenovirus type 12 by evasion of T-cell immunity. *Nature*, **305**, 776–9.

BRADY, J. & KHOURY, G. (1985). *Trans* activation of the Simian virus 40 late transcription unit by T-antigen. *Molecular and Cellular Biology*, **5**, 1391–9.

BRADY, J., BOLEN, J. B., RADONOVICH, M., SALZMAN, N. & KHOURY, G. (1984). Stimulation of Simian virus 40 late gene expression by Simian virus 40 tumor antigen. *Proceedings of The National Academy of Sciences, USA*, **81**, 2040–4.

BRICKELL, P. M., LATCHMAN, D. S., MURPHY, D., WILLISON, K. & RIGBY, P. W. J. (1983). Activation of a *Qa/Tla* class I major histocompatibility antigen gene is a general feature of oncogenesis in the mouse. *Nature*, **306,** 756–60.

BRICKELL, P. M., LATCHMAN, D. S., MURPHY, D., WILLISON, K. & RIGBY, P. W. J. (1985). The class I major histocompatibility gene activated in a line of SV40-transformed mouse cells is *H-2D^d* not *Qa/Tla*. *Nature*, **316,** 162–3.

BURGERT, H-G. & KVIST, S. (1985). An adenovirus type 2 glycoprotein blocks cell surface expression of human histocompatibility class I antigens. *Cell*, **41,** 987–97.

CHIA, W. & RIGBY, P. W. J. (1981). The fate of viral DNA in non-permissive cells infected with Simian virus 40. *Proceedings of The National Academy of Sciences, USA*, **78,** 6638–42.

CLAYTON, C. E. & RIGBY, P. W. J. (1981). Cloning and characterization of the integrated viral DNA from three lines of SV40-transformed mouse cells. *Cell*, **25,** 547–59.

CLAYTON, C. E., LOVETT, M. & RIGBY, P. W. J. 1982*a*. Functional analysis of a Simian virus 40 super T-antigen. *Journal of Virology*, **44,** 974–82.

CLAYTON, C. E., MURPHY, D., LOVETT, M. & RIGBY, P. W. J. (1982*b*). A fragment of the SV40 large T-antigen gene transforms. *Nature*, **299,** 59–61.

COLANTUONI, V., DAILEY, L. & BASILICO, C. (1980). Amplification of integrated viral DNA sequences in polyoma-virus transformed cells. *Proceedings of The National Academy of Sciences, USA*, **77,** 3850–4.

ELIYAHU, D., RAZ, A., GRUSS, P., GIVOL, D. & OREN, M. (1984). Participation of p53 cellular tumour antigen in transformation of normal embryonic cells. *Nature*, **312,** 646–9.

GLAICHENHAUS, N., LEOPOLD, P. & CUZIN, F. (1986). Increased levels of mitochondrial gene expression in rat fibroblast cells immortalized or transformed by viral and cellular oncogenes. *EMBO Journal*, **5,** 1261–5.

HANSEN, U., TENEN, D. G., LIVINGSTON, D. M. & SHARP, P. A. (1981). T-antigen repression of SV40 early transcription from two promoters. *Cell*, **27,** 603–12.

HARTZELL, S. W., BYRNE, B. J. & SUBRAMANIAN, K. N. (1984). The Simian virus 40 minimal origin and the 72-base-pair repeat are required simultaneously for efficient induction of late gene expression with large tumour antigen. *Proceedings of The National Academy of Sciences, USA*, **81,** 6335–9.

HISCOTT, J., MURPHY, D. & DEFENDI, V. (1980). Amplification and rearrangement of integrated SV40 DNA sequences accompany the selection of anchorage-independent transformed mouse cells. *Cell*, **22,** 535–43.

HISCOTT, J. B., MURPHY, D. & DEFENDI, V. (1981). Instability of integrated viral DNA in mouse cells transformed by Simian virus 40. *Proceedings of The National Academy of Sciences, USA*, **78,** 1736–40.

HOEFFLER, W. K. & ROEDER, R. G. (1985). Enhancement of RNA polymerase III transcription by the E1a gene product of adenovirus. *Cell*, **41,** 955–63.

JENKINS, J. R., RUDGE, K. & CURRIE, G. A. (1984). Cellular immortalization by a cDNA clone encoding the transformation-associated phosphoprotein p53. *Nature*, **312,** 651–4.

KELLER, J. M. & ALWINE, J. C. (1984). Activation of the SV40 late promoter: direct effects of T-antigen in the absence of viral DNA replication. *Cell*, **36,** 381–9.

KELLER, J. M. & ALWINE, J. C. (1985). Analysis of an activatable promoter; sequences in the Simian virus 40 late promoter required for T-antigen-mediated *trans* activation. *Molecular and Cellular Biology*, **5,** 1859–69.

KRAYEV, A. S., MARKUSHEVA, T. V., KRAMEROV, D. A., RYSKOV, A. P., SKRYABIN, K. G., BAYEV, A. A. & GEORGIEV, G. P. (1982). Ubiquitous transposon-like repeats B1 and B2 of the mouse genome: B2 sequencing. *Nucleic Acids Research*, **10**, 7461–75.

LANE, D. P., SIMANIS, V., BARTSCH, R., YEWDELL, J., GANNON, J. & MOLE, S. (1985). Cellular targets for SV40 large T-antigen. *Proceedings of The Royal Society of London B*, **226**, 25–42.

LEARNED, R. M., SMALE, S. T., HALTINER, M. M. & TJIAN, R. (1983). Regulation of human ribosomal RNA transcription. *Proceedings of The National Academy of Sciences, USA*, **80**, 3558–62.

MAJELLO, B., LA MANTIA, G., SIMEONE, A., BONCINELLI, E. & LANIA, L. (1985). Activation of major histocompatibility complex class I mRNA containing an *Alu*-like repeat in polyoma virus-transformed rat cells. *Nature*, **314**, 457–9.

MARTIN, R. G. (1981). The transformation of cell growth and transmogrification of DNA synthesis by Simian virus 40. *Advances in Cancer Research*, **34**, 1–68.

MAY, E., LASNE, C., PRIVES, C., BORDE, J. & MAY, P. (1983). Study of the functional activities concomitantly retained by the 115,000 M_r super T-antigen, an evolutionary variant of Simian virus 40 large T-antigen expressed in transformed rat cells. *Journal of Virology*, **45**, 901–13.

MULLIGAN, R. C. & BERG, P. (1981). Selection for animal cells that express the *Escherichia coli* gene coding for xanthine-guanine phosphoribosyl-transferase. *Proceedings of The National Academy of Sciences, USA*, **78**, 2072–6.

PARADA, L. F., LAND, H., WEINBERG, R. A., WOLF, D. & ROTTER, V. (1984). Cooperation between gene encoding p53 tumour antigen and ras in cellular transformation. *Nature*, **312**, 649–51.

POSTEL, E. H. & LEVINE, A. J. (1976). The requirement of Simian virus 40 gene *A* product for the stimulation of cellular thymidine kinase activity after viral infection. *Virology*, **73**, 206–15.

RIGBY, P. W. J. & BERG, P. (1978). Does Simian virus 40 DNA integrate into cellular DNA during productive infection? *Journal of Virology*, **28**, 475–89.

RIGBY, P. W. J. & LANE, D. P. (1983). Structure and Function of Simian virus 40 large T-antigen. *Advances in Viral Oncology*, **3**, 31–57.

RIGBY, P. W. J., BRICKELL, P. M., LATCHMAN, D. S., MURPHY, D., WESTPHAL, K-H. & SCOTT, M. R. D. (1984a). Oncogenic transformation activates cellular genes. In *Genetic Manipulation: Impact on Man and Society*. eds. Arber *et al.*, pp. 227–34. Cambridge: Cambridge University Press.

RIGBY, P. W. J., BRICKELL, P. M., LATCHMAN, D. S., MURPHY, D., WESTPHAL, K-H. & WILLISON, K. (1984b). The activation of cellular genes in transformed cells. *Philosophical Transactions of the Royal Society of London B*, **307**, 347–51.

RIGBY P. W. J., CHIA, W., CLAYTON C. E., & LOVETT, M. (1980). The structure and expression of the integrated viral DNA in mouse cells transformed by simian virus 40. Proceedings of the Royal Society of London B, *210*, 437–50.

RIGBY, P. W. J., LA THANGUE, N. B., MURPHY, D. & SKENE, B. I. (1985). The regulation of cellular transcription by Simian virus 40 large T-antigen. *Proceedings of the Royal Society of London B*, **226**, 15–23.

RIO, D. & TJIAN, R. (1983). SV40 T-antigen binding site mutations that affect autoregulation. *Cell*, **32**, 1227–40.

SAGER, R., ANISOWICZ, A. & HOWELL, N. (1981). Genomic rearrangements in a mouse cell line containing integrated SV40 DNA. *Cell*, **23**, 41–50.

SCHRIER, P. I., BERNARDS, R., VAESSEN, R. T. M. J., HOUWELING, A. & VAN DER EB, A. J. (1983). Expression of class I major histocompatibility antigens switched off by highly oncogenic adenovirus 12 in transformed rat cells. *Nature*, **305**, 771–5.

SCHUTZBANK, T., ROBINSON, R., OREN, M. & LEVINE, A. J. (1982). SV40 large tumor antigen can regulate some cellular transcripts in a positive fashion. *Cell*,

30, 481–90.

Scott, M. R. D. (1982). Activation of cellular genes following transformation by Simian virus 40. PhD Thesis, University of London.

Scott, M. R. D., Brickell, P. M., Latchman, D. S., Murphy, D., Westphal, K-H. & Rigby, P. W. J. (1983a). The use of cDNA cloning techniques to isolate genes activated in tumour cells. In *Modern Approaches to Human Leukaemia* 5, eds Neth *et al*. *Haematology and Blood Transfusion*, **28**, 236–40.

Scott, M. R. D., Westphal, K-H. & Rigby, P. W. J. (1983b). Activation of mouse genes in transformed cells. *Cell*, **34**, 557–67.

Singh, K., Carey, M., Saragosti, S. & Botchan, M. (1985a). Expression of enhanced levels of small RNA polymerase III transcripts encoded by the *B2* repeats in Simian virus 40-transformed mouse cells. *Nature*, **314**, 553–6.

Singh, K., Saragosti, S. & Botchan, M. (1985b). Isolation of cellular genes differentially expressed in mouse NIH3T3 cells and a Simian virus 40-transformed derivative; growth-specific expression of VL30 genes. *Molecular and Cellular Biology*, **5**, 2590–8.

Skene, B. I. (1986). Regulation of cellular gene expression by DNA tumour virus transforming proteins. PhD Thesis, University of London.

Skene, B. I., La Thangue, N. B., Murphy, D. & Rigby, P. W. J. (1986). The regulation of cellular transcription by viral transforming proteins. In *New Avenues in Developmental Cancer Chemotherapy*, eds. Harrap & Connors, Bristol-Myers Cancer Symposium **8**. Orlando, Florida: Academic Press, in press.

Soprano, K. J., Galanti, N., Jonak, G. J., McKercher, S., Pipas, J. M., Peden, K. W. C. & Baserga, R. (1983). Mutational analysis of Simian virus 40 T-antigen: stimulation of cellular DNA synthesis and activation of rRNA genes by mutants with deletions in the T-antigen gene. *Molecular and Cellular Biology*, **3**, 214–19.

Stahl, H., Droge, P. & Knippers, R. (1986). DNA helicase activity of SV40 large tumor antigen. *EMBO Journal*, **5**, 1939–44.

Stringer, J. R. (1982). Mutant of Simian virus 40 large T-antigen that is defective for viral DNA synthesis but competent for transformation of cultured rat cells. *Journal of Virology*, **42**, 854–64.

Tooze, J. (1981). Molecular Biology of Tumor Viruses. *DNA Tumor Viruses*. Revised 2nd ed. New York: Cold Spring Harbor Laboratory.

Williams, J. G., Hoffman, R. & Penman, S. (1977). The extensive homology between mRNA sequences of normal and SV40-transformed human fibroblasts. *Cell*, **11**, 901–7.

Yoshinaga, S., Dean, N., Han, M. & Berk, A. J. (1986). Adenovirus stimulation of transcription by RNA polymerase III: evidence for an Ela-dependent increase in transcription factor IIIC concentration. *EMBO Journal*, **5**, 343–54.

PAPILLOMAVIRUSES AND DISEASE

M. SAVERIA CAMPO* AND WILLIAM F. H. JARRETT†

*The Beatson Institute for Cancer Research,
†Department of Veterinary Pathology, Veterinary School, University
of Glasgow, Garscube Estate, Bearsden, Glasgow, UK

Papillomaviruses (PVs) are oncogenic viruses in their natural hosts, and potentially among the best candidates for furthering our understanding of viral carcinogenesis. They induce hyperproliferation of epithelial cells of the skin or mucosa, although certain types can also infect fibroblasts. The proliferation of the infected cells leads to production of benign tumours, papillomas or warts, some of which can be stimulated towards malignant progression to carcinomas by genetic and/or environmental factors. The best-known examples are: the cottontail rabbit/domestic rabbit system, where the genetic make-up of the host is an important factor determining the extent and the rate of malignant conversion (Kreider, 1980); the progression to carcinomas of the Epidermodysplasia verruciformis (EV) lesions in areas exposed to sunlight in individuals with impaired cell-mediated immunity (Orth et al., 1980); and the transition of papillomas to carcinomas in the upper alimentary canal of cattle feeding on the bracken fern (Jarrett et al., 1978a). The malignant transformation of anogenital warts has been recognized for some time (Oriel, 1971), but the association between cervical cancer and cervical intraepithelial neoplasia (CIN) and papillomavirus infection (Zur Hausen, 1977; Morin et al., 1981; Crum et al., 1984) has only recently attracted world-wide attention and the investigative efforts of several laboratories.

As a rule, the papillomas produce mature virion progeny, whereas the cells of the malignant lesions are not permissive for virus replication. The vegetative virus cycle is tightly linked to the differentiation of epithelial cells (Orth, Jeanteur & Croissant, 1971): viral DNA synthesis can be detected by in situ hybridization in the cells of the granular and corneal layers but not in those of the proliferative basal layers, although it is generally believed that the viral genome is present there and that its expression is responsible for cell transformation and proliferation. The synthesis of the structural capsid proteins takes place in the granular and corneal layers, and virion assembly and virus release occur in the corneal and squamous layer respectively (see Figs 1–4 of Orth *et al.*, 1971). The expression of the viral functions therefore appears to be controlled by the state of differentiation of the epithelial cell itself. The failure to reproduce full keratinocyte differentiation *in vitro* has been partly responsible for the absence of a cell culture system capable of propagating the virus. This has severely hindered detailed analysis of the viral functions for a long time, and only recently, with the availability of molecularly cloned genomes, have the salient structural and functional features of the virus come to light.

MOLECULAR HETEROGENEITY AND LESION SITE SPECIFICITY

The papillomaviruses are remarkably heterogeneous, with over 30 different types described in humans and 6 in cattle. The several viruses have been classified into subgroups, on the basis of both their molecular relationship and of their clinical symptoms. Importantly, the molecular heterogeneity of the viruses appears to underlie their different cytopathological effects.

To date, at least 32 human papillomaviruses (HPV) have been identified, many of them associated with specific lesions. The better-characterized viruses have been divided into seven major groups, A–G (Table 1; Pfister, 1984). However, since new virus types are continuously being isolated, and the relationship between the groups awaits further clarification, it is more convenient for the time being to divide the viruses into subgroups infecting either cutaneous or mucosal tissue (Table 2; for a review, see Smith & Campo, 1985). The virus types found in similar lesions are generally closely related: thus the viruses found in genital lesions display significant nucleic acid homology, and so do the several viruses found in EV lesions.

The six types of bovine papillomaviruses (BPV) are each associated with a specific lesion and fall into two subgroups (Table 3; Campo *et al.*, 1981; Jarrett *et al.*, 1984*b*); subgroup A viruses cause fibropapillomas whereas subgroup B viruses cause true epithelial papillomas. The members of each subgroup are related to each other both genomically and antigenically, but are evolutionary distinct from the members of the other subgroup.

PAPILLOMAVIRUSES AND CANCER

Despite their great multiplicity, only a few papillomavirus types are associated with naturally occurring cancers. Among the several EV

Table 1. *Classification of human papillomaviruses*

Group	Type	Lesion
A	HPV-1	Verruca plantaris
B	HPV-2	Verruca vulgaris
	HPV-3	Verruca plana
	HPV-10	Verruca plana; epidermodysplasia verruciformis (EV)
C	HPV-4	Verruca vulgaris
D	HPV-5	EV in immunosuppressed patients; skin carcinoma
	HPV-8	EV
	HPV-9	EV
	HPV-12	EV
	HPV-14	EV
	HPV-15	EV
	HPV-17	EV; skin carcinoma
E	HPV-6	Condyloma acuminatum
	HPV-11	Condyloma acuminatum; laryngeal papilloma; CIN
	HPV-13	Focal epithelial hyperplasia
F	HPV-7	Verruca vulgaris ('butcher's warts')
G	HPV-16	Condyloma acuminatum; CIN; cervical and penile carcinoma
	HPV-18	Morbus Bowen disease; cervical and penile carcinoma.

From Pfister, 1984.

Table 2. *The relationship between human papillomavirus types and site of lesion*

HPV type	Type of lesion
Skin	
HPV-1, 2, 3, 4, 7, 10, 31, 32	Common cutaneous warts
HPV-3, 5, 8, 9, 10, 12, 14, 15, 17, 19–29	Epidermodysplasia verruciformis
Mucosal tissue	
HPV-13	Buccal cavity
HPV-6, 11, 30	Larynx, genitals
HPV-16, 18	Genitals

From Smith & Campo, 1985.

viruses, only the genomes of HPV-5, 8 and 17 are found in the lesions progressing to carcinomas (Orth *et al.*, 1980; Ostrow *et al.*, 1982; Kremsdorf *et al.*, 1984; Yutsudo, Shimkage & Hakura, 1985) and the DNAs of HPV-16 and 18 are much more often found in the carcinomas of the cervix and the penis than those of HPV-6 or 11 (Durst *et al.*, 1983; Gissmann *et al.*, 1983; Boshart *et al.*, 1984).

Table 3. *Bovine papillomas and their viruses*

Subgroup	Type	Tumour	DNA size (kilobases)	DNA sequence homology with		Immuno cross-reaction with	
				BPV-1 (%)	BPV-4 (%)	BPV-1	BPV-4
	BPV-1	Penile fibropapilloma Teat frond fibropapilloma Adjacent skin fibropapilloma	7.9	100	0	Complete	None
A	BPV-2	Classical skin fibropapilloma Alimentary canal fibropapilloma	7.9	70	0	Complete	None
	BPV-5	'Rice grain' fibropapilloma	7.9	5	0	Partial	None
	BPV-3	Skin true papilloma	7.2	0	50	n.d.	n.d.
B	BPV-4	Alimentary canal true papilloma	7.3	0	100	None	Complete
	BPV-6	Teat true frond papilloma	7.2	0	20	None	n.d.

From Jarrett *et al.*, 1980; Campo *et al.*, 1981; Jarrett *et al.*, 1984*b*.

Likewise, among the BPVs, only BPV-4 is associated with carcinomas of the alimentary canal (Campo *et al.*, 1980), despite abortive infection of the oesophagus by BPV-2 (Jarrett *et al.*, 1984a). However, HPV-10 DNA has been found in some genital cancers (Green *et al.*, 1982) and BPV-2 has been implicated in the aetiology of bladder cancers (Olson *et al.*, 1959; Olson, Pamukcu & Brobst, 1965; Campo, 1984; Campo & Jarrett, 1986).

It is not yet known whether the apparent greater 'malignancy' of some viruses is due to their genetic structure, to a greater propensity of their target cells for neoplastic transformation, or to a combination of both. In an attempt to clarify this point, the genomes of several papillomaviruses have been sequenced and the transcriptional pattern determined.

GENOME ORGANIZATION OF PAPILLOMAVIRUSES AND MOLECULAR ASPECTS OF CELL TRANSFORMATION

The overall genome organization is similar in all virus types, with large overlapping open reading frames (ORFs) contained in only one strand of the genomic DNA, in similar positions and with homologous sequences in different viruses (Fig. 1; Danos *et al.*, 1984). The other strand is blocked by stop codons in all three reading frames and any small ORFs that are present show no conservation of position or apparent coding sequences. The viral mRNAs are generated by complex splicing mechanisms (Fig. 2; Yang, Okayama & Howley, 1985a); and by deletion–insertion mutagenesis of subgenomic DNA fragments and cDNAs, it has been possible to assign several functions to precise regions of the viral genome. In such a way, transcriptional enhancer and promoter sequences have been identified (Campo *et al.*, 1983; Lusky *et al.*, 1983; Nakabayashi, Chattopadhyay & Lowy, 1983), the origin of replication of the DNA has been mapped (Waldek, Rosl & Zentgraf, 1984), sequences responsible for the maintenance of the DNA as a multicopy plasmid have been located (Lusky & Botchan, 1984), and the genes responsible respectively for morphological transformation and trans-activation have been identified (Sarver *et al.*, 1984; Spalhoz, Yang & Howley, 1985; Yang *et al.*, 1985b; Shiller *et al.*, 1986), as have the genes encoding the structural capsid proteins (Engel, Heilman & Howley, 1983) (Fig. 2; Table 4). Although these salient features of the virus cycle have been elucidated in experimental tissue culture systems, it is reasonable to assume that the same, or similar, molecular mechanisms will be operational *in vivo*.

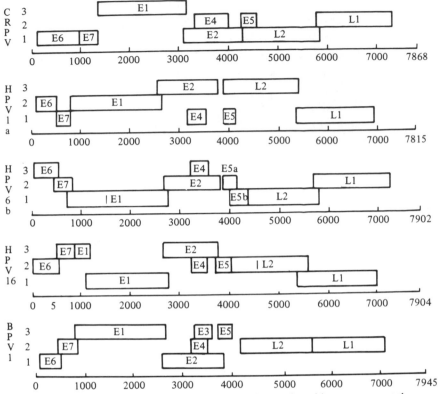

Fig. 1. Genomic organisation of papillomaviruses. The numbered boxes represent the open reading frames (ORFs) and the solid lines represent the linearized viral genomes (from Danos et al., 1984 and Seerdof et al., 1985).

No relevant differences have been found in the genomic organization of the 'more malignant' types of papillomaviruses. However, HPV-5 DNA deleted in the structural region has been found in one case of primary skin carcinoma and in the metastatic deposits of another case (Fig. 3a; Ostrow et al., 1982), and inserts totalling 108 base pairs have been detected in the non-coding region of a subtype of HPV-6 isolated from an invasive vulvar carcinoma (Fig. 3b; Rando et al., 1986). In addition, the genome of HPV-16 and HPV-18 is often deleted in the E1/E2 region and integrated into cellular DNA in the genital cancers (Fig. 3c; Durst et al., 1985; Lehn, Kneg & Sauer, 1985). It is not clear whether these differences are relevant to the malignant state or whether they merely reflect the non-requirement of certain functions for neoplastic transformation.

Little is still known of the interaction between viral and cellular functions in vivo. For instance, it is not known whether the E1 protein

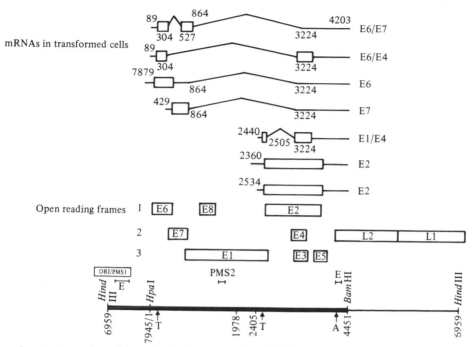

Fig. 2. Genomic and transcriptional organisation of BPV-1. The viral genome, linearized at the *Hind* III site, is represented by the solid line, and the region encoding the early transforming functions by the bold line. The numbers indicate nucleotide positions. E, transcriptional enhancer; T, TATA box; A, polyadenylation site; ORI, origin of DNA replication; PMS, plasmid maintenance sequence. The numbered boxes represent the ORFs. The open boxes represent the coding sequences of the mRNAs and the solid lines their untranslated sequences. The slanted lines represent intervening sequences.

Table 4. *Functions of BPV-1 genes*

ORF	Function
E1	Plasmid maintenance (Lusky & Botchan, 1984)
E2	Trans-activation of transcription (Spalholz *et al.*, 1985)
E3	Unknown
E4	Virion maturation[a] (Doorbar *et al.*, 1986)
E5	Cell transformation (Shiller *et al.*, 1986)
E6	Cell transformation (Yang *et al.*, 1985*b*)
E7	Regulation of copy number (Lusky & Botchan, 1985)
L1	Major capsid protein (Engel *et al.*, 1983)
L2	Minor capsid protein (Engel *et al.*, 1983)

[a] Determined for HPV-1.

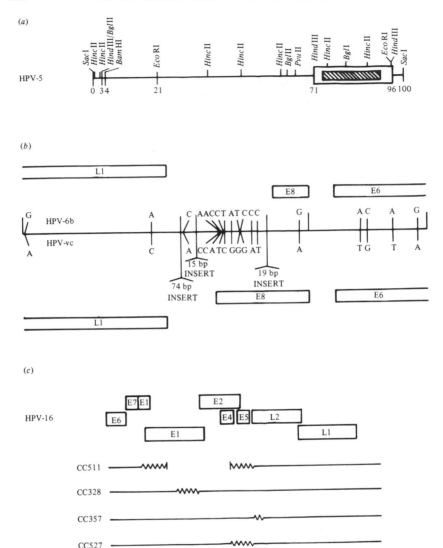

Fig. 3. Genomic organization of HPV DNA found in carcinomas. (a) HPV-5 DNA: the open box represents the deletion affecting the viral DNA in a primary skin squamous cell carcinoma and the shaded box represents the deletion in the viral DNA of a metastatic deposit (Ostrow et al., 1982). (b) HPV-6vc DNA: insertions and single nucleotide changes in the viral DNA of a vulvar carcinoma in relation to HPV-6b (Rando et al., 1986). (c) HPV-16 DNA: organization of viral DNA sequences in four carcinomas of the cervix. The solid lines indicate the viral sequences present in the tumours; the wavy lines indicate the integration sites; the interrupted area in CC511 indicates the deletion affecting the viral DNA in this tumour. Numbered boxes represent the ORFs of HPV-16 (Lehn et al., 1985).

binds to both viral and cellular DNA, or whether the E2 protein trans-activates cellular enhancer sequences as well as the viral ones.

An indication of the possible functions of the viral proteins may be provided by a comparison of the viral nucleotide or amino acid sequences with those of other viral or cellular genes. Thus, the BPV-1 E1 protein is homologous to the polyoma and SV40 large T antigen in the carboxyl terminal half and in the ATPase and nucleotide binding sites (Clertant & Seif, 1984; Seif, 1984); in the cottontail rabbit papillomavirus (CRPV) the carboxyl terminal portion of the E2 protein shows some homology to *c-mos* in the region of a putative phosphotyrosine acceptor site (Danos & Yaniv, 1984), and the E6 protein to the subunit of mitochondrial ATP synthase, in a region involved in ATP binding (Giri, Danos & Yaniv, 1985); and in HPV-8 the E4 protein has some homology with the EBNA 2 antigen of Epstein Barr Virus (EBV) (Fuchs *et al.*, 1986). Although these homologies do not directly shed light on the mechanisms underlying cell transformation, they point to experimentally testable predictions as to the function of the viral gene products.

BOVINE PAPILLOMAVIRUS AND CANCER

Our interest has been focussed for several years on the papilloma-carcinoma system of the upper alimentary canal of cattle, not only because of its intrinsic scientific value and economic importance, but also because it provides a model system more amenable to experimental investigation than the human one.

In the Western Highlands of Scotland, cattle affected by papillomatosis of the upper alimentary canal are at higher risk for the development of squamous cell carcinomas than their lowland counterparts. The alimentary cancers are often accompanied by adenomas and adenocarcinomas of the lower bowels and by cancers of the urinary bladder. Ingestion of bracken fern has been identified as a critical factor in the malignant conversion of the papillomas (Jarrett *et al.*, 1978a) and all the stages between papilloma and carcinoma have been recognized (Jarrett, 1981). Bracken-grazing animals are also heavily immunosuppressed (Evans, Patel & Koohy, 1982).

A papillomavirus, BPV-4, identified as a new entity, has been isolated from the papillomas of the upper alimentary canal, and has been shown to be their causative agent (Campo *et al.*, 1980). It shows a remarkable target specificity for the mucous epithelium of the alimentary canal, and it induces tumours only when injected into this site and not when inoculated into cutaneous epithelia (Jarrett, 1985). It belongs to subgroup B of BPVs inducing proliferation of epithelial cells only (Table 3; Jarrett *et al.*, 1984*b*); it has extensively diverged from the more common fibropapillomaviruses of subgroup A (Campo *et al.*, 1980, 1981; Campo & Coggins, 1982) and its evolutionary remoteness is confirmed by the failure of immune sera from calves bearing alimentary canal papillomas, or from rabbits inoculated with BPV-4, to cross-react with the other BPVs (Jarrett *et al.*, 1980, 1984*b*).

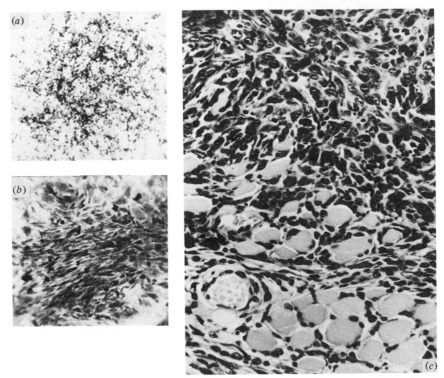

Fig. 4. (*a*) Focus of NIH3T3 mouse fibroblasts transformed by recombinant BPV-4 plasmid (Campo & Spandidos, 1983). (*b*) Focus of C127 mouse fibroblasts transformed by BPV-4 DNA (K. T. Smith & M. S. Campo, unpublished). Cells were fixed in 100% methanol and stained with 10% Giemsa. (*c*) Histological section of an invasive sarcoma induced by inoculation of BPV-4-transformed NIH3T3 cells into nude mice. Cryostat sections were fixed in 3:1 methanol–acetic acid and stained with 6% Giemsa. Figs 4*a* and *b* are reproduced at the same magnification, half as great as that used with Fig. 4*c*.

GENOMIC ORGANIZATION OF BPV-4

According to all the accepted criteria, BPV-4 is a truly oncogenic virus. Its inoculation in hamsters results in the growth of malignant sarcomas (Moar, Jarrett & O'Neil, 1986), and the molecularly cloned viral DNA transforms both NIH3T3 and C127 mouse fibroblasts *in vitro* to the fully malignant phenotype (Campo & Spandidos, 1983; K. T. Smith & M. S. Campo, unpublished). The transformed cells have lost contact inhibition (Fig. 4*a* and *b*), are no longer anchorage dependent, and induce very aggressive sarcomas in nude mice (Fig. 4*c*). In the hamster tumour and in the transformed cells *in vitro* and in the derived nude mouse tumours, the viral DNA persists as multiple episomes with a methylation and restriction fragment pattern indistinguishable from that of viron DNA (Campo & Spandidos, 1983; Moar *et al.*, 1986; K. T. Smith & M. S. Campo, unpublished).

Its genome has been sequenced recently and the identity of the genes has been established by comparison of the nucleotide and

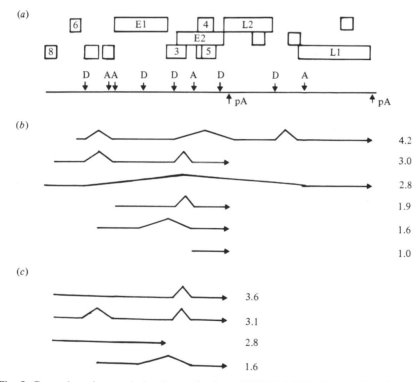

Fig. 5. Genomic and transcriptional organization of BPV-4. (*a*) The linearized viral genome is represented by the solid line; the boxes represent the open reading frames (ORFs); E1, E2 and the numbered boxes are the major early ORFs and L1 and L2 the major late ORFs; the unnumbered ORFs await further analysis. pA, polyadenylation site; D, splice donor site; A, splice acceptor site. (*b*) Viral transcripts in productive papillomas. The horizontal lines represent mRNA sequences, the slanted lines intervening sequences and the arrowheads 3′ termini. The numbers represent transcript sizes in kilobases. (*c*) Viral transcripts in transformed cells *in vitro*. The 3.6 kb and 2.8 kb RNAs are from C127 cells; the 3.1 kb RNA is from NIH3T3 cells, and the 1.6 kb RNA is from both cell types.

amino acid sequences with those of other PVs (K. R. Patel & M. S. Campo, unpublished; Fig. 5a). The DNA is 7.23 kilobases (kb) long; all the major ORFs are found in only one strand and there is extensive overlap between them: for instance, in the E2 region all three reading frames are open. Poly-A addition sites have been found at the 3' end of the E2, L2 and L1 ORFs; the E2 and the L1 sites are used as transcription termination signals, as shown by analysis of viral mRNAs (see below). Splicing donor and acceptor sites have been identified, both within and outside the ORFs, which are shown to be functional by RNA analysis (see below). Putative transcriptional promoter and enhancer sequences have been found 5'to the E region and in between the L2 and L1 ORFs. Therefore the overall genomic organization of BPV-4 is the same as that of other papillomaviruses.

When the amino acid sequence of the major ORFs of BPV-4 are compared with that of the other PVs, maximum homology is found with E1 of BPV-1 (51%), with E2 of CRPV (48%), with HPV-1 L1 (60%), and with HPV-6 L2 (54%) (Table 5). It is worth noting that BPV-4 is more closely related to the purely epitheliotropic papillomaviruses than to BPV-1, confirming earlier hybridization and immunological results. The L2 protein has been implicated in tissue specificity (Danos *et al.*, 1984) and it may be relevant that BPV-4 and HPV-6, both of which infect mucous epithelium of the alimentary and the genital tract respectively, show extensive amino acid homology in this protein.

Table 5. *Amino acid homology between the proteins of BPV-4 and other PVs*

	E1[a]	E2[b]	L1[b]	L2[b]
BPV-1	51	26	56	42
HPV-1	46	36	60	48
HPV-6	42	28	56	54
CRPV	46	48	36	nd[c]

From Campo et al., 1985b.
[a] Percentage of homology in the last 90 C-terminus amino acids of E1; [b] Percentage of homology in the first 50 N-terminus amino acids of L2 and L1 and the last 50 C-terminus amino acids of E2; [c] Not determined.

TRANSCRIPTIONAL ORGANIZATION OF BPV-4

The sequencing data agree with and support the transcriptional data obtained by hybridisation and S1 analysis. Seven virus-specific RNA transcripts have been identified in the papillomas; they measure 4.2, 3.6, 3.0, 2.8, 1.9, 1.6 and 1.0 kb (Fig. 6a; Smith, Patel & Campo, 1986). Hybridization to subgenomic probes revealed that many of these transcripts must be spliced since DNA fragments from widely separate regions of the viral genome hybridized to the same RNAs (Fig. 6a). The splicing pattern has been analysed by S1 digestion of 3'- and 5'-end labelled DNA fragments. With the exception of the 1.0 kb RNA, all the RNA species are internally spliced (Fig. 5b); the 3.0, 1.9, 1.6 and 1.0 kb RNAs utilize the same polyadenylation site at 0.4 map unit (m.u.); and the 4.2 kb RNA and the 2.8 kb RNA terminate at the polyadenylation site at 0.99 m.u.

A comparison of the genomic and transcriptional organization (Fig. 5) allows the assignment of the RNAs to individual ORFs; thus, the 4.2 kb RNA spans both the L2 and L1 ORFs and may be a precursor molecule encoding an L2/L1 fusion polypeptide, as has been suggested for the 4.8 kb RNA of CRPV papillomas (Nasseri & Wettstein, 1984); the 2.8 kb RNA encodes the major structural protein L1; the 1.0 kb RNA probably encodes the E4 protein, and the 3.0, 1.9 and 1.6 kb RNAs span most of the E region and probably encode the E6/E7, the E6/E4 and the E1/E4 fusion polypeptides, as is the case in BPV-1-transformed cells (Yang et al., 1985a).

A similar analysis has been conducted in the transformed mouse fibroblasts in vitro (K. T. Smith & M. S. Campo, unpublished). When transformed cell RNA is hybridized to total viral DNA, only the 1.6 kb RNA is occasionally detected in exceedingly low amounts (barely visible in Fig. 6b, b and d), but if the cells are treated with the tumour promoter TPA or with cycloheximide, additional viral RNA species are detected and the amount of the 1.6 RNA is increased (Fig. 6b, c and e). Interestingly, the transcripts detected in transformed C127 cells are not the same as those detected in NIH3T3 cells; four RNAs of 4.7, 3.6, 2.8 and 1.6 kb are found in the former, whereas only three RNA species, of 4.7, 3.1 and 1.6 kb, are found in the latter (Fig. 6b). Hybridization to subgenomic viral DNA fragments (Fig. 6c and b) reveals that the 1.6 kb RNA found in both cell lines is the same as the 1.6 kb transcript found in papillomas, and that the 3.6 kb RNA of C127 cells and the 3.1 kb RNA of NIH3T3 cells are analogous to the 3.0 kb transcript of papillomas (Fig. 5b

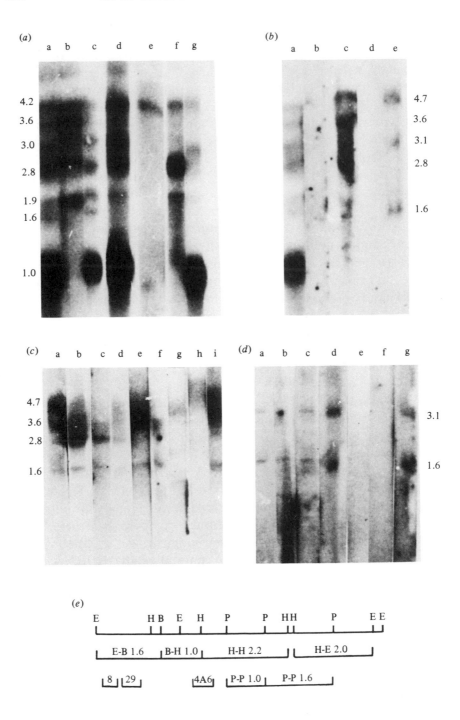

and c). The size difference between the 3.6 and the 3.1 kb transcripts appears to be due to selective use of the splice acceptor/donor sites in the two cells lines. The 2.8 kb RNA found only in C127 cells is different from the structural RNA of the same length found in papillomas. It hybridizes only to the E region of the viral genome and spans the E6/E7 and E1 ORFs (Fig. 6c) and is therefore one of the transforming RNAs. This RNA was not found in papillomas, possibly because it is a rare species in the productive system and its presence there is obscured by the large quantities of the structural RNA. The 4.7 kb RNA is present in very low amounts even after TPA stimulation and could not be mapped. The 1.0 kb RNA which probably encodes the E4 protein and is by far the most abundant species in papillomas, is absent in the transformed cells. A similar RNA of 0.9 kb is found in productive CRPV papilloma only (Nasseri & Wettstein, 1984), raising the possibility that the RNA of both BPV-4 and CRPV is keratinocyte-specific.

TPA has been shown to increase both viral and cellular transcription (Amtman & Sauer, 1982; Greenberg & Ziff, 1984; Campo & Roe, 1986) but the mechanisms for this are still unclear. Our results are interesting in view of the fact that tumour promoters present in bracken have been strongly implicated in carcinogenesis in cattle. The situation observed with the transformed cells *in vitro* may therefore provide a clue as to the interaction between the virus and tumour promoters in naturally occurring cancer.

Fig. 6. BPV-4 transcripts in papillomas and transformed cells *in vitro*. (*a*) Total papilloma RNA (250 μg) was electrophoresed across a 12 cm well. After transfer the membrane was cut into 0.5 cm strips and hybridized to the probes shown diagramatically in (*e*) below the restriction map of BPV-4 DNA. a, pBV4; b, E-B 1.6; c, B-H 1.0; d, H-H 2.2; e, P-P 1.0; f, H-E 2.0; G, 4A6. Sizes are in kilobases. pBV4 is a recombinant plasmid containing the whole viral genome (Campo & Coggins, 1982). (*b*) Transformed cell RNA (40 μg) hybridized to pBV4. a, control oesophageal papilloma RNA (10 μg); b, RNA from partially transformed C127 cells; c, RNA from TPA-treated transformed C127 cells; d, RNA from transformed NIH3T3 cells treated with cycloheximide. (*c*) Total RNA (20 μg) from TPA-treated transformed C127 cells hybridized to the probes shown in E. a, pBV4; b, E-B 1.6; c, 29; d, 8; e, B-H 1.0; f, 4A6; g, P-P 1.0; h, P-P 1.6; i, H-E 2.0. (*d*) Poly-A RNA (5 μg) from transformed NIH3T3 cells treated with cycloheximide hybridized to the probes shown in e. a, E-B 1.6; b, B-H 1.0; c, 4A6; d, H-H 2.2; e, P-P 1.0; f, P-P 1.6; g, H-E 2.0. Hybridization was performed in 50% formamide, 5 × SSC at 42 °C. (*e*) Restriction map and subgenomic probes of BPV-4 DNA. The positions of *Bam*HI, *Eco*RI, *Hind*III and *Pst*I restriction enzyme sites are shown as B, E, H and P respectively, and sizes are given in kilobases. Fragment 8 and 29 were derived from subclone E-B 1.6 and contain open reading frames. Clone 4A6 is a viral cDNA, derived from papilloma RNA, which contains the early polyadenylation site at 0.56 m.u.

THE SEARCH FOR BPV-4 DNA IN CANCERS

Whereas the BPV-4-induced papillomas are highly productive tumours, containing large amounts of mature virus progeny (Jarrett *et al.*, 1978*b*), no virus or viral antigens have been found in the carcinomas (Jarrett *et al.*, 1978*a*). To elucidate the role of the virus in malignant transformation *in vivo*, we have analysed large numbers of squamous cell carcinomas of the upper alimentary canal, associated adenomas and adenocarcinomas of the lower bowels and metastatic deposits for the presence of the viral genome. Out of 70 cases, only 1 transforming papilloma of the oesophagaus and 1 squamous carcinoma of the tongue contained detectable amounts of viral DNA (Campo *et al.*, 1985*a*).

To test whether only a few cells harbouring the viral genome would drive the transformation process, we explanted cell lines from alimentary canal papillomas, transforming papillomas and carcinomas. The cells are epithelial, as shown by the retention of their markers (H. Laird, unpublished), and are fully transformed: they are no longer contact-inhibited, plate in agar and induce tumours in nude mice. None of the cell lines or the nude mouse tumours contain any viral DNA (Fig. 7; Campo *et al.*, 1985*b*; Campo & Jarrett, 1986; B. Watt, W. S. H. Jarret & M. S. Campo, unpublished). These

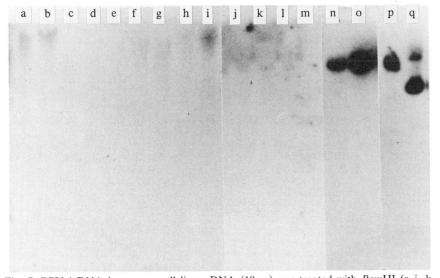

Fig. 7. BPV-4 DNA in tumour cell lines. DNA (10 μg) was treated with *Bam*HI (a–i, k, m–p) or no enzyme (j, l, q). a–c, cell lines explanted from three oesophageal papillomas; d–f, tumours induced in nude mice by the cell line in c; g–i, cell lines explanted from tumours d–f; j,k, cell lines explanted from an oesophageal carcinoma; l,m, cell line explanted from a tumour induced in a nude mouse by cell line j; n,o, BPV-4-transformed NIH3T3 mouse fibroblasts; p,q, oesophageal papilloma. Hybridization was to molecularly cloned BPV-4 DNA, in 50% formamide, 6 × SSC at 42 °C.

results allow us to conclude that the viral genetic information in not required either for the maintenance of, or progression to, the malignant state, and strongly suggest that the execution of virus-mediated cell transformation is an early event, as indicated by the apparently transformed phenotype of the cells derived from 'benign' papillomas.

However, a similar search of urinary bladder cancers revealed the presence of multiple episomal copies of BPV-2 DNA in several cases (Fig. 8a; W. F. H. Jarrett, M. S. Campo, B. W. O'Neil & M. L. Blaxter, unpublished). BPV has already been implicated as one of the causative agents of bladder cancer in cattle (Olson *et al.*, 1959, 1965) and our results confirm at least its association with the disease.

VIRAL, CHEMICAL AND IMMUNOLOGICAL FACTORS IN EXPERIMENTALLY INDUCED TUMOURS

The absence of the viral genome in the alimentary canal cancers contrasts with the permanence and expression of BPV-4 DNA in

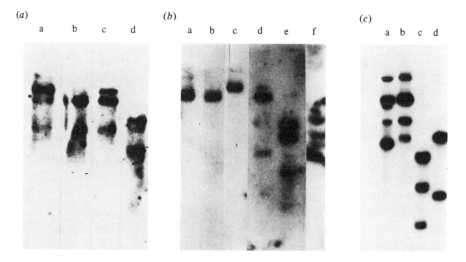

Fig. 8. BPV DNA in urinary bladder cancers and in experimental papillomas. (*a*) DNA (10 μg) from a naturally occurring hemangiosarcoma of the urinary bladder was treated with (a) no enzyme; (b) *Hind*III; (c) *Eco*RI; (d) *Bam*HI. Digestion with *Hind*III was not complete. (*b*) DNA (10 μg) from an experimentally induced carcinoma of the urinary bladder was treated with (a) no enzyme; (b) *Eco*RI; (c) *Hind*III; (d) *Bam*HI; (e) *Hinc*II; (f) *Bgl*II. (*c*) DNA (2 μg) from an experimentally induced papilloma of the palate treated with (a) no enzyme; (b) *Sal*I; (c) *Bam*HI; (d) *Eco*RI; (e) *Hind*III. Hybridization was to molecularly cloned BPV-2 DNA in (*a*) *and* (*b*), and to molecularly cloned BPV-4 DNA in (*c*), in 50% formamide, 6 × SSC at 42 °C.

experimentally transformed cells, and with the persistence of BPV-2 DNA in the urinary bladder cancers. It is also at variance with the presence of papillomavirus genomes in other systems, for instance, genital cancer in humans.

As several other factors (chemical and immunological) appear to be involved in the aetiology of alimentary canal cancer in cattle in addition to the virus, we designed an experiment to dissect the interaction between them, by reproducing the natural conditions that lead to the disease (Campo & Jarret, 1986; W. F. H. Jarret, B. W. O'Neil, M. S. Campo, R. M. C. Blaxter, unpublished).

Calves, born of papillomatosis-free mothers, were obtained at birth and kept in isolation throughout the course of the experiment. They were divided into eight groups, and treated with BPV-4, bracken, azathioprine or quercetin either alone or in combination, or kept untreated as control (Table 6). Azathioprine is an immuno-suppressant, and quercetin is one of the flavonoids present in bracken that is claimed to be carcinogenic (Pamucku, Yalciner & Bryan, 1980).

All the animals treated with a bracken diet or azathioprine were immunosuppressed; all the animals injected in the palate with BPV-4 developed papillomas at the site of injection, but in the immunocom-promised animals (groups 2 and 5) the papillomas spread throughout the palate and the mouth and persisted well beyond their average life span in immunocompetent animals. This is precisely the situation in the field: bracken-eating animals are affected by persistent wide-spread oral papillomatosis (Jarrett et al., 1978a).

Many of the animals that had been immunosuppressed developed urinary bladder cancer (groups 1, 2 and 5), and one animal in group 2 developed carcinomas of the upper alimentary canal and adenomas of the lower bowels. The surviving animals in group 2 are still under observation. As already found in field cases, BPV-4 DNA was pre-sent in the papillomas (Fig. 8c) and BPV-2 DNA in most of the bladder cancers (Fig. 8b), but no viral DNA was found in the cancers of the digestive tract. We have therefore successfully reproduced in experimental conditions both the pathological and molecular aspects of the diseases as they occur in the field.

A comprehensive analysis of all our results suggests the following interpretation of the sequence of events during carcinogenesis in the alimentary canal: BPV-4 executes one of the early events in cell transformation and its genetic information is not required for malig-nant progression; immunosuppression causes the spread and the

Table 6. *Synergism between viral and chemical cofactors*

Group[a]	Treatment[b]	Immunosuppression[c]	Alimentary canal papillomas[d]	Bladder cancers[e]	Cutaneous warts[e]	Lymphocytes[e]
1 (6)	B	+	—	2 (BPV-2)	1 (BPV-2)	4 (BPV-2)
2 (6)	B + V	+	6 (++)	3 (BPV-2)	—	—
3 (6)	V	—	6 (+)	—	—	—
4 (4)	V	—	4 (+)	—	2 (BPV-1)	2 (BPV-1)
5 (6)	A + V	++	6 (++)	3 (BPV-2)	1 (BPV-1)	—
6 (4)	A	++	—	—	2 (BPV-2)	2 (BPV-1 or 2)
7 (4)	Q + V	—	4 (+)	—	2 (BPV-1)	1 (BPB-1 or 2)
8 (4)	—	—	—	—	—	—

[a] The number of animals in each group is in brackets; [b] B, bracken fern; V, BPV-4, inoculated in the palate; A, azathioprine, administered orally; Q, quercetin, administered orally; [c] + indicates the degree of immunosuppression; [d] Number of positive animals in each group; (+) indicates the degree of papillomatosis; [e] Number of positive animals in each group. The virus type is in brackets.

persistence of papillomas, thus increasing the target size for subsequent neoplastic events; and chemicals present in the bracken fern act as promoters upon virally initiated cells, thus achieving full cell transformation.

LATENT AND SUBCLINICAL BPV INFECTIONS

The presence of multiple episomal copies of BPV-2 DNA in the experimentally induced bladder cancers was unexpected as the animals had been papillomatosis-free and had been kept in isolation, and those that had been exposed to virus had received only BPV-4. These observations strongly suggest that BPV-2, or its DNA, is associated with urinary bladder cancers, and it can persist in a latent form.

The occurrence of latent BPV infection was confirmed by the onset, several months after the beginning of the experiment, of cutaneous fibropapillomas on the face and neck at sites where the skin had been damaged. Animals in all groups were affected, except those in groups 2, 3 and 8. The warts contained large amounts of mature virus, identified as BPV-1 in five cases and BPV-2 in three cases (Table 6; Fig. 9a and b). The warts developed in both immunosuppressed and immunocompetent animals, suggesting that skin damage is enough to activate latent virus. It is unlikely that the warts were induced by a contaminant present in the BPV-4 preparation used to infect the palate, rather than by latent virus, because three of the affected animals (groups 1 and 6) had not been exposed to virus at all.

We conclude that the mothers of our animals, although papillomatosis-free, were probably subclinically infected and transmitted latent virus to their offspring, probably at birth. The infection became obvious as a consequence of skin damage and immunosuppression.

SITE OF LATENCY

The presence of latent virus poses the question of the site of latency. We investigated this problem by looking for the presence of viral markers in circulating lymphocytes. We found BPV-2 DNA in the lymphocytes of four animals from group 1 (Fig. 9c), BPV-1 DNA in two animals from group 4 and we were unable to identify the DNA because of the paucity of material, in two animals from group

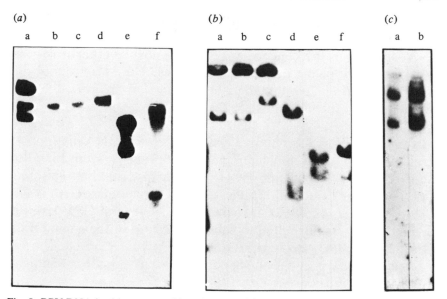

Fig. 9. BPV DNA in skin warts and lymphocytes. (a) and (b) Wart DNA (2 μg) was treated with (a) no enzyme; (b) EcoRI; (c) HindIII; (d) BamHi; (e) HincII; (f) BglII. (c) Non-digested lymphocyte DNA (10 μg) from two animals from group 1. Hybridization was to molecularly cloned BPV-1 DNA in (a), and to molecularly cloned BPV-2 DNA in (b) and (c), in 50% formamide, 6 × SSC at 42 °C.

6 and one from group 7 (Table 6). When we extend this search to field animals, we found both BPV-1 and BPV-2 DNA in some of the lymphocytes preparations, although the amount of DNA was much lower than in the lymphocytes of experimental animals.

The significance of the presence of BPV genomes in lymphocytes is not clear. Although we have no definitive proof, the possibility that the lymphocytes are indeed the site of latent infection is very attractive: when the skin is damaged, an inflammation reaction occurs and the lymphocytes move to the wound; the viral DNA can infect the keratinocytes, and, once in a permissive cellular environment, could induce cell proliferation and papillomatosis.

DISCUSSION

Although we have identified six different types of BPV, we have found only two associated with cancer in the natural host: BPV-4 with cancer of the alimentary canal and BPV-2 with cancers of the urinary bladder. However, the role of the two BPVs in malignant transformation is still obscure; BPV-4 is present in the premalignant lesions but not in the frankly malignant ones, and, although BPV-2 DNA is found in the cancers, we do not know whether it (or the virus) is present in the premalignant lesions. Therefore, either the two viruses bring about malignant progression by two different mechanisms, or the mechanism is the same but the phenotypic manifestation different.

In other papilloma–carcinoma systems, the viral DNA may persist through the phase of malignant progression and is often found in the carcinomas (Gissmann, 1984). In humans, multiple episomal copies of HPV-5 are found in the EV skin carcinomas (Orth et al., 1980; Ostrow et al., 1982), and the DNA of several HPV types is found in the premalignant and malignant lesions of the genital tract (Durst et al., 1983; Boshart et al., 1984; Millan et al., 1986); the genome of CRPV is also found in papilloma-derived skin carcinomas of rabbits (Favre, Jibard & Orth, 1982).

Also, the genome of several BPV types persists in equine sarcoids (Lancaster, 1981), in experimental tumours in hamster (Moar et al., 1981a, 1986) and in transformed cells in vitro (Moar et al., 1981b; Law et al., 1981; Campo & Spandidos, 1983). Nevertheless, the genome of BPV-4 is absent in the alimentary canal cancers. The possibility that BPV-4 is not one of the factors involved in carcinogenesis is unlikely, as there is complete concurrence of time, geographical distribution and anatomical site between the benign and the malignant tumours (Jarrett et al., 1978a), all stages of transformation from papilloma to carcinoma have been observed and recognized (Jarrett, 1981), and viral DNA has been found in one case of transforming papilloma (Campo et al., 1985a).

The absence of viral DNA in the cancers argues against the requirement for its retention and expression for the maintenance of the transformed state. This conclusion is strengthened by several observations: one, cell lines explanted from alimentary canal carcinomas are fully transformed but do not contain any viral DNA (Campo et al., 1985b; Campo & Jarrett, 1986; B. Watt, W. S. H. Jarrett & M. S. Campo, unpublished); two, several lines of rat fibroblasts transformed in vitro by BPV-1 have lost the viral genome (Cuzin et al., 1985); three, no viral transcripts have been found in a number of cancers of the genital tract, despite the presence of HPV DNA (Lehn et al., 1985).

The role of BPV-4 in malignant transformation seems to be an indirect one. The virus may activate cellular oncogenes, and/or it may induce increased production of growth factors with ensuing cell proliferation; bracken would supply a promoter, a mutagen or an immunosuppressant as one of several factors involved in later events.

Experiments are in progress to identify somatically heritable changes in the cellular genetic make-up which may result from viral infection, and which may constitute early events in the neoplastic process. Whatever the mechanism underlying cell transformation, the genetic constitution of the virus seems to be an important factor, since the BPV-2-induced fibropapillomas of the alimentary canal do not undergo malignant transformation and no connection with cancer has been observed (Jarrett *et al.*, 1984*a*). There are similarities between the oesophageal fibropapillomas and the urinary bladder cancers: both contain the genome of BPV-2 but infection is abortive, and both fibroblasts and epithelial cells are transformed. However, the fibropapillomas are not malignant.

Therefore, either BPV-2 is not involved in carcinogenesis, and the presence of its DNA in the tumours is opportunistic and irrelevant to neoplastic progression, or the full expression of its oncogenic potential is linked to its cellular environment. The results of Olson and collaborators (1959, 1965) suggest an active role for BPV-2 in bladder neoplasia. In this case, the expression of the BPV-2 functions is either continuously required for the maintenance of the transformed state, or is needed only at some early stage of transformation, but the viral DNA would still be allowed to replicate in the permissive cells of the epithelium of the urinary bladder.

The presence of BPV-2 DNA in the bladder cancers and of BPV-1 and BPV-2 in skin warts in animals that had been exposed only to BPV-4 or to no virus indicates that these viruses can persist in a latent form. Possible latency of papillomaviruses is an important problem. The DNA of HPV-11 has been found in histologically normal epithelium of the larynx, as late as 2 years after remission of laryngeal warts (B. Steinberg, personal communication); HPV-5 has been found in the sperm of both an adult male affected by EV and of his unaffected young son (R. Ostrow, personal communication); HPV-6 DNA was present in the foreskin of a newborn boy (A. Roman, personal communication); and both HPV-16 and HPV-18 DNA was found in histologically normal cervices (Millan *et al.*, 1986). Presumably each of these viruses can be reactivated by changes in the intracellular environment, in the immune or hormonal systems, or by external factors such as trauma, ultraviolet (UV) rays or chemicals. Immunosuppression and skin damage appeared to activate BPV-1 and BPV-2 in our animals.

The BPV genome was also present in lymphocytes, and it is tempting to speculate that they are (one of) the site(s) of latent infection. DNA 'transfection' would take place *in vivo* between lymphocytes and epithelial cells and, once the viral DNA was in a permissive cell, the expression of its function would lead either to overt infection and papillomatosis or to cellular transformation.

These considerations have obvious implications for our understanding of the role of papillomavirus in human cancer.

It has already been mentioned that HPV is associated with lesions of the uterine cervix. Whether it operates as an initiator, as would appear to be the case in the bovine system, or as a promoter, as circumstantial evidence from other systems would suggest, viral infection is an important factor in the aetiology of cervical carcinoma. Successful management and treatment of viral cervical lesions, possible precursors to cancer, require knowledge of the mode of viral spread, of latency and reactivation, of the involvement of the immune system, and of the molecular mechanisms that bring about cell transformation.

The bovine system, being both 'natural' and amenable to experimentation, should provide at least some of the answers needed to control an increasingly common and increasingly aggressive human cancer.

ACKNOWLEDGMENTS

Thanks are due to all our colleagues who took part in this work and provided their data for this review. The work described here has been supported by the Cancer Research Campaign, the Agricultural and Food Research Council and the National Institutes of Health. MSC is a Fellow of the Cancer Research Campaign.

REFERENCES

AMTMANN, E. & SAUER, G. (1982). Activation of non-expressed bovine papilloma virus genomes by tumour promoters. *Nature*, **296**, 675–7.

BOSHART, M., GISSMANN, L., IKENBERG, H., KLEINHEINZ, A., SCHEURLEN, W. & ZUR HAUSEN, H. (1984). A new type of papillovavirus DNA, its presence in genital cancer biopsies and in cell lines derived from cervical cancer. *EMBO Journal*, **3**, 1151–7.

CAMPO, M. S. (1984). Detection and identification of papillomaviruses in benign and malignant tumours of cattle. In *Recent Advances in Viral Diagnosis*, ed.

M. S. McNulty & J. B. McFerran, pp. 72–94. Dordrecht Martinus Nijhoff Publishers.

CAMPO, M. S. & COGGINS, L. W. (1982). Molecular cloning of bovine papillomavirus genomes and comparison of their sequence homologies by heteroduplex mapping. *Journal of General Virology*, **63**, 255–64.

CAMPO, M. S. & JARRETT, W. F. H. (1986). Papillomavirus infection in cattle: viral and chemical cofactors in naturally occurring and experimentally induced tumours. In *Papillomaviruses*, CIBA Foundation Symposium, **120**, ed. D. Evered & S. Clark, pp. 117–30. Chichester: John Wiley & Sons.

CAMPO, M. S., MOAR, M. H., JARRETT, W. F. H. & LAIRD, H. M. (1980). A new papillomavirus associated with alimentary tract cancer in cattle. *Nature*, **286**, 180–2.

CAMPO, M. S., MOAR, M. H., LAIRD, H. M. & JARRETT, W. F. H. (1981). Molecular heterogeneity and lesion site specificity of cutaneous bovine papillomaviruses. *Virology*, **113**, 323–35.

CAMPO, M. S., MOAR, M. H., SATIRANA, M. L., KENNEDY, I. M. & JARRETT, W. F. H. (1985a). The presence of bovine papillomavirusm type 4 DNA is not required for the progression to, or the maintenance of, the malignant state in cancers of the alimentary canal in cattle. *EMBO Journal*, **4**, 1819–25.

CAMPO, M. S. & ROE, F. A. (1986). The tumour promoter TPA enhances the transformation to the TK$^+$ phenotype of LA cells transfected with chimeric plasmids without interaction with the bovine papillomavirus regulatory sequences. *Annales de l'Institut Pasteur*, **137**, 27–35.

CAMPO, M. S., SMITH, K. T., JARRETT, W. F. H. & MOAR, M. H. (1985b). Presence and expression of bovine papillomavirus type 4 in tumours of the alimentary canal of cattle and its possible role in transformation and malignant progression. In *Papillomaviruses: Molecular and Clinical Aspects*, ed. P. M. Howley & T. R. Broker, pp. 305–326. New York: Alan R. Liss.

CAMPO, M. S. & SPANDIDOS, D. (1983). Molecularly cloned bovine papillomavirus DNA transforms mouse fibroblasts *in vitro*. *Journal of General Virology*, **64**, 549–57.

CAMPO, M. S., SPANDIDOS, D., LANG, J. & WILKIE, N. M. (1983). Transcriptional control signals in the genome of bovine papillomavirus type 1. *Nature*, **303**, 77–80.

CLERTANT, P. & SEIF, I. (1984). A common function for polyoma virus large-T and papillomavirus E1 proteins? *Nature*, **311**, 276–9.

CRUM, C. P., IKENBERG, H., RICHART, R. M. & GISSMANN, L. (1984). Human papillomavirus type 16 and early cervical neoplasia. *New England Journal of Medicine*, **310**, 880–3.

CUZIN, F., MENEGUZZI, G., BINETRUY, B., CERNI, C., CONNAN, G., GRISONI, M. & DE LAPEYRIÈRE, O. (1985). Stepwise tumoural progression in rodent fibroblasts transformed with bovine papillomavirus type 1 DNA. In *Papillomaviruses: Molecular and Clinical Aspects*, ed. P. M. Howley and T. R. Broker, pp. 473–87. New York: Alan R. Liss.

DANOS, O., GIRI, I., THIERY, F. & YANIV, M. (1984). Papillomavirus genomes: sequences and consequences. *Journal of Investigative Dermatology*, **83**, 7s–11s.

DANOS, O. & YANIV, M. (1984). An homologous domain between the C-mos gene product and a papilloma virus polypeptide with a putative role in cellular transformation. In *Cancer Cells*, 2, *Oncogenes and Viral Genes*. ed. G. F. Vande Woude, A. J. Levine, W. C. Topp & J. D. Watson, pp. 291–4. New York: Cold Spring Harbor Laboratory.

DOORBAR, J., CAMPBELL, D., GRAND, R. J. A. & GALLIMORE, P. H. (1986). Identification of the human papillomavirus-1a E4 gene product. *EMBO Journal*, **5**, 355–62.

DURST, M., GISSMANN, L., IKENBERG, H. & ZUR HAUSEN, H. (1983). A papilloma-virus DNA from a cervical carcinoma and its prevalence in cancer biopsy samples from different geographic regions. *Proceedings of the National Academy of Sciences, USA*, **80**, 3812–15.

DURST, M., KLEINHEINZ, A., HOTZ, M. & GISSMANN, L. (1985). The physical state of human papillomavirus type 16 DNA in benign and malignant genital tumours. *Journal of General Virology*, **66**, 1515–22.

ENGEL, L. W., HEILMAN, C. A. & HOWLEY, P. M. (1983). Transcriptional organiza-tion of bovine papillomavirus type 1. *Journal of Virology*, **47**, 516–28.

EVANS, W. C., PATEL, M. C. & KOOHY, Y. (1982). Acute bracken poisoning in homogastric and ruminant animals. *Proceedings of the Royal Society of Edin-burgh, B (Biol. Sci,)*, **81**, 29–64.

FAVRE, M., JIBARD, N. & ORTH, G. (1982). Restriction mapping and physical charac-terization of the domestic rabbit papillomavirus genome in transplantable V × 2 and V × 7 domestic rabbit carcinomas. *Virology*, **119**, 298–309.

FUCHS, P. G., IFTNER, T., WENINGER, J. & PFISTER, H. (1986). Epidermodysplasia verruciformis-associated human papillomavirus type 8: genomic sequence and comparative analysis. *Journal of Virology*, **58**, 626–34.

GIRI, I., DANOS, O. & YANIV, M. (1985). Genomic structure of the cottontail rabbit (Shope) papillomavirus. *Proceedings of the National Academy of Sciences, USA*, **82**, 1580–4.

GISSMANN, L. (1984). Papillomaviruses and their association with cancer in animals and man. *Cancer Surveys*, **3**, 161–81.

GISSMAN, L., WOLNICK, L., IKENBERG, H., KOLDOVSKY, U., SCHNURCH, H. G. & ZUR HAUSEN, H. (1983). Human papillomavirus type 6 and 11 DNA sequences in genital and laryngeal papillomas and in some cervical cancers. *Proceedings of the National Academy of Sciences, USA*, **80**, 560–3.

GREEN, M., BRACKMANN, K. H., SANDERS, P. R., LOEWENSTEIN, P. M., FREEL, J. H., EISINGER, M. & SWITLYK, S. A. (1982). Isolation of a human papillomavirus from a patient with epidermodysplasia verruciformis: Presence of related viral DNA genomes in human urogenital tumours. *Proceedings of the National Acad-emy of Sciences, USA*, **79**, 4437–41.

GREENBERG, M. E. & ZIFF, E. G. (1984). Stimulation of 3T3 cells induces transcrip-tion of the c-fos proto-oncogene. *Nature*, **311**, 433–8.

JARRETT, W. F. H. (1981). Papillomaviruses and cancer. In *Recent Advances in Histopathology*, ed. P. P. Anthony & R. N. M. Macsween, pp. 35–48. Edinburgh: Churchill & Livingstone.

JARRETT, W. F. H. (1985). The natural history of bovine papillomavirus infections. In Advances in Viral Oncology, ed. G. Klein, **5**, 83–102.

JARRETT, W. F. H., CAMPO, M. S., BLAXTER, M. L., O'NEIL, B. W., LAIRD, H. M., MOAR, M. H. & SARTIRANA, M. L. (1984a). Alimentary fibropapilloma in cattle: a spontaneous tumour, non-permissive for papillomavirus replication. *Journal of the National Cancer Institute*, **73**, 499–504.

JARRETT, W. F. H., CAMPO, M. S., O'NEIL, B. W., LAIRD, H. M. & COGGINS, L. W. (1984b). A novel bovine papillomavirus (BPV-6) causing true epithelial papillomas of the mammary gland skin: a member of a proposed new BPV sub-group. *Virology*, **136**, 255–64.

JARRETT, W. F. H., McNEIL, P. E., GRIMSHAW, T. R., SELMAN, I. E. & McINTYRE, W. I. M. (1978a). High incidence area of cattle cancer with a possible interaction between an environmental carcinogen and a papillomavirus. *Nature*, **274**, 215–17.

JARRETT, W. F. H., McNEIL, P. E., LAIRD, H. M., O'NEIL, B. W., MURPHY, J., CAMPO, M. S. & MOAR, M. H. (1980). Papillomaviruses in benign and malignant tumours of cattle. In *Viruses in Naturally Occurring Cancers*, eds M. Essex,

G. Todaro & H. Zur Hausen, pp. 215–22. New York: Cold Spring Harbor Laboratory.

JARRETT, W. F. H., MURPHY, J., O'NEIL, B. W. & LAIRD, H. M. (1978b). Virus-induced papillomas of the alimentary tract of cattle. *International Journal of Cancer*, **22**, 323–8.

KREIDER, J. W. (1980). Neoplastic progression of the Shope rabbit papilloma. In *Viruses in Naturally Occurring Cancers*, Cold Spring Harbor Conference on Cell Proliferation, **7**, 283–300, ed. M. Essex, G. Todaro & H. Zur Hausen. New York: Cold Spring Harbor Laboratory.

KREMSDORF, D., FAVRE, M., JABLONSKA, S., OBALEK, S., RUEDA, L. A., LUTZNER, M. A., BLANCHET-BARDON, C., VAN VOORST VADER, P. C. & ORTH, G. (1984). Molecular cloning and characterization of the genomes of nine newly recognised human papillomavirus types associated with epidermodysplasia verruciformis. *Journal of Virology*, **52**, 1013–18.

LANCASTER, W. D. (1981). Apparent lack of integration of bovine papillomavirus DNA in virus-induced equine and bovine tumors and virus transformed mouse cells. *Virology*, **108**, 251–5.

LAW, M.-F., FOWY, D. R., DVORETZKY, J. & HOWLEY, P. (1981). Mouse cells transformed by bovine papilloma virus contain only extrachromosomal viral DNA sequences. *Proceedings of the National Academy of Sciences, USA*, **78**, 2727–31.

LEHN, H., KNEG, P. & SAUER, G. (1985). Papillomavirus genomes in human cervical tumours; Analysis of their transcriptional activity. *Proceedings of the National Academy of Sciences, USA*, **82**, 5540–4.

LUSKY, M., BERG, L., WEIHER, H. & BOTCHAN, M. (1983). Bovine papilloma virus contains an activator of gene expression at the distal end of the transcriptional unit. *Molecular and Cellular Biology*, **3**, 1108–22.

LUSKY, M. & BOTCHAN, M. R. (1984). Characterization of the bovine papilloma virus plasmid maintenance sequences. *Cell*, **36**, 391–401.

LUSKY, M. & BOTCHAN, M. R. (1985). Genetic analysis of the bovine papillomavirus type 1 transacting replication factors. *Journal of Virology*, **53**, 955–65.

MILLAN, D. W. H., DAVIS, J. A., TORBETT, T. E. & CAMPO, M. S. (1986). DNA sequences of human papillomavirus types 11, 16 and 18 in lesions of the uterine cervix in the West of Scotland. *British Medical Journal*, **293**, 93–6.

MOAR, M. H., CAMPO, M. S., LAIRD, H. M. & JARRETT, W. F. H. (1981a). Persistence of nonintegrated viral DNA in bovine cells transformed *in vitro* by bovine papillomavirus type 2. *Nature*, **293**, 749–51.

MOAR, M. H., CAMPO, M. S., LAIRD, H. M. & JARRETT, W. F. H. (1981b). Unintegrated viral DNA sequences in a hamster tumour induced by bovine papillomavirus. *Journal of Virology*, **39**, 945–9.

MOAR, M. H., JARRETT, W. F. H. & O'NEIL, B. W. (1986). Viral DNA sequences detected in a hamster liposarcoma induced by bovine papillomavirus type 4. *Journal of General Virology*, **67**, 187–90.

MORIN, C., BRAUN. L., CASAS-CORDERO, M., SHAH, K. V., ROY, M., FORTIER, M. & MEISELS, A. (1981). Confirmation of the papillomavirus etiology of condylomatous cervix lesions by the peroxidase-antiperoxidase technique. *Journal of the National Cancer Institute*, **66**, 831–4.

NAKABAYASHI, Y., CHATTOPADHYAY, S. K. & LOWY, D. R. (1983). The transforming function of bovine papillomavirus DNA. *Proceedings of the National Academy of Sciences, USA*, **80**, 5832–6.

NASSERI, M. & WETTSTEIN, F. O. (1984). Differences exist between viral transcripts in cottontail rabbit papillomavirus-induced benign and malignant tumours as well as non-virus-producing and virus producing tumours. *Journal of Virology*, **51**, 706–12.

OLSON, C., PAMUKCU, A. M. & BROBST, D, F. (1965). Papilloma-like virus from bovine urinary bladder tumours. *Cancer Research*, **25**, 840–9.

OLSON, C., PAMUKCU, A. M., BROBST, D. F., KOWALCZYK, T., SATTER, E. J. & PRICE, J. M. (1959). A urinary bladder tumour induced by a bovine cutaneous papilloma agent. *Cancer Research*, **19**, 779–83.

ORIEL, J. D. (1971). Natural history of genital warts. *British Journal of Venereal Disease*, **47**, 1–8.

ORTH, G., FAVRE, M., BREITBURD, F., CROISSANT, O., JABLONSKA, S., OBALEK, S., JARZABEK-CHORZELSKA & RZESA, G. (1980). Epidermodysplasia verruciformis: a model for the role of papillomaviruses in human cancers. In *Viruses in Naturally Occurring Cancers*, ed. M. Essex, G. Todaro & H. Zur Hausen, pp. 259–82. New York: Cold Spring Harbor Laboratory.

ORTH, G., JEANTEUR, P. & CROISSANT, O. (1971). Evidence for localization of vegetative viral DNA replication by autoradiographic detection of RNA-DNA hybrids in sections of tumours induced by Shape papillomavirusa. *Proceedings of the National Academy of Sciences, USA*, **68**, 1876–80.

OSTROW, R. S., BENDER, M., NIIMURA, M., SEKI, T., KAWASHIMA, N., PASS, F. & FARAS, A. J. (1982). Human papillomavirus DNA in cutaneous primary and metastasized squamous cell carcinomas from patients with epidermodysplasia verruciformis, *Proceedings of the National Academy of Sciences, USA*, **79**, 1634–8.

PAMUCKU, A. M., YALCINER, J. F. & BRYAN, G. T. (1980). Quercetin, a rat intestinal and bladder carinogen present in bracken fern (*Pteridium aquilinium*). *Cancer Research*, **40**, 1468–72.

PFISTER, H. (1984). Biology and biochemistry of papillomaviruses. *Reviews of Physiological and Biochemical Pharmacology*, **99**, 111–81.

RANDO, R. F., GROFF, D. E., CHINKJIAN, J. G. & LANCASTER, W. D. (1986). Isolation and characterization of a novel human papillomavirus type 6 DNA from an invasive vulvar carcinoma. *Journal of Virology*, **57**, 353–6.

SARVER, N., RABSON, M. S., YANG, Y., BYRNE, J. C. & HOWLEY, P. M. (1984). Localization and analysis of bovine papillomavirus type 1 transforming functions. *Journal of Virology*, **52**, 377–88.

SCHILLER, J. T., VOSS, W. C., VOUSDEN, K. H. & LOWY, D. R. (1986). E5 open reading frame of bovine papillomavirus type 1 encodes a transforming gene. *Journal of Virology*, **57**, 1–6.

SEERDOF, K., KRAMMER, G., DURST, M., SUHAI, S. & ROWEKAMP, W. G. (1985). Human papillomavirus type 16 DNA sequence. *Virology*, **145**, 181–5.

SEIF, I. (1984). Sequence homology between the large tumor antigen of polyoma viruses and the putative E1 protein of papilloma viruses. *Virology*, **138**, 347–52.

SMITH, K. T. & CAMPO, M. S. (1985). Papillomaviruses and their involvement in oncogenesis. *Biomedicine and Pharmacotherapy*, **39**, 405–14.

SMITH, K. T., PATEL, K. R. & CAMPO M. S. (1986). Transcriptional organisation of bovine papillomavirus type 4. *Journal of General Virology*, **67**, 2381–93.

SPALHOLZ, B. A., YANG, Y. & HOWLEY, P. M. (1985). Transactivation of a bovine papilloma virus transcriptional regulatory element by the E2 gene product. *Cell*, **42**, 183–91.

WALDECK, W., ROSL, F. & ZENTGRAF, H. (1984). Origin of replication in episomal bovine papilloma virus type 1 DNA isolated from transformed cells. *EMBO Journal*, **33**, 2173–8.

YANG, Y. C., OKAYAMA, H. & HOWLEY, P. M. (1985a). Bovine papillomavirus contains multiple transforming genes. *Proceedings of the National Academy of Sciences, USA*, **82**, 1030–4.

YANG, Y., SPALHOLZ, B. A., RABSON, M. S. & HOWLEY, P. M. (1985b). Dissociation of transforming and trans-activation functions for bovine papillomavirus type 1. *Nature*, **318**, 575–7.

YUTSUDO, M., SHIMKAGE, T. & HAKURA, A. (1985). Human papillomavirus type 17 DNA in skin carcinoma tissue of a patient with epidermodysplasia verruciformis. *Virology*, **144**, 295–8.

ZUR HAUSEN, H. (1977). Human papilloma viruses and their possible role in squamous cell carcinomas. *Current Topics in Microbiology and Immunology*, **78**, 1–30.

ADENOVIRUS AS A MODEL OF DISEASE

HAROLD S. GINSBERG*†, ULLA LUNDHOLM—BEAUCHAMP*, AND GREGORY PRINCE†

Department of Microbiology and Medicine, College of Physicians and Surgeons of Columbia University, and †Laboratory of Infectious Diseases, National Institutes of Health, Bethesda, MD 20892, USA

INTRODUCTION

The independent and almost simultaneous discoveries of adenoviruses by Rowe *et al.* (1953) and by Hilleman & Werner (1954) immediately gave hints for the multifaceted roles these viruses would prove to play in humans and in many other species of animals. Rowe, Huebner and their co-workers found adenoviruses emerging from cultured explants of human tonsils and adenoids when they attempted to use these tissues as substrates for the isolation of viruses causing the common cold (Rowe *et al.*, 1953). During an investigation of an influenza-like epidemic in US Army recruits, Hilleman & Werner (1954) isolated agents, later shown to be immunologically related to the viruses Rowe *et al.* isolated (Ginsberg *et al.*, 1954). This epidemic disease was studied intensively by the Commission on Acute Respiratory Diseases during World War II and termed 'Acute Respiratory Disease of Recruits' or 'ARD'. Thus, adenoviruses at their discovery were seen to be both latent viruses and aetiological agents of acute respiratory disease (the first such virus isolated since the initial discovery of influenza virus in 1933, Smith, Andrews & Laidlaw, 1933). This beginning initially stimulated the interest of a few virologists who saw their potential in the studies of latency (a field whose significance was just beginning to be recognized), the important role of adenoviruses in disease, and the need to understand the fundamental characteristics of a new family of viruses which infected humans.

Forty-four distinct types of adenoviruses are now known to infect humans and to produce a wide variety of acute infections including an influenza-like illness, acute pharyngitis, pneumonia, conjunctivitis, keratoconjunctivitis, cystitis and severe enteritis (Strauss, 1984). Moreover, all adenoviruses tested have been shown to transform non-permissive rodent cells and some types are even oncogenic in newborn hamsters and mice (Graham, 1984). The numerous roles adenoviruses can play in pathogenesis of disease are striking!

The early cytochemical, biochemical, and morphological studies of adenoviruses revealed it to be a seemingly simple, DNA-containing virus that could be readily investigated. Such proved to be the case; but as the investigations became more sophisticated neither its genome nor its replication proved to be uncomplicated: the DNA genome was found to encode a number of early and late viral functions, and the expression of the gene products was tightly regulated. Isolation of a number of conditionally lethal, temperature-sensitive mutants (Williams *et al.*, 1971; Ensinger & Ginsberg, 1972; Begin & Weber, 1975; Young, Shenk & Ginsberg, 1984), a variety of naturally occurring deletion and substitution mutants (Young, Shenk & Ginsberg, 1984), and laboratory constructed mutants (Young, Shenk & Ginsberg, 1984) have permitted an increasing understanding of gene regulation of viral replication. It is the objective of this paper to review beginning studies, using a variety of mutants, to gain insight into the molecular mechanisms involved in the pathogenic processes that adenoviruses induce. Emphasis will be placed on the mechanisms of: 1. latent infections of tonsils and adenoids, using a persistent infection of human T cells as the model; and 2. production of pneumonia in cotton rats, an experimental animal that responds with lesions closest to the disease produced in humans (Pacini, Dubrovi & Clyde, 1984).

ADENOVIRUS LATENT INFECTION

After the initial isolation of adenoviruses, several investigators confirmed the original finding that adenoviruses, usually types 1, 2, 5, and 6, emerged from cultured explants of tonsils and adenoids. In the several studies it was shown that these viruses were present in 75 to 85 per cent of children who underwent tonsillectomy. These findings, however, did not present any evidence as to whether the latent state was dependent upon integration of the viral genome, similar to a prophage, whether the viral DNA replicated as an episome, or whether normal viral replication occurred but at a restrained pace so that few, if any, cells were killed. Study of DNA from tonsils yielded Southern blot hybridization patterns that were too complex to permit a determination of whether the DNA was integrated (Green *et al.*, 1979).

Table 1. *Adenovirus DNA sequences in human lymphocytes*

Lymphocyte origin	Adenovirus virus DNA present*
Normal blood	7/11
Leukemia patients†	3/3

* Positive spot-blot hybridization with Ad5 DNA nick-translated probes.
† Samples from peripheral blood and bone marrow.

Since lymphocytes from tonsils, adenoids, and lymph nodes (another site from which viruses had been isolated by culture explant) are continually shed into the lymphatics, and thus into the blood stream, peripheral lymphocytes were studied to determine whether they contained adenovirus DNA. Dot blot hybridization, using a type 5 adenovirus (Ad5) DNA probe ([^{32}P]-labelled), showed that total DNA extracted from peripheral lymphocytes contained adenovirus DNA sequences in 7 out of 11 apparently normal individuals (Table 1). The viral DNA was predominantly, but not exclusively, in DNA from T-lymphocytes. Horvath, Palkonyay & Weber (1986) reported similar findings but their report indicated a higher percentage of cells contained viral DNA and many more copies were present. Adenovirus DNA sequences were also found to be present in DNAs from three out of three lymphocytes or bone marrow cells from patients with lymphocytic leukaemia.

Southern blot hybridizations were then carried out in an attempt to determine whether the viral DNA was integrated into that of the host chromosome. The bands, however, were very light and only the largest band(s) could be detected. When *in situ* hybridizations were carried out with lymphoyctes from two of the normal volunteer donors, DNA could only be detected in 1 in 10^4–10^6 cells, and the intensity of the signal as compared with quantitative markers, implied that each positive cell contained only 1–3 copies. These data suggest the basis for the inconclusive findings obtained with Southern blot hybridization on electrophoresed cellular DNA restriction fragments. It is important to note, however, that Horvath, Palkonyay & Weber (1986) found adenovirus DNA sequences in higher copy number and in a much higher percentage of cells, therefore they successfully carried out Southern blot hybridization analysis but did not find evidence of viral DNA integration (Horvath, Palkonyay & Weber, 1986).

To investigate further the intimate relationship of adenoviruses with human lymphocytes, efforts were directed toward the establishment of a persistent infection in a lymphocyte cell line. The Molt 4 T-cell line, derived from a human lymphoma, was used because its characteristics had been described, and we showed them to be free of detectable adenovirus DNA. B-cell lines were not deemed suitable for these studies since they all had been immortalized by Epstein–Barr virus, which potentially could be a complicating feature since the replicating and functioning viral DNA persisted. Adenovirus infection of lymphocyte cultures and umbilical cord lymphocytes, with or without EB virus infection, have been initiated and persistent infections were established for several months (Lambriex & van der Veen, 1976; Andiman & Miller, 1982; Faucon & Desgranges, 1980; Horvath et al., 1983).

The studies were initiated with a well-characterized wild-type adenovirus type 5 (WtAd5) using a varying number of infectious virions per cell. When greater than 1 PFU per cell was employed, infection could be maintained for only one to two weeks (about two to six cell passages) before cell death occurred. When a multiplicity of infection (MOI) of 0.1 to 1 PFU per cell was used, a more persistent infection could be established. However, under the most favorable conditions the infected cell line could be maintained for only 10 to 12 weeks (Fig. 1) even though less than 20 per cent of the cells were infected, as assayed by indirect immunofluorescence.

To determine whether this inability to establish a latent or persistent infection in T-cells cultured in vitro was under the control of a particular viral gene, a variety of Ad5 mutants that replicated in Molt 4 lymphocytes were employed. Most emphasis was placed upon an early region 3 mutant, (H5sub304), isolated by Jones & Shenk (1978). The E3 region was termed 'nonessential' since this mutant had a part of the E3 region deleted (83.2–85.1 map units) as well as a DNA substitution, and it could replicate like Wt Ad5 in HeLa and KB cells. However, it seemed unlikely that an infectious agent with a relatively small genome could afford to develop with excess baggage of almost 10 per cent of its DNA. The E3 region encodes two early proteins, a 19 K glycoprotein and a 14 K protein. The latter protein is encoded in the region deleted in H5sub304. Challberg & Ketner (1981) recently isolated H2d1801, a mutant in

Infection of T-Lymphocytes Line (Molt 4) with Wild-Type 5 Adenovirus

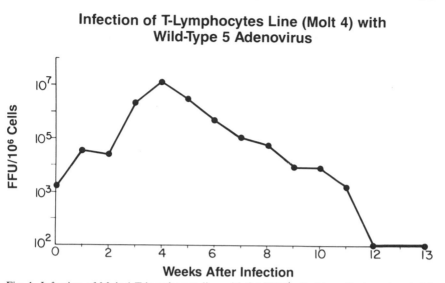

Fig. 1. Infection of Molt 4 T-lymphocyte line with 0.1 PFU/cell of type 5 adenovirus (wild-type). Cells were divided and fed three times per week. The indirect immunofluorescence assay was used to determine the infectious virus titre; FFU = fluorescent focus units.

Infection of Molt 4 T-Cell Line with H5 sub 304

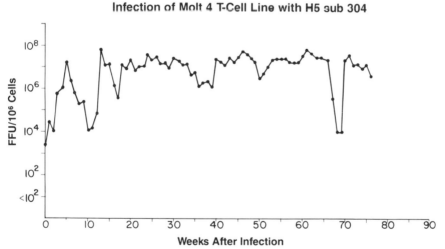

Fig. 2. Infection of Molt 4 T-lymphocyte line with 0.1 PFU/cell of H5sub304. Conditions of cell passage and assay of infectious virus were the same as described for Fig. 1.

the E3 region that encodes the 19 K glycoprotein; H5d1801 is presently being investigated in this model.

A persistent infection was readily established in Molt 4 lymphocytes with H5sub304 (Fig. 2); it seems likely that this infection is stable and can be maintained indefinitely through serial passage of infected cells, since it has now been maintained for over three years

Effect of Specific Antibodies on H5 sub 304 Replication in T-Lymphocytes

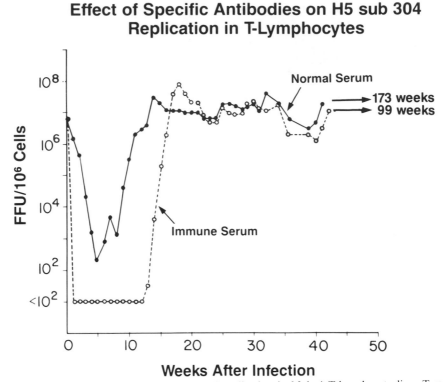

Fig. 3. Effect of neutralizing Abs on viral replication in Molt 4 T-lymphocyte line. Two aliquots of cells were obtained from persistent cell culture (shown in Fig. 2) 578 days after it was initiated. To one culture, neutralizing Abs were added in 10-fold excess of that required to neutralize completely the virus present; to the other culture, the same dilution of normal rabbit serum was added to the culture fluid. Serum was replenished with each cell passage (three times weekly) for 12 weeks, at which time the cells in both cultures were washed three times, and the culture continued under the same conditions as in Fig. 2.

(Fig. 3). During this period, from 2 to almost 20 per cent of the cells were infected at varying times, and the intracellular content of infectious virus varied as noted (Fig. 2). Virus was always detectable in the culture supernatant, indicating that extracellular viral spread could occur.

To investigate the state of the virus, or its genome, during the persistent infection, neutralizing antibodies were added to the culture medium during continuous passage of infected cells. Antibodies were added in 10-fold excess of that required to neutralize the quantity of virus present in the culture 578 days after the initial infection (Fig. 2). Normal, preimmune rabbit serum was similarly added to the medium of a sample from the same culture. Infectious virus could not be isolated from cultures three days after neutralizing

Abs had been added, and the infectivity titre gradually decreased in cultures receiving normal serum (Fig. 3). After five weeks, serum was removed from both sets of cultures, the cells were washed and the cultures were continued. In the culture to which immune serum had been added, infectious virus could not be detected for an additional seven weeks, but thereafter the viral titre rapidly increased. However, the infectivity titre immediately increased after normal serum serum was deleted from the culture medium (Fig. 3).

These data suggested that either infectious virus was present but could not be detected or that the viral DNA was present, perhaps in an integrated state, and was then induced to produce viral particles after the inhibiting factor (i.e. neutralizing Abs) was removed. To test these possibilities DNA was extracted from a sample of cells removed from culture during the period Abs were present and Southern blot hybridization carried out on enzyme-restricted DNA fragments separated by electrophoresis in agarose gels. This experiment clearly showed that viral DNA was present during the period that infectious virus was not detectable, and that the viral DNA was present in amounts of at least one-half that in cells bathed in normal serum. It was striking that there was not any detectable integrated viral DNA. These findings indicated that the viral DNA was present in episomal form throughout the period of Ab treatment. The data also suggested that the DNA could have been in replicating viral particles, and that it was neutralized when cells were disrupted by Abs that had been engulfed into cytoplasmic vacuoles. It is important to note, however, that, in experiments in which Abs were left in the culture medium for longer periods, 8 to 12 weeks, DNA gradually decreased and in some experiments disappeared. These findings imply that infected cells probably cannot divide, which would be predicted from experiments done in KB and HeLa cells. Thus, if intracellular transmission is prohibited, virus is eventually eradicated even though it can persist and replicate intracellularly for long periods.

These experiments revealed another striking phenomen. During this long period of persistent adenovirus infection, viral mutants continuously appeared. The mutations detected occurred in the E3 or E1a region of the viral genome. Most of the mutants were present in no more than 5 per cent of the total population of infectious viral particles at any single time of testing. One E3 mutant, presently termed NRS-1, appeared in the culture treated with preimmune, normal rabbit serum (NRS) and it rapidly outgrew the parent Wt Ad5 virus. The nucleotide changes and the phenotypic expression of these mutations are being investigated.

APPROACH TO UNDERSTANDING THE MOLECULAR BASIS OF ADENOVIRUS PNEUMONIA

Adenoviruses produce a variety of diseases in humans, among which respiratory infections, including pneumonia, are the most prominent (Straus, 1984). Until recently, however, it has been impossible to investigate the mechanisms of pathogenesis of these diseases since there was not an animal model available for such investigations. Pacini, Dubovi & Clyde (1984) recently satisfied this need demonstrating that intranasal inoculation of cotton rats with type 5 adenovirus produced a pneumonia that pathologically mimicked that described in humans. Based upon these observations an investigation was initiated to attempt to determine which gene products of adenovirus are essential to produce the cell injury leading to pneumonia. This approach was possible because there are now extant a series of type 5 adenovirus mutants with genetic defects throughout the genome. We have available conditionally lethal, temperature-sensitive mutants (Williams et al., 1971; Ensinger & Ginsberg, 1972) as well as deletion and substitution mutants either constructed in our laboratory (Babbis, Fisher & Ginsberg, 1984a, b; Babbis & Ginsberg, 1984; P. Fremuth & H. S. Ginsberg, unpublished data) or generously supplied by Drs T. Shenk and G. Ketner.

Baseline studies, which initiated this investigation, showed that two species of cotton rats available at the Laboratory of Infectious Diseases, National Institutes of Health, *Sigmadon hispidus* and *Fulviventer*, were susceptible to intranasal infection with only 10^3–10^4 PFU/animal. Virus replicated rapidly in both nares and lungs, producing lesions at both sites. When 10^8 PFU/cotton rat was used as the infecting inoculum, infectious virus attained maximum titre in two to three days and began to decline rapidly after five days. By seven to ten days after infection, virus had fallen to titres of 10^2 to 10^3 PFU/gram of tissue, as assayed by a modified, indirect immunofluorescence–serial dilution titration. Similar kinetics of viral replication were noted in the nares. Gross and histological changes in the lung were observed by one day after infection, and maximum pathology developed five to seven days after infection. Thus, the pathological lesions followed viral replication by 2–4 days, which is a common observation in the relationship between viral multiplication and development of pneumonia in laboratory models. Antibodies measured by ELISA assay appeared by seven days after intranasal infection, reached maximum titre 14–21 days after infection, and maintained the maximum titre until at least two months after infection (the longest period measured).

Pathological lesions were first seen in the bronchioles, where cells began to shed, and infected cells were detected by indirect immunofluorescent assay of frozen sections of lungs. A rare bronchiolar, epithelial cell contained a basophilic, intranuclear inclusion body surrounded by a vacuolar region. Monocytic infiltration of the peribronchiolar region also developed at that time, and increased alveolar involvement became evident. The infiltrating cells appeared to be a mixture of lymphocytes and macrophages; an occasional polymorphonuclear leucocyte was also present.

Studies have begun using a series of mutants, and initial results appear to offer promise that at least some of the goals of this research will be attained: i.e. an understanding of the genes required for the development of adenovirus disease. These preliminary results suggest that development of viral pneumonia, at least in the cotton rat, only requires early gene functions. These preliminary findings will be summarized.

Intranasal infection with the temperature-sensitive mutant H5ts125, which cannot replicate its DNA at the non-permissive temperature and does not manifest any late functions (Ensinger & Ginsberg, 1972; Young, Shenk & Ginsberg, 1984), produced pneumonia as extensive as that induced by wild-type virus. This pneumonia developed with a large viral inoculum (10^8 PFU/cotton rat) although virus did not replicate in the animals' lungs where the body temperature is about 39 °C (viral multiplication occurred normally in the nares where the temperature is close to 32 °C, the permissive temperature). The above data suggested that the pneumonia developed because all of the viral early proteins were expressed normally except the 72 K DNA-binding protein, which gene is mutated in H5ts125 (Ensinger & Ginsberg, 1972).

To test the above hypothesis, experiments were carried out using a different DNA-minus, temperature-sensitive mutant, H5ts149 (Ensinger & Ginsberg, 1972), which is mutated in the viral DNA polymerase gene (Stillman, Tamanoi & Mathews, 1982). H5ts149, like H5ts125, also produced pneumonia as extensive as wild-type virus.

The above hypothesis is supported by studies using a deletion mutant in early region 1b(E1b), H5dl110, which produces a truncated 55 K protein. This mutant, which in KB cell cultures cannot shut-off host cell protein synthesis (Babbis & Ginsberg, 1984), produced little or no pulmonary infiltration, although viral replication was only minimally reduced. The E1a deletion and insertion mutants tested (Babbis, Fisher & Ginsberg, 1984) also did not produce pneumonia; these mutants were very defective in viral multiplication owing to decreased transcription of early genes.

DISCUSSION

Latency or persistent infection and respiratory disease are two of the prominent features of adenovirus infections in humans. The mechanisms, by which adenoviruses effect this pathogenic process, are the goals of research in progress and our recent findings are the basis of this communication.

It has not been possible to isolate adenoviruses directly from tonsils or adenoids; but virus eventually produces cytopathic changes in explants of human tonsils and adenoids, and infectious virus can then be isolated (Rowe et al., 1953; H. S. Ginsberg, unpublished data). It is on the basis of the inability to isolate virus from the original extracts of tonsils and adenoids that the infection has been termed a *latent infection*. It must be realized, however, that in those individuals whose tonsils and adenoids eventually yield adenoviruses from their cultured explants, relatively high titres of neutralizing Abs are present in their blood and the extracellular fluids bathing their tonsils and adenoids. The inability to isolate infectious virus directly from primary extracts of the lymphoid organs may be effected by the presence of neutralizing Abs. According to this hypothesis the persistence of virus in cultured lymphocytes could be akin to the so-called latent virus in tonsils and adenoids. This possible phenomenon requires direct proof. If the proof is forthcoming it would be similar to the latent state of herpes simplex virus in nerve ganglia.

The cotton rat has once more proved to be a valuable animal for the study of viral pathogenesis. Type 5 adenovirus (Ad5) was shown to replicate in the lungs of the *Sigmadon hispidus* cotton rat and produce a pneumonia which histologically resembled that produced in humans (Pacini, Dubovri & Clyde, 1984). This important observation permitted us to initiate an investigation to determine the molecular mechanism(s) required for pneumonia production, taking advantage of a series of Ad5 mutants, each with a single defect, so that the group of mutants available encompass the entire genome. Since pneumonia followed adenovirus replication according to the kinetics shown for other viruses, and therefore the lesions are not a toxic effect of the viral inoculum, it was apparent pathogenesis of the pneumonia was a direct effect of one or more viral gene products.

These preliminary studies suggest that cell damage and subsequent host response (i.e. cellular infiltration producing pneumonia) result from effects of early gene functions. Thus, H5ts125 and H5ts149, neither of which can replicate at the non-permissive temperature (39.5 °C) owing to a single-base pair change in the DNA-binding protein gene (H5ts125) or the DNA polymerase gene (H5ts149), produced pneumonia to an extent comparable to that of Ad5 wild-type virus. The pneumonia was produced even though virus did not multiply in the lungs because viral DNA cannot replicate at the lung temperature, and therefore viral structural proteins (i.e. late gene products) were also not synthesized. It appears critical, however, that all early gene products be made at the non-permissive temperature, even though the mutated gene product is not functional. These findings imply that sufficient cells were infected by the viral inoculum, 10^8 PFU/cotton rat, so that the number of cells initially injured was sufficient to induce the host mononuclear cell infiltrative response.

The possibility that early gene functions supply the critical protein(s) required to effect adenovirus penumonia is supported by the finding that infections with H5dl110, which contains an E1b mutation affecting the production of the early 55 K protein, produces only minimal pneumonia. Viral replication is significantly less than in lungs infected with wild-type virus, but the viral yield is more than 5 logs greater than in the lungs of cotton rats infected with the same amounts of H5ts125. Therefore, the markedly reduced pneumonia in H5dl110-infected cotton rats cannot be the result of viral replication that was decreased ten-fold or less. The failure to produce pulmonary lesions can best be explained by the finding that although viral DNA replication and viral transcription, early and late, are comparable to Ad5 wild-type virus, host protein synthesis is not blocked and synthesis of late viral proteins is significantly reduced (Babbiss & Ginsberg, 1984).

Initial experiments with other E1b mutants, which affect either the 55 K protein (like H5dl110, but in different regions) or the 19 K protein, and several E1a deletion mutants (data not presented) support the above conclusion that adenovirus early gene functions are primarily responsible for the production of pneumonia in cotton rats. The preliminary data available imply that more than one early gene product is involved in the process. The identification of which gene products are responsible and clarification of the mechanisms by which each viral protein acts should emerge from this approach described.

SUMMARY

Latent infection and pneumonia, two pathogenic processes that result from adenovirus infections in humans, were investigated. Viral DNA was detected in seven samples of peripheral lymphocytes from eleven apparently normal individuals. The state of the adenovirus DNA relative to cellular DNA could not be determined, however, owing to the low copy number of viral DNA present to only a small number of lymphocytes. As an experimental approach to the problem, a persistent infection was established with H5sub304, an early region 3 mutant of type 5 adenovirus (Ad5). An infection with Ad5 wild-type virus could be maintained for only 10 to 12 weeks; but with H5sub304, an infection persisting for more than three years was established, and it appears that this infection can be maintained indefinitely. Addition of excess neutralizing antibodies to this culture medium made it impossible to isolate virus from the washed lymphocytes, but intracellular viral DNA persisted, and detectable infectious virus rapidly increased after Abs were removed. Of particular interest was the finding that mutants with nucleotide changes in the E1a or E3 region continuously appeared during the period of persistent infection.

An excellent animal model of type 5 adenovirus pneumonia was developed in two species of cotton rats, confirming the findings of Pacini et al. Virus rapidly multiplied in the lungs and nares, and the development of pneumonia followed by about two to four days: virus reached maximal infectivity titre in the lungs about two to three days after intranasal infection with 10^8 PFU/cotton rat, and the greatest pulmonary infiltration developed after five to seven days. The production of pneumonia is being investigated using a series of mutants having single-base changes, deletions, or insertions in the early and late transcription regions throughout the genome.

The DNA-minus, temperature-sensitive mutants H5ts125 and H5ts149, which have mutations in the DNA-binding protein gene and the DNA-polymerase gene, respectively, produce pneumonia to the same extent as Ad5 wild-type virus, although virus did not replicate at the temperature of cotton-rat lungs, 39 °C. Under this condition all early genes are expressed except that mutated, viral DNA cannot replicate and late viral proteins are not produced. These experiments indicate that cell damage and cellular replication are caused by early gene functions. The early 1b 55 K protein appears to be one gene product responsible for production of adenovirus pneumonia.

REFERENCES

ANDIMAN, W. E. & MILLER, G. (1982). Persistent infections with adenovirus types 5 and 6 in lymphoid cells from humans and woolly monkeys. *Journal of Infectious Diseases*, **145**, 83–8.

BABBISS, L. E., FISHER, P. B. & GINSBERG H. S. (1984a). Effect on transformation of mutations in the early region 1b-encoded 21- and 55-Kilodalton protein of adenovirus 5. *Journal of Virology*, **52**, 389–95.

BABBIS, L. E. & GINSBERG, H. S. (1984). Adenovirus type 5 early region 1b gene product is required for efficient shut-off of host protein synthesis. *Journal of Virology*, **50**, 202–12.

BABBISS, L. E., FISHER, P. B. & GINSBERG, H. S. (1984b). Deletion and insertion mutations in early region 1a of type 5 adenoviruses producing cold-sensitive or defective phenotypes for transformation. *Journal of Virology*, **49**, 731–40.

BEGIN, M. & WEBER, J. (1975). Genetic analysis of adenovirus type 2. I. Isolation and genetic characterization of temperature-sensitive mutants. *Journal of Virology*, **15**, 1–7.

CHALLBERG, S. S. & KETNER, G. (1981). Deletion mutants of adenovirus 2: isolation and initial characterization of virus carrying mutations near the right end of the viral genome. *Virology*, **114**, 196–209.

ENSINGER, M. J. & GINSBERG, H. S. (1972). Selection and preliminary characterization of temperature-sensitive mutants of type 5 adenoviruses. *Journal of Virology*, **10**, 328–39.

FAUCON, N. & DESGRANGES, C. (1980). Persistence of human adenovirus 5 in human cord blood lymphoblastoid cell lines transformed by Epstein–Barr virus. *Infection and Immunity*, **29**, 1180–4.

GINSBERG, H. S., BADGER, G. F., DINGLE, J. H., JORDAN, W. S. JR & KATZ, S. (1954). Etiologic relationship of the RI-67 agent to acute respiratory disease (ARD). *Journal of Clinical Investigations*, **34**, 1077–86.

GRAHAM, F. L. (1984). Transformation by and oncogenicity of human adenoviruses. In *The Adenoviruses*, ed. H. S. Ginsberg, pp. 339–98. New York: Plenum Press.

GREEN, M., WOLD, W. S. M., MACKEY, J. K. & RIGDENS, P. (1979). Analysis of human tonsil and cancer DNAs and RNAs for human sequences of Group C (serotypes 1, 2, 5, and 6) human adenoviruses. *Proceedings of the National Academy of Sciences, USA.*

HILLEMAN, M. R. & WERNER, J. R. (1954). Recovery of a new agent from patients with acute respiratory illness. *Proceedings of the Society for Experimental Biology and Medicine*, **85**, 183–8.

HORVATH, J., KULCISAIR, G., UGRYUEMOV, J. P., DAN, P., NASZ, I., BARINSKY, I. F., SIMON, G. & ONGRADI, J. (1983). Effect of adenovirus infection on human peripheral lymphocytes. *Acta Microbiologica Hungarica*, **30**, 203–9.

HORVATH, J., PALKONYAY, L. & WEBER, J. (1986). Group C adenovirus DNA sequences in human lymphoid cells. *Journal of Virology*, **59**, 189–92.

JONES, N. & SHENK, T. (1978). Isolation of deletion and substitution mutants of adenovirus type 5. *Cell* **13**, 181–8.

LAMBRIEX, M. & VAN DER VEEN, J. (1976). Comparison of replication of adenoviruses type 2 and type 4 in human lymphocyte cultures. *Infection and Immunity*, **14**, 619–22.

PACINI, D. L., DUBROVI, E. J. & CLYDE, W. A. JR (1984). A new animal model for human respiratory tract disease due to adenoviruses. *Journal of Infectious Diseases*, **150**, 92–7.

ROWE, W. P., HUEBNER, R. J., GILMORE, L. K., PARROTT, R. N. & WARD, T. G. (1953). Isolation of a cytopathogenic agent from human adenoids undergoing spontaneous degeneration in tissue culture. *Proceedings of the Society for Experimental Biology and Medicine*, **84**, 570–3.

SMITH, W., ANDREWS, C. H. & LAIDLAW, P. P. (1933). A virus obtained from influenza patients. *Lancet*, **2**, 66–8.

STILLMAN, B. W., TAMANOI, F. & MATHEWS, M. B. (1982). Purification of an adenovirus-coded DNA polymerase that is required for initiation of DNA replication, *Cell*, **31**, 613–23.

STRAUS, S. E. (1984). Adenovirus Infections in Humans. In *The Adenoviruses*, ed. H. S. Ginsberg, pp. 451–96. New York: Plenum Press.

VAN DER VLEIT, P. C., LEVINE, A. H., ENSINGER, M. J. & GINSBERG, H. S. (1975). Thermolabile DNA binding proteins from cells infected with a temperature-sensitive mutant of adenoviruses defective in viral DNA synthesis. *Journal of Virology*, **15**, 348–84.

WILLIAMS, J. F., GHARPURE, M., USTACELEBI & McDONALD, S. (1971). Isolation of temperature-sensitive mutants of adenovirus type 5. *Journal of General Virology*, **11**, 95–101.

YOUNG, C. S. H., SHENK, T. & GINSBERG, H. S. (1984). The genetic system. *The Adenoviruses*, ed. H. S. Ginsberg, pp. 125–72. New York: Plenum Press.

HERPES SIMPLEX VIRUS GLYCOPROTEINS AND PATHOGENESIS

HOWARD S. MARSDEN

MRC Virology Unit, Institute of Virology, University of Glasgow, Church Street, Glasgow G11 5JR, Scotland

INTRODUCTION

Herpesviruses are large viruses. Over 80 members of the group have been isolated, which between them infect a wide range of animals including fish, frogs, rodents, birds, marsupials and mammals, including man. Following a primary infection herpesviruses can produce latent infections lasting throughout the life of the host and may periodically be reactivated to produce distressing or life threatening symptoms and diseases of economic importance. Herpesvirus genomes comprise double stranded DNA of approximate size 100–200 kilobase pairs. The virion consists of at least four morphologically distinct substructures: core, capsid, tegument and envelope. Glycoproteins are located in the envelope. For reviews see Stevens (1975), Honess (1984), Hill (1985) Spear (1985) and Dargan (1986).

It is perhaps trivial to say that virus-induced pathogenesis reflects the expression of virus functions. Nevertheless, our attempts to understand the disease process requires us to understand the process of infection and the mechanisms of host resistance to infection (Mims & White, 1984). This means understanding how viruses are transmitted and enter the host and then individual cells, what determines tissue tropism, how expression of virus genes is regulated and how viruses are assembled and emerge from cells.

Herpesvirus glycoproteins have been the subject of intense interest over the last decade because they are involved in the primary interactions of the virus with the host and hence are thought to play an important role in pathogenicity and to be likely candidates for subunit vaccines. Best studied of the human herpesviruses are herpes simplex virus types 1 and 2 (HSV-1 and HSV-2) which cause oral/facial and genital lesions. Our knowledge of the glycoproteins of these viruses is presently greater than that of the other herpesviruses.

In this chapter I have summarised our present knowledge of the glycoproteins of herpes simplex virus and where possible identified known functions and properties. Later sections deal with the host's immune response to them and the contribution this makes to host resistance. Pathogenesis is clearly multifactorial and evidence for involvement of other virus genes is discussed. However, our understanding of the molecular basis of herpes simplex virus pathogenicity is meagre and the number of virus genes recognised to be involved will undoubtedly increase.

IDENTIFICATION OF GENETICALLY DISTINCT HSV GLYCOPROTEIN SPECIES

Six glycoproteins have been positively identified in HSV-infected cells: gB, gC, gD (Spear, 1976), gE (Baucke & Spear, 1979), gG (Marsden et al., 1978, 1984; Roizman et al., 1984) and gH (Buckmaster et al., 1984). The glycoproteins designated g92K by Marsden et al. (1978, 1984) and gG by Roizman et al. (1984) have now been shown using monoclonal antibodies to be identical (Balachandran & Hutt-Fletcher, 1985; Olofsson et al., 1986). Glycoproteins B, C, D (Spear, 1976), E (Para et al., 1980), G (Frame et al., 1986) and H (Richman et al., 1986) are found in virions.

Five of the HSV glycoproteins (gB, gC, gD, gE and gG) have counterparts in both HSV-1 and HSV-2 infected cells based on biochemical or immunological criteria or analysis of DNA sequence data (Balachandran et al., 1981, 1982a; Baucke & Spear, 1979; Dowbenko & Lasky, 1984; Eisenberg et al., 1982b; Frink et al., 1983; Hope & Marsden, 1983; Hope et al., 1982; Lee et al., 1982a,b; Para et al., 1982b, 1983; Pereira et al., 1980, 1981; Swain et al., 1985; Zezulak & Spear, 1983; Zweig et al., 1983; McGeoch et al., 1987). No HSV-2 counterpart of gH-1 has yet been identified. Experiments concerning gB, gC, gD, gE and gG-2 have been reviewed (Spear, 1985) and are not discussed in detail here. Map locations of the genes encoding the glycoproteins are shown in Fig. 1.

Recently the HSV-1 equivalent of gG-2 has been identified. Three groups, independently, detected a novel HSV-1 glycoprotein mapping to the left of gD in a position consistent with the product of HSV-1 gene U_S4 (McGeoch et al., 1985) and all speculated that this was gG-1. Ackermann et al. (1986) used HSV-1/HSV-2 intertypic recombinants to map the target antigen of a type-1 specific monoclonal antibody, H1379, to the left of gD and overlapping gene U_S4. A similar approach with the type-1 specific monoclonal antibody LP10 was used by Richman et al. (1986) who showed, in addition, that the target antigen of LP10 mapped to the U_S4 open

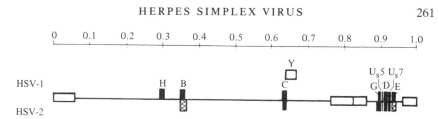

Fig. 1. Location of the genes encoding HSV glycoproteins. The HSV genome consists of long and short unique regions (thin horizontal lines) each bounded by inverted repeats (open rectangles). The repeat regions allow recombination leading to inversion of both long and short regions so that four isomeric forms are possible.

The HSV genome is represented in the prototype orientation together with the approximate map locations of the genes encoding the viral glycoproteins. The scale represents the fractional genome length (mu). Solid boxes represent those genes which have been sequenced. Hatched boxes represent those genes which have not been sequenced. The open box shows the location of gY (Marsden *et al.*, unpublished data). The glycoproteins predicted to be encoded by genes U_S5 and U_S7 have not yet been identified.

reading frame. A different approach was used by Frame *et al.* (1986): they raised an antiserum to a synthetic oligopeptide corresponding to a region of the predicted U_S4 protein and used the antiserum in immunoprecipitation experiments to identify the U_S4 gene product. The U_S4 product was shown to be a glycoprotein by mannose labelling experiments, hence a new glycoprotein was identified.

Definitive evidence showing that U_S4 encodes the HSV-1 equivalent of gG-2 was provided by McGeoch *et al.* (1987). They obtained the DNA sequence of the entire HSV-2 *Hind*IIIl fragment (located within U_S) and identified six potential genes one of which, HSV-2 U_S4, showed homology with HSV-1 U_S4. Using an antiserum raised against a dodecapeptide representing a stretch of amino acids near the C-terminus of the predicted HSV-2 U_S4 polypeptide this gene was shown to encode gG-2. The dodecapeptide was identical, except at one position, to the corresponding part of the HSV-1 U_S4 protein so that induced antiserum might be type common. The antiserum immunoprecipitated the HSV-1 U_S4 gene products previously recognised by Frame *et al.* (1986) and so unambiguously identified U_S4 as encoding the HSV-1 equivalent of gG-2. Accordingly, the HSV-1 U_S4 product was designated gG-1.

Comparison of the HSV-1 and HSV-2 U_S4 DNA sequences (McGeoch *et al.*, 1985; McGeoch *et al.*, 1987) shows they have diverged considerably: more than any equivalent HSV-1/HSV-2 genes so far studied (Fig. 2). The HSV-2 gene contains an insertion of about 1460 base pairs in the coding sequence. Accordingly, the predicted unprocessed mol. wt. of gG-1 is much smaller (25 237) than that of gG-2 (72 239) and, as might be predicted, this is reflected

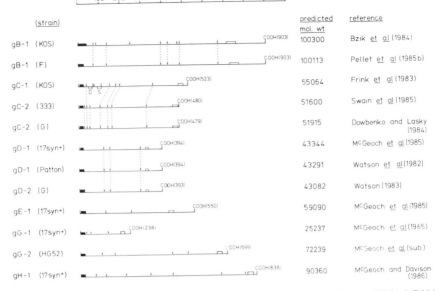

Fig. 2. Summary of the HSV glycoprotein primary structures based on published DNA sequences (references are given in the right-hand column). Each horizontal line represents the length of the polypeptide chain, the number of amino acids in the protein is given in parenthesis beside the carboxy (COOH) terminus. The amino (NH₂) terminus is taken as amino acid (aa) number 1. The glycoprotein represented and the strain from which the DNA was sequenced is indicated on the left hand side. Essential features predicted for each glycoprotein are: the signal sequence (solid boxes), membrane anchoring sequences (open boxes) and potential sites for the addition of N-linked oligosaccharides (arrows). For glycoproteins which have been sequenced from more than one virus strain, conserved potential oligosaccharide addition sites are shown by dotted lines. The molecular weight of the predicted primary amino acid sequence is also indicated. The stippled boxes below the line representing gC-1 indicate sequences missing from gC-2 as predicted by Dowbenko & Lasky (1984) (D) or Swain *et al.* (1985) (S).

in the apparent sizes of the glycoproteins on SDS polyacrylamide gels.

Glycoprotein G-1, labelled over several hours, has been variously described to migrate as a single species close to gD (Richman *et al.*, 1986), two species of apparent mol. wt. 60–68 K and 44–48 K (Ackermann *et al.*, 1986) or three species of apparent mol. wt. 56 K, 48 K and 37 K (Frame *et al.*, 1986). The possibility that these multiple forms were not all products of HSV-1 U$_S$4 but represented other proteins complexed to U$_S$4 was made less likely by the experiments of Ackerman *et al.* (1986) who used protein blotting to identify the species. The basis for these multiple forms remains obscure but does appear to be strain dependent (Ackermann *et al.*, 1986). All forms were found in virions purified from HSV-1 strain 17 (Frame *et al.*, 1986).

The amount of gG-1 in strain 17 virions has been estimated to be about half that of gB, gC and gD, thus gG-1 should not be considered a minor species (Frame *et al.*, 1986). Failure to have been detected earlier was probably due to its relatively low molar methionine content and comigration with other herpes virus glycoproteins.

A particularly interesting feature of the analysis by McGeoch *et al.* (1986) is the identification of a local homology between gG-2 and gD which they interpret as suggesting that the evolution of the genes for gG and gD may have proceeded through a duplication event.

The DNA sequencing studies of McGeoch *et al.* (1985, 1986) encompassing the whole of the short unique region of the HSV-1 genome, enabled them to predict the existence of two glycoproteins in addition to $gD(U_S6)$, $gE(U_S8)$ and $gG(U_S4)$; these represent the products of genes U_S5 and U_S7 but their presence remains to be verified. If detected, five HSV glycoproteins would be encoded by adjacent genes. The 42 K and 55 K polypeptides translated *in vitro* from mRNA hybridising to the HSV-1 *Bam*H1 *j* fragment (Lee *et al.*, 1982*a*) may correspond to gG-1 and the product of gene U_S7 respectively. Additional glycoproteins may be encoded in U_L.

A glycoprotein, designated gY (Palfreyman *et al.*, 1983), is induced in cells infected with HSV-1. Glycoprotein Y has the same apparent mol. wt. as gC-1 but has a more basic isoelectric point. It is not known if gY is antigenically related to any previously identified glycoproteins but it was not precipitated by a number of monoclonal antibodies directed against gB, gC, and gD. However, the gene encoding it has been mapped by HSV-1/HSV-2 intertypic recombinants and lies between coordinates 0.64 and 0.67. These span the region encoding gC-1 (Fig. 1), suggesting that gY and gC-1 may be related (Marsden *et al.*, in preparation).

GLYCOPROTEIN HOMOLOGUES IN OTHER HERPESVIRUSES

Considerable progress is being made in identifying equivalent glycoproteins in different herpesviruses. This has been made possible in part by adoption of common nomenclatures for the glycoproteins of individual herpesviruses, by the availability of increasing numbers of monospecific antisera, by the rapidly accumulating DNA sequence data and by development of predictive methods to identify possible glycoproteins from analysis of DNA sequences (Perlman & Halvorson, 1983; McGeoch, 1985). Table 1 lists the counterparts of the

Table 1. *Herpesvirus glycoprotein homologues*

Virus	HSV-1[1]	HSV-2[1]	VZV[5,6]	PRV	EBV[18]
	gB-1	gB-2	gpII[7,8]		BALF4[18,19]
	gC-1	gC-2	(gpV)[7,9]	gIII[14,15]	
	gD-1	gD-2	_[10]		_[20]
	gE-1	gE-2	gpI[7]	gI	_[20]
	gG-1	gG-2	_[10]	gX[16,17]	_[20]
	gH-1	ni[2]	gpIII[7,11,12]		BXLF2[11,12]
	(U_S5)[3]	(U_S5)[4]	_[10]		_[20]
	(U_S7)[3]	(U_S7)[4]	gpIV[7,13]		_[20]

1. Identification of HSV-1 glycoproteins and the HSV-2 homologues has been discussed in the text.
2. No homologue yet identified.
3. Protein not yet identified but existence predicted from DNA sequence (McGeoch *et al.*, 1985; McGeoch, 1985).
4. Protein not yet identified but existence predicted from DNA sequence (McGeoch *et al.*, 1987).
5. Complete DNA sequence available (Davison & Scott, 1986).
6. New common glycoprotein nomenclature (Davison *et al.*, 1986).
7. Davison & Scott (1986).
8. Kitamura *et al.* (1986).
9. Protein not yet identified but existence predicted from DNA sequence (Davison & Scott, 1986).
10. No predicted homologue in VZV (Davison & McGeoch, 1986).
11. McGeoch & Davison (1986).
12. U. Gompels & A. C. Minson, personal communication.
13. Davison & McGeoch (1986).
14. Robbins *et al.* (1986).
15. Not essential in tissue culture. Wathen & Wathen (1986).
16. Rea *et al.* (1985).
17. Homology with gG-2. McGeoch *et al.*, (1987).
18. Complete DNA sequence available (Baer *et al.*, 1984).
19. Pellett *et al.* (1985a).
20. No predicted homologue in EBV (Davison & Taylor, unpublished data; McGeoch, unpublished data cited in McGeoch & Davison, 1986).

HSV-1 glycoproteins B, C, D, E, G and H for two alphaherpesviruses (VZV and PRV) and one gammaherpesvirus (EBV).

The EBV genome has been completely sequenced (Baer *et al.*, 1984). The protein predicted to be encoded by the BXLF2 reading frame shows homology with HSV-1 gH-1 (McGeoch & Davison, 1986; U. Gompels & A. C. Minson, personal communication) and the glycoprotein predicted to be encoded by the EBV BALF4 reading frame shows homology with HSV gB (Pellett *et al.*, 1985a). The related EBV glycoproteins gp350 and gp220 are encoded by differently spliced transcripts from the BLLF1 reading frame (Baer *et al.*, 1984; Biggin *et al.*, 1984; Beisel *et al.*, 1985) but HSV homologues have not been identified. The EBV glycoproteins predicted to be encoded by BDLF3 and BILF2 have not yet been detected; whether these genes will lack HSV counterparts remains to be determined.

Four HCMV glycoproteins have been identified using monoclonal antibodies and designated gA, gB, gC and gD (reviewed by Pereira, 1985). So far no sequence data has been available for comparison with HSV. The equine herpesvirus type 1 138 K glycoprotein and bovine mammillitis virus 120 K and 130 K glycoproteins are antigenically and structurally related to HSV gB (Snowden *et al.*, 1985).

PROCESSING OF HSV GLYCOPROTEINS

The processing of HSV glycoproteins B, C, D and E has recently been reviewed (Spear, 1985; Campadelli-Fiume & Serafini-Cessi, 1985). Briefly, there is evidence for both N-linked and O-linked oligosaccharides on all these glycoproteins. They are also sulphated, gE being the most heavily labelled (Hope *et al.*, 1982; Hope & Marsden, 1983). Glycoprotein E is modified by the addition of fatty acid (Johnson & Spear, 1983).

Glycoprotein G-2 labels heavily with glucosamine, only moderately with mannose and poorly, if at all, with inorganic sulphate (Marsden *et al.*, 1984). gG-2 is modified by addition of both N- and O-linked oligosaccharides (Balachandran & Hutt-Fletcher, 1985; Serafini-Cessi *et al.*, 1985). Serafini-Cessi *et al.* (1985) estimated the ratio of O- to N-linked oligosaccharides to be higher than that of any other HSV glycoprotein so far studied including gC-1 which itself was judged to have a large number of O-linked oligosaccharides (Dall'Olio *et al.*, 1985). Processing of mature gG-2 appears to involve a major proteolytic cleavage step (Balachandran & Hutt-Fletcher, 1985).

Glycoprotein G-2 is unique amongst HSV-2 glycoproteins in possessing affinity for *Helix pomatia* lectin (Olofsson *et al.*, 1986), a property shared by gC-1 but by no other HSV-1 glycoproteins (Olofsson *et al.*, 1981, 1983).

Glycosylation plays some role in infectivity as virions produced in the presence of tunicamycin, a drug which effectively inhibits N-linked glycosylation, are not infectious (Kousoulas *et al.*, 1983). However, complete glycosylation is not necessary as virions produced in the presence of ammonium chloride ions contain partially

Table 2. *Functions of the HSV glycoproteins*

Function	HSV glycoproteins					
	gB	gC	gD	gE	gG	gH
Essential for replication in tissue culture	Yes	No		No		Yes
Present in virions	+	+	+	+	+	+
Adsorption	+	+	+	+		
Penetration	+	+				
Cell fusion (syn)	+	+	+			+
Egress						+
Fc-receptor	−	−	−	+	−	−
C3b-receptor	−	gC-1	−	−	−	−
NK specificity	+	+				
Neutralisation	+	+	+	+	+	+
Cell-mediated immunity	+	+	+	+	+	

+ evidence has been presented (see text) indicating that the glycoprotein is involved in a particular function.
− tested and found not to be involved.
gC-1 indicates that the C3b receptor is specific for gC of HSV-1 (and not HSV-2).
no symbol indicates that the glycoprotein has not been implicated in the function but could possibly be.

glycosylated glycoproteins and are infectious (Kousoulas *et al.*, 1983) as are virus grown in cells with defective glycosyltransferases (Campadelli-Fiume *et al.*, 1982).

FUNCTIONS AND ACTIVITIES OF HSV GLYCOPROTEINS

The properties associated with the HSV glycoproteins have recently been reviewed (Spear, 1985). Here I have emphasised new developments and summarised the properties of the glycoproteins (Table 2). Unless specifically stated otherwise, failure to implicate a glycoprotein in a particular function should not be interpreted to mean that the glycoprotein does not play a role in that function. Until all the HSV glycoproteins have been identified, monoclonal antibodies against them prepared and the genes saturated with mutations the information in Table 2 will remain incomplete.

Mutants have been isolated which do not express gC-1 (Hoggan & Roizman, 1959; Holland *et al.*, 1984; Heine *et al.*, 1974) or gC-2 (Cassai *et al.*, 1975/76; Zezulak & Spear, 1984) and these viruses grow in tissue culture. More recently, a deletion mutant lacking gE has been isolated (Longnecker & Roizman, 1986). This deletion mutant grows in Vero, 143TK⁻ and rabbit skin cells demonstrating that gE also is not necessary for growth in tissue culture.

Only two glycoproteins, gB and gH, have been shown to be essential for virus growth in tissue culture. The lesion responsible for the temperature-sensitive (*ts*) phenotype of the mutant *ts*B5 which fails to induce gB at non-permissive temperature (Manservigi *et al.*, 1977; Little *et al.*, 1981) maps within gB (DeLuca *et al.*, 1982). The mutant *ts*Q26 maps to a small region of the genome adjacent to the TK gene (Weller *et al.*, 1983) within the gH coding sequence (McGeoch & Davison, 1986; Gompels & Minson, 1986).

Adsorption

Glycoprotein B and to a lesser extent gD have been implicated in the adsorption of virions to cells by experiments involving virosomes (liposomes into which proteins extracted from purified virions have been inserted; Johnson *et al.*, 1984). The authors found the cell-associated fraction of virosomes was enriched for gB (and the non-glycosylated protein VP16). Further, antibodies against gB inhibited binding of virosomes to cells and antibodies against gC and gD also inhibited binding though less efficiently. Consistently, virosomes depleted of gB and gD bound less well to cells, the effect being more pronounced in the case of gB.

Fuller & Spear (1985) have suggested the term viral attachment component (VAC) for the viral structure which mediates attachment to cells. Antibodies specific for gD, gC and high concentrations of Fc fragments inhibited HSV adsorption thus suggesting that gD, gC and possibly gE form part of the VAC which might be either a single component comprising several proteins or several separated structures made up of different proteins. Fuller & Spear (1985) pointed out that since gC is not required for infectivity its presence in the VAC cannot be essential; recently isolated deletion mutants (Longnecker & Roizman, 1986) extend this conclusion to gE.

The VAC may contain other important constituents apart from glycoproteins. The presence of the non-glycosylated protein VP16 in the cell-associated fraction of virosomes (Fuller & Spear, 1985) raises the possibility that VP16 might in some way be involved. Finally, the cellular receptors to which HSV binds are specific for the serotype (Vahlne *et al.*, 1979, 1980; Addison *et al.*, 1984).

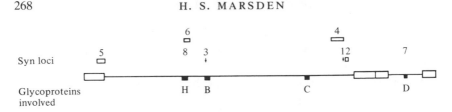

Fig. 3. Location of identified syn loci on the HSV-1 genome. The syn1, 2 and 3 loci were originally identified and designated by Ruyechan et al. (1979). Other loci discussed in the text have been given numbers here for the purpose of identification. Glycoproteins involved in the syn phenotype are indicated.

Penetration

At least two separate complementation groups affecting penetration have been identified using *ts* mutants. One, represented by *ts*B5 (Sarmiento *et al.*, 1979) and *ts*J12 (Little *et al.*, 1981) is in gB-1 and the other, represented by *ts*1204 (Addison *et al.*, 1984), has been mapped to within a 400 bp region laying to the left of gB between map coordinates 0.322 and 0.325, however, from DNA sequencing this region appears not to encode polypeptides with the characteristics of glycoproteins (D. J. McGeoch, personal communication). The mutation in gB-1 affecting penetration involves a substitution of alanine for valine at amino acid 552 (Bzik *et al.*, 1984; DeLuca *et al.*, 1984).

Epstein *et al.* (1984) mapped a locus affecting penetration to between coordinates 0.70 and 0.83 and found that a low gC/gB ratio favoured penetration of HSV-1 into XC cells.

Cell fusion

There has been considerable interest in HSV-1-induced syncytium (syn) formation since it was first described (Hoggan & Roizman, 1959) in part because the syn plaque morphology could be used as a convenient non-selectable marker (Brown *et al.*, 1973) and in part due to the importance of understanding the mechanism(s) which regulate the fusion of two membranes. However, the problem appears to be very complex – evidence for the existence of seven viral loci involved in syncytium formation has been recently reviewed (Spear, 1985) and their genome locations are shown in Fig. 3. Ruyechan *et al.* (1979) originally mapped and designated the syn1, 2 and 3 loci. The other four syn loci have been termed syn4–syn7 here for the purpose of identification on the figure. Syn4 is represented by mutant KOS-78R, it overlaps but is not the same as syn1, (Little & Schaffer, 1981), syn5 is represented by mutant KOS-804 (Little & Schaffer, 1981), syn6 involves a deletion at the 5′ end of the

TK gene (Sanders *et al.*, 1982) and syn7 involves gD (see below). An eighth locus has been suggested by the involvement of gH (Gompels & Minson, 1986).

The region closely defining the syn1 locus (Pogue-Geile *et al.*, 1984; Bond & Person, 1984) has now been completely sequenced and contains an open reading frame which potentially encodes a transmembrane protein of 338 amino acids (Debroy *et al.*, 1985). The nucleotide sequences of this same region from the syncytium-forming mutants MP and syn20 were also determined. Both sequences have amino acid substitutions at position 40 of the putative fusion protein: from alanine to valine for MP and alanine to threonine for syn20. A late transcript from the putative fusion gene has been detected (Debroy *et al.*, 1985) but the protein itself remains to be identified.

Glycoprotein B plays a role in cell fusion and the mutation responsible for the syn3 phenotype lies within it. This follows from experiments showing that temperature-sensitivity for production of functional gB-1 correlates with temperature-sensitivity in the ability of the mutant to induce cell fusion (Manservigi *et al.*, 1977; Haffey & Spear, 1980), from marker rescue experiments (DeLuca *et al.*, 1982) and DNA sequencing of the syn3 mutation (Bzik *et al.*, 1984). The syn determinant in *ts*B5 is due to replacement of an arginine at amino acid position 857 with a histidine. Amino acid 857 is predicted to lie within the cellular cytoplasmic domain. This syn mutation is in a different part of gB from that affecting penetration and suggests that penetration is not simply due to membrane fusion.

The relationship between the synthesis of gC-1 and lack of virus-induced cell fusion (syn$^+$) remains obscure. Early studies noted a correlation between the gC-1 negative (gC-1$^-$) and syn phenotypes (Manservigi *et al.*, 1977) but exceptions were found demonstrating that the two phenotypes can be genetically separated (Honess *et al.*, 1980; Lee *et al.*, 1982*b*). More recently, a strong correlation was found by Machuca *et al.* (1986) who co-transfected Vero cells with intact DNA from strain F (syn$^+$) and DNA fragments from strain MP (syn) then selected 40 recombinant viruses which produced syn plaques on Vero cells. The DNA fragments used have map coordinates 0.706–0.745, 0.745–0.810 and 0.710–0.761 and lie to the right of the gene encoding gC-1. Each of the transfected fragments gave rise to three different types of recombinants as judged by the ability to form syn plaques on Hep2 cells: 15 were syn$^+$ and gC$^+$, 17 were syn and gC$^-$ (with one exception which produced low levels of gC), and the remainder produced intermediate plaque types and either normal or reduced levels of gC. These results emphasise the strong correlation between the gC⁻ and syn phenotypes and provide support for the existence of syn loci on either side of 0.745 map units.

Glycoprotein D also has been implicated in cell fusion since anti-gD monoclonal antibodies blocked fusion whereas anti-gB, anti-gC and anti-gE antibodies did not (Noble *et al.*, 1983). Similar experiments have implicated gH in cell fusion and identified an anti-gD monoclonal antibody, LP2, which does not block fusion (Gompels & Minson, 1986; Minson *et al.*, 1986). Why anti-gB monoclonal antibodies fail to block cell fusion when gB is clearly involved in cell fusion is unclear but this feature may be important in evaluating the implied involvement of gD and gH.

A correlation was observed between the anti-fusion and neutralising activities of gD monoclonal antibodies (Fuller & Spear, 1985) which led the authors to speculate that the neutralising activity was due to blocking of the membrane fusion required for penetration. Identification of the anti-gD monoclonal antibody, LP2, which neutralises but does not block fusion and AP12, which does not neutralise but blocks fusion (Minson *et al.*, 1986) appears to conflict with that speculation. However, Minson *et al.* consider that the conflict may arise from the different methods of quantitating neutralisation and conclude that while studies with monoclonal antibodies implicate gD in cell fusion it has at least one other function required for infectivity.

The syn6 locus involving deletions at the 5′ end of the TK gene and the syn8 locus (gH) may be one and the same. TK and gH are adjacent genes transcribed from right to left, in the prototype genome orientation, with the promoter region for gH lying within the 3′ terminus of the TK gene (Sharpe *et al.*, 1983; McGeoch & Davison, 1986). The TK-containing plasmid used to generate deletions within TK (Sanders *et al.*, 1982) also contains about half of the gH gene. It is possible that recombination of the deleted plasmid back into the virus genome produced rearrangements affecting gH expression. Indeed the recombinant virus did not grow as well as the parent virus which led the authors to suggest that the recombinant virus was affected in a second function, as well as the virus TK.

Egress

Envelopment of the virus appears to occur at the inner nuclear membrane. Arguments favouring egress of virions via transport vesicles operating between the rough endoplasmic reticulum, the Golgi apparatus and the cell surface (Johnson & Spear, 1983) have been presented by Spear (1985).

Only gH has been implicated to play a role in egress. Buckmaster *et al.* (1984) observed that the anti-gH monoclonal antibody LP11, when added to infected cell monolayers after infection, inhibited plaque formation. This property is not displayed by antibodies against other HSV glycoproteins suggesting that gH may play a role in cell-to-cell spread of infectious virus.

C3b receptor

Human endothelial cells infected with HSV-1 develop receptors for the C3b component of complement (Cines *et al.*, 1982). Friedman *et al.* (1984) demonstrated that gC-1, but not gC-2 functions as the receptor. Comparison of the DNA sequence data for the genes encoding gC-1 (Frink *et al.*, 1983) and gC-2 (Dowbenko & Lasky, 1984; Swain *et al.*, 1985) identified a sequence of 28 amino acids in gC-1 which is not present in gC-2 and this region may be involved in binding of C3b. Both gC-1 and C3b have now been purified to homogeneity (as judged by silver staining of purified proteins) and a direct interaction between them has been demonstrated (Cohen *et al.*, Abstracts 10th International Herpesvirus Workshop, 1985, p. 158).

Fc receptor

Early work identifying an Fc receptor on the surface of HSV-infected cells has been reviewed (Spear, 1985): the HSV protein with affinity for the Fc region of IgG was identified and designated gE (Baucke & Spear, 1979; Para *et al.*, 1982a) and was shown to be present on the surface of virions as well as infected cells (Para *et al.*, 1980). Similar Fc-receptors are induced by HCMV (Keller *et al.*, 1976; Rahman *et al.*, 1976; Westmoreland *et al.*, 1976) and VZV (Ogata & Shigeta, 1979).

Johansson *et al.* (1984) have examined the specificity of the HSV-1 Fc receptor for human immunoglobulins. IgG4 binds more strongly than does IgG1 or IgG2. IgG3 does not bind to the Fc receptor nor do the other human immunoglobulins IgA, IgM or IgD or the structurally related beta$_2$-microglobulin.

In the presence of serum but not in its absence, gE on HSV-1-infected cells is modified, probably by proteolytic cleavage, and gE-related proteins of lower molecular weight (55 K/57 K and 32 K/34 K/35 K) are secreted into the medium. Unlike gE, the secreted proteins do not possess affinity for the Fc end of IgG (Cross, Hope & Marsden in preparation).

ANTIGENIC STRUCTURE OF HSV GLYCOPROTEINS

Knowledge of the physical and antigenic structure of the HSV glycoproteins is limited but most progress has been made on glycoproteins C and D.

Epitopes of gC-1 were identified using a panel of gC-specific virus-neutralising monoclonal antibodies and a series of antigenic variants selected for resistance to neutralisation with individual members of the antibody panel (Marlin *et al.*, 1985). Nine epitopes were identified clustered into two distinct antigenic domains. Domain I is located in the carboxy terminal portion of the molecule and domain II in the amino terminal half within the first 248 amino acids. Domain II is composed of three overlapping but distinct subdomains.

Eight separate antigenic domains on gD have been identified (Eisenberg *et al.*, 1985). This was achieved using a battery of anti-gD monoclonal antibodies. The antibodies were grouped according to neutralising activity, HSV type-specificity, reactivity with native, denatured and truncated forms of gD and reactivity with fragments of gD. A series of studies, made possible by the DNA sequence of the gD-1 and gD-2 genes and involving synthesis of peptides corresponding to regions of the gD-1 and gD-2 proteins, allowed construction of an epitope map and investigation of the fine structure of some antigenic determinants (Cohen *et al.*, 1984; Dietzschold *et al.*, 1984; Eisenberg *et al.*, 1985).

These studies showed three of the epitopes defined by group II, V and VII monoclonal antibodies were sequential determinants located between amino acids 268–287, 340–356 and 11–19 respectively of the mature glycoprotein. Four additional groups of monoclonal antibodies (I, III, IV and VI) reacted with discontinuous epitopes since antibody reactivity was lost when the glycoproteins were denatured; truncated forms of gD were used to localise these four epitopes to the first 260 amino acids of the mature protein.

On the epitope map of gD, the group V epitope is on the carboxy terminal side of the putative transmembrane sequence suggesting it lies within the cytoplasmic domain which provides a ready explanation for the failure of antibodies recognising this epitope to neutralise virus (Eisenberg *et al.*, 1982*a*) or confer protective immunity in passive immunisation studies (Rector *et al.*, 1984).

Additional mutants in gD resistant to neutralisation by four monoclonal antibodies have been reported (Minson *et al.*, 1986). Of these, two were assigned to two of the groupings defined by Eisenberg *et al.* (1982*a*). The others had properties not entirely consistent with any of the groups and one, AP7, had the novel and interesting property of enhancing three-fold to five-fold the infectivity of HSV-2 but not HSV-1 (Minson *et al.*, 1986).

INVOLVEMENT OF HSV GLYCOPROTEINS IN THE HOST IMMUNE RESPONSE

The host response to HSV infection involves both natural resistance (Lopez, 1985) and the adaptive humoral and cell-mediated responses (Norrild, 1985; Nash *et al.*, 1985). The studies described below indicate that the viral glycoproteins are the important targets of the host's immune response.

Natural resistance

Evidence implicating natural killer (NK) cells in resistance to HSV has been reviewed (Lopez, 1985). Recent experiments (Bishop *et al.*, 1986) suggest that human cells with NK activity possess clonal specificity for HSV-1-infected target cells. Experiments of this type are difficult because NK cells cannot be cloned and propogated *in vitro*. However, various dilutions of cells highly enriched for NK activity were made and the cells grown in short term cultures to obtain just sufficient cells to test for cell-mediated cytolysis of infected and uninfected cells. By using a monoclonal antibody resistant (mar) mutant (marB3.1) with an altered gB epitope and a mutant lacking gC-1 (syn LD70) the authors demonstrated that cell cultures having a high statistical probability of containing NK cells derived from a single ancestor recognise specific viral glycoprotein determinants on HSV-1-infected target cells.

Humoral response

By immunising animals with preparations of HSV virions or infected cells neutralising antibodies can be elicited by all the HSV glycoproteins (Balachandran *et al.*, 1982*b*; Balachandran & Hutt-Fletcher, 1985; Buckmaster *et al.*, 1984; Cohen *et al.*, 1972; Eberle & Courtney, 1980; Eisenberg *et al.*, 1982*a,b*; Holland *et al.*, 1983; Honess & Watson, 1974; Para *et al.*, 1982*a*; Pereira *et al.*, 1980; Powell *et al.*, 1974; Showalter *et al.*, 1981; Watson & Wildy, 1969; Zezulak & Spear, 1983) including the recently identified gG-1 (V. Sullivan & G. L. Smith – personal communication). Purified glycoproteins B, C, D, and E can individually elicit neutralising antibodies (Eisenberg *et al.*, 1982*b*; Long *et al.*, 1984; Para *et al.*, 1982*a*; Roberts *et al.*, 1985; Schrier *et al.*, 1983).

In experiments in which monoclonal antibodies were prepared by injecting mice with purified virions or extracts of infected cells, gD elicited the most potent neutralising antibodies (Vestergaard & Norrild, 1978; Showalter *et al.*, 1981; Balachandran *et al.*, 1982*a,b*; Rector *et al.*, 1982; Holland *et al.*, 1983; Para *et al.*, 1985). It is the potency of anti-gD antibodies which makes gD an attractive candidate for a subunit vaccine. As discussed by Para *et al.* (1985), the formal possibility that potent neutralising antibodies of other specificities were selected against in these studies seems most unlikely.

The neutralising titres of four monoclonal antibodies against gH-1 (designated 110 000 MW in the study) have been reported and were all quite low (Showalter *et al.*, 1981). However, one was recently identified having a very high titre (Gompels & Minson, 1986), raising the possibility that gH may be as effective as gD in eliciting potent neutralising antibodies. If this possibility turns out to be true then gH also becomes a candidate for a subunit vaccine.

Monoclonal antibodies directed against individual glycoproteins can protect animals from an otherwise lethal HSV infection. The specificities of monoclonal antibodies passively conferring protection include gB-1, gB-2, gC-1, gD-1, gD-2, gE-1, gE-2 and gH-1 (Balachandran *et al.*, 1982*a*; Dix *et al.*, 1981; Kapoor *et al.*, 1982; Para *et al.*, 1985; Rector *et al.*, 1982; Roberts *et al.*, 1985; Schrier, 1983; Sethi *et al.*, 1983; Simmons & Nash, 1985).

The question of whether antibodies which recognise different epitopes on the same or different glycoproteins are equivalent in their protective capacity has been investigated (Kumel *et al.*, 1985). Protection was found to be highly epitope specific. These experiments utilised mutant virus with specific epitope alterations in gB-1, gC-1 and gD-1 which were obtained by selection for virus resistant to neutralisation by individual monoclonal antibodies (Holland *et al.*, 1983; Marlin *et al.*, 1985). In general, monoclonal antibodies specific for gC provided protection at low doses, antibodies specific for gB were less protective while those specific for gD were least protective although there were exceptions to these generalisations. No correlation was found between the *in vitro* neutralisation titre and their protective capacity *in vivo*. Of particular interest, since the experiments were performed in a strain of mice deficient in the complement cascade, was the finding with three monoclonal antibodies, two against gD and one against gB: these antibodies neutralised virus in the absence of complement yet conferred no greater protection than antibodies which required complement for neutralisation. This finding suggests that neutralisation is not the major protective mechanism against virus infection (Kumel *et al.*, 1985).

Cell-mediated response

Cell-mediated immunity is an important part of the host's response to HSV infection and is required for the clearance of virus from the sites of infection (reviews, Nash *et al.*, 1985; Nash & Wildy, 1983). Anti-HSV cytotoxic T lymphocytes confer protection on mice against a lethal intraperitoneal infection with HSV-1 (Sethi *et al.*, 1983).

HSV-1 glycoproteins appear to be involved in cell-mediated immunity. Target cells in which synthesis of mature HSV glycoproteins was impaired by 2-deoxy-D-glucose, tunicamycin or use of a *ts* mutant were reduced in the susceptibility to T-cell-mediated lysis (Lawman *et al.*, 1980; Carter *et al.*, 1981). The observation that non-neutralising antibodies to glycoproteins B, C, D, E and G can passively confer protection against a lethal HSV challenge shows that these glycoproteins function as targets for antibody-dependent cell-mediated cytoxicity (Balachandran *et al.*, 1982*b*; Eberle *et al.*, 1981; Schrier *et al.*, 1983). A similar conclusion for gB and gD was reached by Rector *et al.* (1984), who showed that antibodies which did not recognise cell-free virus could passively confer protection.

VIRAL GENES AFFECTING THE PATHOGENICITY OF HSV IN MICE

Pathogenicity has been measured by inoculation with virus of animals such as guinea pigs, rabbits, tree shrews and mice. Various doses of virus are used and mortality or different states of disease are recorded. Sites of inoculation have included the brain, cornea, footpad, peritoneal cavity and vagina. The route of inoculation also appears to be an important determinant (reviewed by Dix *et al.*,

1983; Lopez, 1985) and different strains of virus differ in their pathogenicity (Dix *et al.*, 1983; Kaerner *et al.*, 1983). The pathogenicity of HSV for mice varies between different strains of mice (reviewed by Lopez, 1985), and resistance of mice to lethal HSV-1 infections is a dominant trait involving the host's haemopoietic defence system. With this number of variables, care must be exercised when making comparisons between experiments.

Thymidine kinase

The first viral function to be identified which affects the pathogenicity of HSV was thymidine kinase (TK) (Field & Wildy, 1978; Tenser *et al.*, 1979).

Glycoproteins

Recently, the pathogenicity of both gC-1$^-$ and gC-2$^-$ mutants has been measured. Intravaginal inoculations of mice with a gC-2$^-$ mutant produced local inflammation, spread of virus infection in the nervous system and lethality in a manner identical to that shown by the parental HSV-2 strain 333 (Johnson *et al.*, 1986). Intracerebral inoculation of mice with strain 333 or gC-2$^-$ virus produced similar LD$_{50}$s (the mutant was six-fold lower than the parent in one experiment and three-fold lower in another) which supports the impression that gC-2 does not appear to be involved in pathogenicity.

Evidence is accumulating, using mutants, that gC-1 also may not be involved in pathogenicity. Although one HSV-1 mutant (gC$^-$39), which does not produce any detectable form of gC-1, was more than 10 000-fold less pathogenic than the parent strain (KOS-321) following intracerebral inoculation (Kumel *et al.*, 1985), other gC-1 mutants are apparently pathogenic (unpublished data quoted in Kumel *et al.*, 1985). As pointed out by the authors, there may be other undetected mutations in gC-39 which give rise to the non-pathogenic phenotype. Moreover, strain MP which is gC-1$^-$ (Hoggan & Roizman, 1959; Holland *et al.*, 1984) was found to be highly virulent following intracranial inoculation (Dix *et al.*, 1983).

Variations in the antigenic structure of glycoproteins B, C and D appear to affect virus pathogenicity. Kumel *et al.* (1985) screened 39 monoclonal antibody resistant mutants and found that four *mar*C mutants, as well as the gC$^-$39 mutant, were significantly reduced in pathogenicity (1000–100 000 fold). Pathogenicity was not affected by *mar*B or *mar*D mutants with single epitope changes but multi *mar*B mutants showed significantly reduced pathogenicity. As pointed out by the authors, firm conclusions will depend on excluding the possibilities that unrecognised mutations might account for altered pathogenicities or that the temperature sensitive phenotype of a multi *mar*B mutant (unpublished work quoted in Kumel *et al.*, 1985) might account for the reduced pathogenicity.

Other loci

Three groups have independently identified a region on the HSV genome located between 0.70 and 0.83 map units which is involved in pathogenesis. Thompson & Stevens (1983) attempted to use HSV-1/HSV-2 intertypic recombinants to map neurovirulence in mice following intracranial inoculation but found that most recombinants were not useful as reduced virulence could be attributed to inefficient replication in any cell type at 38.5 °C (the normal temperature of the mouse). However, one recombinant RE6 was useful, it was completely non-virulent yet replicated as well as the parent virus at 38.5 °C in cultured cells, expressed a functional TK and grew in mouse tissue *in vivo*. The authors also excluded the possibility that a 'hidden' non-*ts* mutation in *ts*E (Brown *et al.*, 1973), the parent virus from which RE6 was generated (Marsden *et al.*, 1978), accounted for the lack of neurovirulence. To do this they selected a *ts*+ revertant of *ts*E and showed it to be equally neurovirulent as the wild-type parent from which *ts*E was isolated. The physical location of the function(s) responsible for non-virulence of RE6 was mapped by cotransfection to between 0.71 and 0.83 map units (Thompson *et al.*, 1983, 1985). Cleavage of the fragment which restored virulence (HSV-1 *Hind*III *c*) with *Eco*RI destroyed that property and virulence was restored with the *Bam*HI *l* fragment (0.71–0.745) suggesting the defect resides near the *Eco*RI site at 0.72 map units.

This same region of DNA (0.71–0.83 map units) was involved in patterns of occular disease. Centifanto-Fitzgerald *et al.* (1982) recorded (*a*) the morphology of dendritic lesions (*b*) the severity of epithelial disease and (*c*) the incidence and duration of stromal disease, and found that transfer of the 0.71–0.83 region of the stromal disease-producing strain (MP) to the genome of the epithelial disease-producing strain (F) gave recombinants with the disease-producing disease characteristic of the parent MP virus. They did not report the growth properties of their recombinants at 38.5 °C but, since disease patterns rather than virulence was used to map the locus, the analysis seems valid. Whether the defect is the same as that studied by Thompson and colleagues is not known.

A third study showing involvement of this same genome region in pathogenicity was reported by Rosen et al., (1985). They used HSV-1 strain HFEM, previously shown to contain a deletion of about 4 kbp between 0.7 and 0.8 map units (Halliburton et al., 1980) and found it to be apathogenic in tree shrews and mice. Pathogenicity could be restored to HFEM by cotransfection from strain F (a pathogenic strain) of the BamHI b fragment thus mapping the pathogenicity determinant to between 0.753 and 0.809 map units. To further define this position the deletion in HFEM was mapped to between 0.762 and 0.790 map units (Rosen & Darai, 1985) who found that the deletion was substituted in the pathogenic recombinant by about 3.45 kbp of DNA from strain F4 BamHI b.

More recently, another region of the genome has been implicated in pathogenesis (Thompson et al., 1986). Cotransfection of the avirulent HSV-1 KOS strain with the HindIII a fragment of the virulent HSV-1 strain 17 restored the pathogenicity thus demonstrating the importance of the region 0.25–0.53 map units: this region of the KOS genome affected virus replication in the sacral ganglia and brain.

Thus it is apparent that pathogenesis involves a number of loci in addition to TK. Rosen and Darai's results indicates involvement of functions mapping between 0.762 and 0.790 map units. Thompson and coworkers suggested involvement of one region near 0.72 and another between 0.25 and 0.53. The 0.25–0.53 region is large and several proteins have been mapped there including gB and gH. The locus identified by Centifanto-Fitzgerald et al., between 0.71 and 0.83, could correspond to that identified by Rosen and Darai or represent another region. Introduction of specific mutations in proteins mapping in the various regions should help elucidate their involvement in pathogenesis.

ACKNOWLEDGEMENTS

I wish to thank Dr Duncan McGeoch and Dr Graham Hope for extensive discussions, Dr Graham Hope for preparation of the figures, Dr Duncan McGeoch and Dr Andrew Davison for access to unpublished data and Miss Linda Shaw for typing the manuscript. I am grateful to Professor Subak-Sharpe, Dr Barklie Clements, Dr Margaret Frame, Dr Graham Hope and Dr Duncan McGeoch for critically reading the manuscript.

REFERENCES

ACKERMANN, M., LONGNECKER, R., ROIZMAN, B. & PEREIRA, L. (1986). Identification, properties and gene location of a novel glycoprotein specified by herpes simplex virus 1. *Virology*, **150**, 207–20.

ADDISON, C., RIXON, F. J., PALFREYMAN, J. W., O'HARA, M. & PRESTON, V. G. (1984). Characterisation of a herpes simplex virus type 1 mutant which has a temperature-sensitive defect in penetration of cells and assembly of capsids. *Virology*, **138**, 246–57.

BAER, R., BANKIER, A. T., BIGGIN, M. D., DEININGER, P. L., FARRELL, P. J., GIBSON, T. J., HATFULL, G., HUDSON, G. S., SATCHWELL, S. C., SEGUIN, C., TUFFNELL, P. S. & BARRELL, B. G. (1984). DNA sequence and expression of the B95-8 Epstein–Barr virus genome. *Nature*, **310**, 207–11.

BALACHANDRAN, N., BACCHETTI, S. & RAWLS, W. E. (1982*a*). Protection against lethal challenge of BALB/c mice by passive transfer of monoclonal antibodies to five glycoproteins of herpes simplex virus type 2. *Infection and Immunity*, **37**, 1132–7.

BALACHANDRAN, N., HARNISH, D., KILLINGTON, R. A., BACCHETTI, S. & RAWLS, W. E. (1981). Monoclonal antibodies to two glycoproteins of herpes simplex virus. *Journal of Virology*, **39**, 438–46.

BALACHANDRAN, N., HARNISH, D., RAWLS, W. E. & BACCHETTI, S. (1982*b*). Glycoproteins of herpes simplex virus type 2 as defined by monoclonal antibodies. *Journal of Virology*, **44**, 344–55.

BALACHANDRAN, N. & HUTT-FLETCHER, L. M. (1985). Synthesis and processing of glycoprotein gG of herpes simplex virus type 2. *Journal of Virology*, **54**, 825–32.

BAUCKE, R. B. & SPEAR, P. G. (1979). Membrane proteins specified by herpes simplex viruses. V. Identification of an Fc-binding glycoprotein. *Journal of Virology*, **32**, 779–89.

BEISEL, C., TANNER, J., MATSUO, T., THORLEY-LAWSON, D., KEZDY, F. & KIEFF, E. (1985). Two major outer envelope glycoproteins of Epstein–Barr virus are encoded by the same gene. *Journal of Virology*, **54**, 665–74.

BIGGIN, M., FARRELL, P. & BARRELL, B. (1984). Transcription and DNA sequence of the *Bam*HI L fragment of Epstein–Barr virus. *EMBO Journal*, **3**, 1083–90.

BISHOP, G. A., KÜMEL, G., SCHWARTZ, S. A. & GLORIOSO, J. C. (1986). Specificity of human natural killer cells in limiting dilution culture for determinants of herpes simplex virus type 1 glycoproteins. *Journal of Virology*, **57**, 294–300.

BOND, V. C. & PERSON, S. (1984). Fine structure physical map locations of alterations that affect cell fusion in herpes simplex type 1. *Virology*, **132**, 368–76.

BROWN, S. M., RITCHIE, D. A. & SUBAK-SHARPE, J. H. (1973). Genetic studies with herpes simplex virus type 1. The isolation of temperature sensitive mutants, their arrangement into complementation groups and recombination analysis leading to a linkage map. *Journal of General Virology*, **18**, 329–46.

BUCKMASTER, E. A., GOMPELS, U. & MINSON, A. (1984). Characterisation and physical mapping of an HSV-1 glycoprotein of approximately 115×10^3 molecular weight. *Virology*, **139**, 408–13.

BZIK, D. J., FOX, B. A., DELUCA, N. A. & PERSON, S. (1984). Nucleotide sequence of a region of the herpes simplex virus type 1 gB glycoprotein gene: Mutations affecting rate of virus entry and cell fusion. *Virology*, **137**, 185–90.

CAMPADELLI-FIUME, G., POLETTI, L., DALL'OLIO, F. & SERAFINI-CESSI, F. (1982). Infectivity and glycoprotein processing of herpes simplex virus type 1 grown in a ricin-resistant cell line deficient in N-acetylglucosaminyl transferase I. *Journal of Virology*, **43**, 1061–71.

CAMPADELLI-FIUME, G. & SERAFINI-CESSI, F. (1985). Processing of the oligo-saccharide chains of herpes simplex virus type 1 glycoproteins. In: *The Herpes-viruses*, Vol. 3, ed. B. Roizman, pp. 357–82. New York: Plenum Press.

CARTER, V. C., SCHAFFER, P. A. & TREVETHIA, S. S. (1981). The involvement of herpes simplex virus type 1 glycoproteins in cell-mediated immunity. *Journal of Immunology*, **126**, 1655–65.

CASSAI, G., MANSERVIGI, R., CORALLINI, A. & TERNI, M. (1975/76). Plaque disso-ciation of herpes simplex viruses: Biochemical and biological characters of the viral variants. *Intervirology*, **6**, 212–23.

CENTIFANO-FITZGERALD,Y. M., YAMAGUCHI, T., KAUFMAN, H. E., TOGNON, M. & ROIZMAN, B. (1982). Occular disease pattern induced by herpes simplex virus is genetically determined by a specific region of the viral DNA. *Journal of Experi-mental Medicine*, **155**, 475–89.

CINES, D. B., LYSS, A. P., BINA, M., CORKEY, R., KEFALIDES, N. A. & FRIEDMAN, H. M. (1982). Fc and C3 receptors induced by herpes simplex virus on cultured human endothelial cells. *Journal of Clinical Investigation*, **69**, 123–8.

COHEN, G. H., DIETZCHOLD, B., PONCE DE LEON, M., LONG, D., GOLUB, E., VARRICHIO, A., PEREIRA, L. & EISENBERG, R. J. (1984). Localisation and syn-thesis of an antigenic determinant of herpes simplex virus glycoprotein D that stimulates the production of neutralising antibody. *Journal of Virology*, **49**, 102–8.

COHEN, G. H., PONCE DE LEON, M. & NICHOLS, C. (1972). Isolation of a herpes simplex virus specific antigenic fraction which stimulates the production of neutra-lising antibody. *Journal of Virology*, **10**, 1021–30.

DALL'OLIO, F., MALAGOLINI, N., SPEZIALI, V., CAMPADELLI-FIUME, G. & SERAFINI-CESSI, F. (1985). Sialylated oligosaccharides O-glycosidically linked to glycopro-tein C from herpes simplex virus type 1. *Journal of Virology*, **56**, 127–34.

DARGAN, D. J. (1986). The structure and assembly of herpesviruses. In: *Electron Micro-scopy of Proteins*, Vol. 5, *Viral Structure*, pp. 359–437. London: Academic Press.

DAVISON, A. J. (1983). DNA sequence of the U_s component of the varicella-zoster virus genome. *EMBO Journal*, **2**, 2203–9.

DAVISON, A. J., EDSON, C. M., ELLIS, R. W., FORGHANI, B., GILDEN, D., GROSE, C., KELLER, P. M., VAFAI, A., WROBLEWSKA, Z. & YAMANISHI, K. (1986). New common nomenclature for glycoprotein genes of varicella-zoster virus and their glycosylated products. *Journal of Virology*, **57**, 1195–7.

DAVISON, A. J. & MCGEOCH, D. J. (1986). Evolutionary comparisons of the S segments in the genomes of herpes simplex virus type 1 and varicella-zoster virus. *Journal of General Virology*, **67**, 597–611.

DAVISON, A. J. & SCOTT, J. E. (1986). The complete DNA sequence of varicella-zoster virus. *Journal of General Virology*, **68**, 1–57.

DAVISON, A. J., WATERS, D. J. & EDSON, C. M. (1985). Identification of the products of a varicella-zoster virus glycoprotein gene. *Journal of General Virology*, **66**, 2237–42.

DEBROY, C., PEDERSON, N. & PERSON, S. (1985). Nucleotide sequence of a herpes simplex virus type 1 gene that causes cell fusion. *Virology*, **145**, 36–48.

DELUCA, N., BZIK, D. J., BOND, V. C., PERSON, S. & SNIPES, W. (1982). Nucleotide sequences of herpes simplex virus type 1 (HSV-1) affecting virus entry, cell fusion and production of glycoprotein gB (VP7). *Virology*, **122**, 411–23.

DeLuca, N., Bzik, D., Person, S. & Snipes, W. (1981). Early events in herpes simplex virus type 1 infection: Photosensitivity of fluorescein isothiocyanate treated virions. *Proceedings of the National Academy of Science (USA)*, **78**, 912–16.

DeLuca, N., Person, S., Bzik, D. J. & Snipes, W. (1984). Genome locations of temperature sensitive mutants in glycoprotein gB of herpes simplex virus type 1. *Virology*, **137**, 382–9.

Dietzschold, B., Eisenberg, R. J., Ponce de Leon, M., Golub, E., Hudecz, F., Varrichio, A. & Cohen, G. H. (1984). Fine structure analysis of type-specific and type-common antigenic sites of herpes simplex virus glycoprotein D. *Journal of Virology*, **52**, 431–5.

Dix, R. D., McKendall, R. R. & Baringer, J. R. (1983). Comparative neurovirulence of herpes simplex virus type 1 strains after peripheral or intracerebral inoculation of BALB/c mice. *Infection and Immunity*, **40**, 103–12.

Dix, R. D., Pereira, L. & Baringer, R. (1981). Use of monoclonal antibodies directed against herpes simplex virus glycoproteins to protect mice against acute virus-induced neurological disease. *Infection and Immunity*, **34**, 192–9.

Dowbenko, D. J. & Lasky, L. A. (1984). Extensive homology between the herpes simplex virus type 2 glycoprotein F gene and the herpes simplex virus type 1 glycoprotein C gene. *Journal of Virology*, **52**, 154–63.

Eberle, R. & Courtney, R. J. (1980). Preparation and characterisation of specific antisera to individual glycoprotein antigens comprising the major glycoprotein region of herpes simplex virus type 1. *Journal of Virology*, **35**, 902–17.

Eberle, R., Russell, R. G. & Rouse, B. T. (1981). Cell-mediated immunity to herpes simplex virus: Recognition of type-specific antigens by cytotoxic T cell populations. *Infection and Immunity*, **34**, 795–803.

Eisenberg, R. J., Long, D., Pereira, L., Hampar, B., Zweig, M. & Cohen, G. H. (1982a). Effect of monoclonal antibodies on limited proteolysis of native glycoprotein gD of herpes simplex virus type 1. *Journal of Virology*, **41**, 478–88.

Eisenberg, R. J., Long, D., Ponce de Leon, M., Matthews, J. T., Spear, P. G., Gibson, M. G., Lasky, L. A., Berman, P., Golu, B. E. & Cohen, G. H. (1985). Localisation of epitopes of herpes simplex virus type 1 glycoprotein D. *Journal of Virology*, **53**, 634–44.

eisenberg, R. J., Ponce de Leon, M., Pereira, L., Long, D. & Cohen, G. H. (1982b). Purification of glycoprotein gD of herpes simplex virus types 1 and 2 by use of monoclonal antibody. *Journal of Virology*, **41**, 1099–104.

Epstein, A. L., Jacquemont, B. & Machuca, I. (1984). Infection of a restrictive cell line (XC) cells by intratypic recombinants of HSV-1: Relationship between penetration of the virus and relative amounts of glycoprotein C. *Virology*, **132**, 315–24.

Field, H. J. & Wildy, P. (1978). The pathogenicity of thymidine kinase-deficient mutants of herpes simplex virus in mice. *Journal of Hygiene (Camb.)*, **81**, 261–77.

Frame, M. C., Marsden, H. S. & McGeoch, D. J. (1986). Novel herpes simplex virus type 1 glycoproteins identified by antiserum against a synthetic oligopeptide from the predicted product of gene US4. *Journal of General Virology*, **67**, 745–51.

Friedman, H. M., Cohen, G. H., Eisenberg, R. J., Seidel, C. A. & Cines, D. B. (1984). Glycoprotein C of herpes simplex virus 1 acts as a receptor for the C3b complement component on infected cells. *Nature*, **309**, 633–5.

Frink, R. J., Eisenberg, R., Cohen, G. & Wagner, E. K. (1983). Detailed analysis of the portion of the herpes simplex virus type 1 genome encoding glycoprotein C. *Journal of Virology*, **45**, 634–47.

FULLER, A. O. & SPEAR, P. G. (1985). Specificities of monoclonal and polyclonal antibodies that inhibit adsorption of herpes simplex virus to cells and lack of inhibition by potent neutralising antibodies. *Journal of Virology*, **55**, 475–82.

GLORIOSO, J., SCHRÖDER, C. H., KÜMEL, G., SZCZESIUL, M. & LEVINE, M. (1984). Immunogenicity of herpes simplex virus glycoproteins gC and gB and their role in protective immunity. *Journal of Virology*, **50**, 805–12.

GOMPELS, U. & MINSON, A. (1986). The properties and sequence of glycoprotein H of herpes simplex virus type 1. *Virology*, **153**, 230–47.

HAFFEY, M. L. & SPEAR, P. G. (1980). Alterations in glycoprotein gB specified by mutants and their partial revertants in herpes simplex virus type 1 and relationship to other mutant phenotypes. *Journal of Virology*, **35**, 114–28.

HALLIBURTON, I. W., MORSE, L. S., ROIZMAN, B. & QUINN, K. E. (1980). Mapping of the thymidine kinase genes of type 1 and type 2 herpes simplex virus using intertypic recombinants. *Journal of General Virology*, **49**, 235–53.

HEINE, J. W., HONESS, R. W., CASSAI, E. & ROIZMAN, B. (1974). Proteins specified by herpes simplex virus. XII. The virion polypeptides of type 1 strains. *Journal of Virology*, **14**, 640–51.

HILL, T. J. (1985). Herpes simplex virus latency. In: *The Herpesviruses*, Vol. 3, B. Roizman, pp. 175–240. New York: Plenum Press.

HOGGAN, M. D. & ROIZMAN, B. (1959). The isolation and properties of a variant of herpes simplex producing multinucleated giant cells in monolayer culture in the presence of antibody. *American Journal of Hygiene*, **70**, 208–19.

HOLLAND, T. C., HOMA, F. L., MARLIN, S. D., LEVINE, M. & GLORIOSO, J. (1984). Herpes simplex virus type 1 glycoprotein C-negative mutants exhibit multiple phenotypes, including secretion of truncated glycoproteins. *Journal of Virology*, **52**, 566–74.

HOLLAND, T. C., MARLIN, S. D., LEVINE, M. & GLORIOSO, J. (1983). Antigenic variants of herpes simplex virus selected with glycoprotein-specific monoclonal antibodies. *Journal of Virology* **45**, 672–82.

HONESS, R. W. (1984). Herpes simplex and 'the herpes complex' diverse observations and a unifying hypothesis. *Journal of General Virology*, **65**, 2077–107.

HONESS, R. W., BUCHAN, A., HALLIBURTON, I. W. & WATSON, D. H. (1980). Recombination and linkage between structural and regulatory genes of herpes simplex virus type 1: Study of the functional organisation of the genome. *Journal of Virology*, **34**, 716–42.

HONESS, R. W. & WATSON, D. H. (1974). Herpes simplex virus-specific polypeptides studied by polyacrylamide gel electrophoresis of immune precipitates. *Journal of General Virology*, **22**, 171–85.

HOPE, R. G. & MARSDEN, H. S. (1983). Processing of glycoproteins induced by herpes simplex virus type 1: Sulphation and nature of the oligosaccharide linkages. *Journal of General Virology*, **64**, 1943–53.

HOPE, R. G., PALFREYMAN, J. W., SUH, M. & MARSDEN, H. S. (1982). Sulphated glycoproteins induced by herpes simplex virus. *Journal of General Virology*, **58**, 399–415.

JOHANSSON, P. J. H., HALLBERG, T., OXELIUS, V-A., GRUBB, A. & BLOMBERG, J. (1984). Human immunoglobulin class and subclass specificity of Fc receptors induced by herpes simplex virus type 1. *Journal of Virology* **50**, 796–804.

JOHNSON, D. C., McDERMOTT, M. R., CHRISP, C. & GLORIOSO, J. C. (1986). Pathogenicity in mice of herpes simplex virus type 2 mutants unable to express glycoprotein C. *Journal of General Virology*, **58**, 36–42.

JOHNSON, D. C. & SPEAR, P. G. (1983). O-linked oligosaccharides are acquired by herpes simplex virus glycoproteins in the golgi apparatus. *Cell* **32**, 987–97.

JOHNSON, D. C., WITTELS, M. & SPEAR, P. G. (1984). Binding to cells of virosomes containing herpes simplex virus type 1 glycoproteins and evidence for fusion. *Journal of Virology*, **52**, 238–47.

KAERNER, H. C., SCHRÖDER, C. H., OTT-HARTMANN, A., KÜMEL, G. & KIRCHNER, H. (1983). Genetic variability of herpes simplex virus: development of a pathogenic variant during passaging of a nonpathogenic herpes simplex virus type 1 virus strain in mouse brain. *Journal of Virology*, **46**, 83–93.

KAPOOR, A. K., NASH, A. A., WILDY, P., PHELAN, J., McLEAN, C. S. & FIELD, H. J. (1982). Pathogenesis of herpes simplex virus in congenitally athymic mice: the relative roles of cell-mediated and humoral immunity. *Journal of General Virology*, **60**, 225–33.

KELLER, R., PEITCHER, R., GOLDMAN, J. M. & GOLDMAN, M. (1976). An IgG-Fc receptor induced in cytomegalovirus infected human fibroblasts. *Journal of Immunology*, **116**, 772–7.

KITAMURA, K., NAMAZUE, J., CAMPO-VERA, H., OGINO, T. & YAMANISHI, K. (1986). Induction of neutralising antibody against varicella-zoster virus (VZV) by VZV gp3 and cross-reactivity between VZV gp3 and herpes simplex viruses gB. *Virology*, **149**, 74–82.

KOUSOULAS, K. G., BZIK, D. J., DeLUCA, N., & PERSON, S. (1983). The effect of ammonium chloride and tunicamycin on the glycoprotein content and infectivity of herpes simplex virus type 1. *Virology*, **125**, 468–74.

KÜMEL, G., KAERNER, H. C., LEVINE, M., SCHRÖDER, C. H. & GLORIOSO, J. C. (1985). Passive immune protection by herpes simplex virus-specific monoclonal antibodies and monoclonal antibody-resistant mutants altered in pathogenicity. *Journal of Virology*, **56**, 930–7.

LAWMAN, M. J. P., COURTNEY, R. J., EBERLE, R., SCHAFFER, P. A., O'HARA, M. K. & ROUSE, B. T. (1980). Cell-mediated immunity to herpes simplex virus: Specificity of cytotoxic T cells. *Infection and Immunity*, **30**, 451–61.

LEE, G. T-Y., PARA, M. F. & SPEAR, P. G. (1982a). Location of the structural genes for glycoproteins D and E and for other polypeptides in the S component of herpes simplex type 1 DNA. *Journal of Virology*, **43**, 41–9.

LEE, G. T-Y., POGUE-GEILE, K. L., PEREIRA, L. & SPEAR, P. G. (1982b). Expression of herpes simplex virus glycoprotein C from a DNA fragment inserted into the thymidine kinase gene of this virus. *Proceedings of the National Academy of Science (USA)*, **79**, 6612–16.

LITTLE, S. P., JOFRE, J. T., COURTNEY, R. J. & SCHAFFER, P. A. (1981). A virion associated glycoprotein essential for infectivity of herpes simplex virus type 1. *Virology*, **115**, 149–60.

LITTLE, S. P. & SCHAFFER, P. A. (1981). Expression of the syncytial (syn) phenotype in HSV-1, strain KOS: Genetic and phenotypic studies of mutants in two syn loci. *Virology*, **112**, 686–702.

LONG, D., MADARA, T. J., PONCE DE LEON, M., COHEN, G. H., MONTGOMERY, P. C. & EISENBERG, R. J. (1984). Glycoprotein D protects mice against lethal challenge with herpes simplex virus types 1 and 2. *Infection and Immunity*, **37**, 761–4.

LONGNECKER, R. & ROIZMAN, B. (1986). Generation of an inverting herpes simplex virus 1 mutant lacking the L-S junction *a* sequences, an origin of DNA synthesis, and several genes including those specifying glycoprotein E and the α47 gene. *Journal of Virology*, **58**, 583–91.

LOPEZ, C. (1985). Natural resistance mechanisms in herpes simplex virus infections. In: *The Herpesviruses*, Vol. 4, ed. B. Roizman & C. Lopez, pp. 1–36. New York: Plenum Press.

McGEOCH, D. J. (1985). On the predictive recognition of signal peptide sequences. *Virus Research*, **3**, 271–86.

McGEOCH, D. J. & DAVISON, A. J. (1986). DNA sequence of the herpes simplex virus type 1 gene encoding glycoprotein gH, and the identification of homologues in the genomes of varicella-zoster virus and Epstein–Barr virus. *Nucleic Acids Research*, **14**, 4281–92.

McGEOCH, D. J., DOLAN, A., DONALD, S. & RIXON, F. J. (1985). Sequence determination and genetic content of the short unique region in the genome of herpes simplex virus type-1. *Journal of Molecular Biology*, **181**, 1–13.

McGEOCH, D. J., MOSS, H. W. M., McNAB, D. & FRAME, M. C. (1987). DNA sequence and genetic content of the *Hind*III*l* region in the short unique component of the herpes simplex virus type 2 genome: identification of the gene encoding glycoprotein G, and evolutionary comparisons. *Journal of General Virology*, (in press).

MACHUCA, I., JACQUEMONT, B. & EPSTEIN, A. (1986). Multiple adjacent or overlapping loci affecting the level of gC and cell fusion mapped by intratypic recombinants of HSV-1. *Virology*, **150**, 117–25.

MANSERVIGI, R., SPEAR, P. G. & BUCHAN, A. (1977). Cell fusion induced by herpes simplex virus is promoted and suppressed by different viral glycoproteins. *Proceedings of the National Academy of Science (USA)*, **74**, 3913–17.

MARLIN, S. D., HOLLAND, T. C., LEVINE, M. & GLORIOSO, J. C. (1985). Epitopes of herpes simplex virus type 1 glycoprotein gC are clustered in two distinct antigenic sites. *Journal of Virology*, **53**, 128–36.

MARSDEN, H. S., BUCKMASTER, A., PALFREYMAN, J. W., HOPE, R. G. & MINSON, A. C. (1984). Characterisation of the 92000-dalton glycoprotein induced by herpes simplex virus type 2. *Journal of Virology*, **50**, 547–54.

MARSDEN, H. S., STOW, N. D., PRESTON, V. G., TIMBURY, M. C. & WILKIE, N. M. (1978). Physical mapping of herpes simplex virus induced polypeptides. *Journal of Virology*, **28**, 624–42.

MIMS, C. A. & WHITE, D. O. (1984). *Viral Pathogenesis and Immunology*. Oxford: Blackwell.

MINSON, A. C., HODGMAN, T. C., DIGARD, P., HANCOCK, D. C., BELL, S. E. & BUCKMASTER, E. A. (1986). An analysis of the biological properties of monoclonal antibodies against glycoprotein D of herpes simplex virus and identification of amino acid substitutions that confer resistance to neutralisation. *Journal of General Virology*, **67**, 1001–13.

NASH, A. A., LEUNG, K-N. & WILDY, P. (1985). The T-cell mediated immune response of mice to herpes simplex virus. In: *The Herpesviruses*, Vol. 4, ed. B. Roizman & C. Lopez, pp. 87–102. New York: Plenum Press.

NASH, A. A. & WILDY, P. (1983). Immunity in relation to the pathogenesis of herpes simplex virus. In: *Human Immunity to Viruses*, ed. F. A. Ennis, pp. 179–92. New York: Academic Press.

NOBLE, A. G., LEE, G. T-Y., & SPEAR, P. G. (1983). Anti-gD monoclonal antibodies inhibit cell fusion induced by herpes simplex virus type 1. *Virology*, **129**, 218–24.

NORRILD, B. (1985). Humoral response to herpes simplex virus infections. In: *The Herpesviruses*, Vol. 4, ed. B. Roizman & C. Lopez, pp. 69–86. New York: Plenum Press.

OGATA, M. & SHIGETA, S. (1979). Appearance of immunoglobulin G Fc receptors on cultured human cells infected with varicella-zoster virus. *Infection and Immunity*, 26, 770–74.

OLOFSSON, S., JEANSSON, S. & LYCKE, E. (1981). Unusual lectin-binding properties of a herpes simplex virus type 1-specific glycoprotein. *Journal of Virology*, **38**, 564–70.

OLOFSSON, S., LUNDSTRÖM M., MARSDEN, H., JEANSSON, S & VAHLNE, A. (1986). Characterisation of a herpes simplex type 2-specified glycoprotein with affinity for N-acetylgalactosamine-specific lectins and its identification as g92K or gG. *Journal of General Virology*, 737–44.

OLOFSSON, S., SJÖBLOM, I., LUNDSTRÖM, M., JEANSSON, S. & LYCKE, E. (1983). Glycoprotein C of herpes simplex virus type 1: characterisation of O-linked oligo-saccharides. *Journal of General Virology*, **64**, 2735–47.

PALFREYMAN, J. W., HAARR, L., CROSS, A., HOPE, R. G. & MARSDEN, H. S. (1983). Processing of herpes simplex type 1 glycoproteins: two-dimensional gel analysis using monoclonal antibodies. *Journal of General Virology*, **64**, 873–86.

PARA, M. F., BAUCKE, R. B. & SPEAR, P. G. (1980). Immunoglobulin G (Fc) binding receptors on virions of herpes simplex virus type 1 and transfer of these receptors to the cell surface by infection. *Journal of Virology*, **34**, 512–20.

PARA, M. F., BAUCKE, R. B. & SPEAR, P. G. (1982a). Glycoprotein gE of herpes simplex virus type 1: effect of anti-gE on virion infectivity and on virus-induced Fc-binding properties. *Journal of Virology*, **41**, 129–36.

PARA, M. F., GOLDSTEIN, L. & SPEAR, P. G. (1982b). Similarities and differences in the Fc-binding glycoprotein (gE) of herpes simplex virus types 1 and 2 and tentative mapping of the viral gene for this glycoprotein. *Journal of Virology*, **41**, 137–44.

PARA, M. F., PARISH, M. L., NOBLE, A. G. & SPEAR, P. G. (1985). Potent neutralis-ing activity associated with anti-glycoprotein D specificity among monoclonal antibodies selected for binding to herpes simplex virions. *Journal of Virology*, **55**, 483–8.

PARA, M. F., ZEZULAK, K. M., CONLEY, A. J., WEINBERGER, M., SNITZER, K. & SPEAR, P. G. (1983). Use of monoclonal antibodies against low 75 000 molecular weight glycoproteins specified by herpes simplex virus type 2 in glycoprotein identification and gene mapping. *Journal of Virology*, **45**, 1223–7.

PELLETT, P. E., BIGGIN, M. D., BARRELL, B. & ROIZMAN, B. (1985a). Epstein–Barr virus genome may encode a protein showing significant amino acid and predicted secondary structure homology with glycoprotein B of herpes simplex virus 1. *Journal of Virology*, **56**, 807–13.

PELLETT, P. E., KOUSOULAS, K. G., PEREIRA, L. & ROIZMAN, B. (1985b). Anatomy of the herpes simplex virus 1 strain F glycoprotein B gene: Primary sequence and predicted protein structure of the wild type and of monoclonal antibody-resistant mutants. *Journal of Virology* **53**, 243–53.

PEREIRA, L. (1985). Glycoproteins specified by human cytomegalovirus. In: *The Herpesviruses*, Vol. 3, ed. B. Roizman, pp. 383–404. New York: Plenum Press.

PEREIRA, L., DONDERO, D., NORRILD, B. & ROIZMAN, B. (1981). Differential immu-nologic reactivity and processing of glycoproteins gA and gB of herpes simplex virus types 1 and 2 made in Vero and HEp-2 cells. *Proceedings of the National Academy of Science (USA)*, **78**, 5202–6.

PEREIRA, L., KLASSEN, T. & BARRINGER, R. J. (1980). Type-common and type-specific monoclonal antibodies to herpes simplex virus type 1. *Infection and Immu-nity*, **29**, 724–32.

PERLMAN, D. & HALVORSON, H. O. (1983). A putative signal peptidase recognition and sequence in prokaryotic signal peptides. *Journal of Molecular Biology*, **167**, 391–409.

POGUE-GEILE, K. L., LEE, G. T-Y., SHAPIRA, S. K. & SPEAR, P. G. (1984). Fine mapping of mutations in the fusion-inducing MP strain of herpes simplex virus type 1. *Virology*, **136**, 100–9.

POWELL, K. L., BUCHAN, A., SIM, C. & WATSON, D. H. (1974). Type-specific protein in herpes simplex virus envelope reacts with neutralising antibody. *Nature*, **249**, 360–1.

RAHMAN, A. A., TESCHNER, M., SETHI, K. K. & BRANDIS, H. (1976). Appearance of IgG (Fc) receptor(s) on cultured human fibroblasts infected with human cytomegalovirus. *Journal of Immunology*, **117**, 253–8.

REA, T. J., TIMMINS, J. G., LONG, G. W. & POST, L. E. (1985). Mapping and sequencing of the gene for the pseudorabies virus glycoprotein which accumulates in the medium of infected cells. *Journal of Virology*, **54**, 21–9.

RECTOR, J. T., LAUSCH, R. N. & OAKES, J. E. (1982). Use of monoclonal antibodies for analysis of antibody-dependent immunity to ocular herpes simplex virus type 1 infection. *Infection and Immunity*, **38**, 168–74.

RECTOR, J. T., LAUSCH, R. N. & OAKES, J. E. (1984). Identification of infected cell-specific monoclonal antibodies and their role in host resistance to ocular herpes simplex virus type 1 infection. *Journal of General Virology*, **65**, 657–61.

RICHMAN, D. D., BUCKMASTER, A., BELL, S., HODGMAN, C. & MINSON, A. C. (1986). Identification of a new glycoprotein of herpes simplex virus type 1 and genetic mapping of the gene that codes for it. *Journal of Virology*, **57**, 647–55.

ROBBINS, A. K., WATSON, R. J., WHEALY, M. E., HAYS, W. W. & ENQUIST, L. W. (1986). Characterisation of a pseudorabies virus glycoprotein gene with homology to herpes simplex virus type 1 and type 2 glycoprotein C. *Journal of Virology*, **58**, 339–47.

ROBERTS, P. L., DUNCAN, B. E., RAYBOULD, J. G. & WATSON, D. H. (1985). Purification of herpes simplex virus glycoproteins B and C using monoclonal antibodies and their ability to protect mice against lethal challenge. *Journal of General Virology*, **66**, 1073–85.

ROIZMAN, B., NORRILD, B., CHAN, C. & PEREIRA, L. (1984). Identification and preliminary mapping with monoclonal antibodies of a herpes simplex virus 2 glycoprotein lacking a known type-1 counterpart. *Virology*, **133**, 242–7.

RÖSEN, A., & DARAI, G. (1985). Mapping of the deletion in the genome of HSV-1 strain HFEM responsible for its avirulent phenotype. *Medical Microbiology and Immunology*, **173**, 329–43.

ROSEN, A., GELDERBLOM, H. & DARAI, G. (1985). Transduction of virulence in herpes simplex virus type 1 from a pathogenic to an apathogenic strain by a cloned viral DNA fragment. *Medical Microbiology and Immunology*, **173**, 257–78.

RUYECHAN, W. T., MORSE, L. S., KNIPE, D. M. & ROIZMAN, B. (1979). Molecular genetics of herpes simplex virus. II. Mapping of the major viral glycoproteins and of the genetic loci specifying the social behaviour of infected cells. *Journal of Virology*, **29**, 677–87.

SANDERS, P. G., WILKIE, N. M. & DAVISON, A. J. (1982). Thymidine kinase deletion mutants of herpes simplex virus type 1. *Journal of General Virology*, **63**, 277–95.

SARMIENTO, M., HAFFEY, M. & SPEAR, P. G. (1979). Membrane proteins specified by herpes simplex virus. III. Role of glycoprotein VP7 (B2) in virion infectivity. *Journal of Virology*, **29**, 1149–58.

SCHRIER, R. D., PIZER, L. I. & MOORHEAD, J. W. (1983). Type-specific delayed hypersensitivity and protective immunity induced by isolated herpes simplex virus glycoprotein. *Journal of Immunology*, **130**, 1413–18.

SERAFINI-CESSI, F., MALAGOLINI, N., DALL'OLIO, F., PEREIRA, L. & CAMPADELLI-FIUME, G. (1985). Oliogsaccharide chains of herpes simplex virus type 2 glycoprotein gG.2. *Archives of Biochemistry and Biophysics,* **240**, 866–76.

SETHI, K. K., OMATA, Y. & SCHNEWEIS, K. E. (1983). Protection of mice from fatal herpes simplex virus type 1 infection by adoptive transfer of cloned virus-specific and H-2 restricted cytotoxic T Lymphocytes. *Journal of General Virology*, **64**, 443–7.

SHARPE, J. A., WAGNER, M. J. & SUMMERS, W. C. (1983). Transcription of herpes simplex virus genes *in vivo*: Overlap of a late promoter with the 3' end of the early thymidine kinase gene. *Journal of Virology*, **45**, 10–17.

SHOWALTER, S. D., ZWEIG, M. & HAMPAR, B. (1981). Monoclonal antibodies to herpes simplex virus type 1 proteins, including the immediate early protein ICP4. *Infection and Immunity*, **34**, 684–92.

SIMMONS, A. & NASH, A. A. (1985). Role of antibody in primary and recurrent herpes simplex virus infection. *Journal of Virology*, **53**, 944–8.

SNOWDEN, B. W., KINCHINGTON, P. R., POWELL, K. L. & HALLIBURTON, I. W. (1985) Antigenic and biochemical analysis of gB of herpes simplex virus type 1 and type 2 and of cross reacting glycoproteins induced by bovine mammillitis virus and equine herpesvirus type 1. *Journal of General Virology*, **66**, 231–47.

SPEAR, P. G. (1976). Membrane proteins specified by herpes simplex viruses. I. Identification of four glycoprotein precursors and their products in type-1 infected cells. *Journal of Virology*, **17**, 911–1008.

SPEAR, P. G. (1985). Glycoproteins specified by herpes simplex virus. In: *The Herpesviruses*, Vol. 3, ed. B. Roizman, pp. 315–56. New York: Plenum Press.

STEVENS, J. G. (1975). Latent herpes simplex virus and the nervous system. *Current Topics in Microbiology and Immunology*, **70**, 31–50.

SWAIN, M. A., PEET, R. W. & GALLOWAY, D. A. (1985). Characterisation of the gene encoding herpes simplex virus type 2 glycoprotein C and comparison with the type 1 counterpart. *Journal of Virology*, **53**, 561–9.

TENSER, R. B. & DUNSTAN, M. E. (1979). Herpes simplex virus thymidine kinase expression in infection of the trigeminal ganglion. *Virology*, **99**, 417–22.

TENSER, R. B., MILLER, R. L. & RAPP, F. (1979). Trigeminal ganglion infection of thymidine kinase-negative mutants of herpes simplex virus. *Science*, **205**, 915–18.

THOMPSON, R. L., COOK, M. L., DEVI-RAO, G. B., WAGNER, E. K. & STEVENS, J. G. (1986). Functional and molecular analyses of the avirulent wild-type herpes simplex virus type 1 strain KOS. *Journal of Virology*, **58**, 203–11.

THOMPSON, R. L., DEVI-RAO, G. B., STEVENS, J. G. & WAGNER, E. K. (1985). Rescue of a herpes simplex virus type 1 neurovirulence function with a cloned DNA fragment. *Journal of Virology*, **55**, 504–8.

THOMPSON, R. L. & STEVENS, J. G. (1983). Biological characterisation of a herpes simplex virus intertypic recombinant which is completely and specifically non-neurovirulent. *Virology*, **131**, 171–9.

THOMPSON, R. L., WAGNER, E. K. & STEVENS, J. G. (1983). Physical location of a herpes simplex virus type 1 gene function(s) specifically associated with a 10 million-fold increase in HSV neurovirulence. *Virology*, **131**, 180–92.

VAHLNE, A., SVENNERHOLM, B. & LYCKE, E. (1979). Evidence for herpes simplex virus type-selective receptors on cellular plasma membranes. *Journal of General Virology*, **44**, 217–25.

VAHLNE, A., SVENNERHOLM, B., SANDBERG, M., HAMBERGER, A. & LYCKE, E. (1980). Differences in attachment between herpes simplex type 1 and type 2 viruses to neurons and glial cells. *Infection and Immunity*, **28**, 675–80.

VESTERGAARD, B. F. & NORRILD, B. (1978). Crossed immunoelectrophoretic analysis and viral neutralising activity of five monospecific antisera against five different herpes simplex virus glycoproteins. In *Oncogenesis and Herpesviruses III*, part I, pp. 225–34. IARC Scientific Publications No. 24.

WATHEN, M. W. & WATHEN, L. M. K. (1986). Characterisation and mapping of a nonessential pseudorabies virus glycoprotein. *Journal of Virology*, **58**, 173–8.

WATSON, R. J. (1983). DNA Sequence of the herpes simplex virus type 2 glycoprotein D gene. *Gene*, **26**, 307–12.

WATSON, R. J., WEIS, J. H., SALSTROM, J. S. & ENQUIST, L. W. (1982). Herpes simplex virus type-1 glycoprotein D gene: nucleotide sequence and expression in *Escherichia coli. Science*, **218**, 381–4.

WATSON *et al.* (1982)

WATSON, D. H. & WILDY, P. (1969). The preparation of 'monoprecipitin' antisera to herpes virus specific antigens. *Journal of General Virology*, **4**, 163–8.

WEIS, J. H., ENQUIST, L. W., SALSTROM, J. S. & WATSON, R. J. (1983). An immunologically active chimeric protein containing herpes simplex virus type 1 glycoprotein D. *Nature*, **302**, 72–4.

WELLER, S. K., ASCHMAN, D. P., SACKS, W. R., COEN, D. M. & SCHAFFER, P. A. (1983). Genetic analysis of temperature-sensitive mutants of HSV-1: The combined use of complementation and physical mapping for cistron assignment. *Virology*, **130**, 290–305.

WESTMORELAND, D., ST JEOR, S. S. & RAPP, F. (1976). The development by cytomegalovirus infected cells of binding affinity for normal human immunoglobulin. *Journal of Immunology*, **116**, 1566–70.

ZEZULAK, K. M. & SPEAR, P. G. (1983). Characterisation of a herpes simplex virus type 2 75 000 molecular weight glycoprotein antigenically related to herpes simplex virus type 1 glycoprotein C. *Journal of Virology*, **7**, 553–62.

ZEZULAK, K. M. & SPEAR, P. G. (1984). Mapping of the structural gene for the herpes simplex virus type 2 counterpart of herpes simplex virus type 1 glycoprotein C and identification of a type 2 mutant which does not express this glycoprotein. *Journal of Virology*, **49**, 741–7.

ZWEIG, M., SHOWALTER, S. D., BLADEN, S. V., HEILMAN, J. C. (JR) & HAMPAR, B. (1983). Herpes simplex virus type 2 glycoprotein gF and type 1 glycoprotein gC have related antigenic determinants. *Journal of Virology*, **47**, 185–92.

BACULOVIRUS PATHOGENESIS

D. C. KELLY

Chemical Defence Establishment, Porton Down, Salisbury, Wiltshire

Baculoviruses cause disease exclusively in invertebrates especially insects. The disease caused is spectacular – on occasion resulting in the complete disruption of the internal organs of the animal. The high efficiency of baculoviruses in killing insect larvae has led to the use of the viruses as biological insecticides although there is no generally acclaimed use of such pathogens in agriculture or horticulure.

An understanding of the pathogenesis of baculoviruses at the molecular level should lead to the design of efficient microbial insecticides, and, in certain instances, antidotes to prevent baculovirus disease in commercially important insects (silkmoths and honey bees). Insects are the most abundant animal species and man's main competitor for natural resources including foodstuffs. Microbial insecticides which have defined host specificity and virulence are the ideal biotechnological insecticide product.

THE VIRUSES

Baculoviruses include nuclear polyhedrosis, granulosis, and the so-called non-occluded baculoviruses. The basic virus particle is rod-shaped (hence the name from the Latin *baculum* – a rod or stick) and comprises a rod-shaped nucleocapsid surrounded by an envelope and this is the basic infectious unit. Some baculoviruses possess these particles 'occluded' within a large proteinaceous crystal and the crystals are variously known as polyhedra, granules, whole inclusion bodies, or polyhedral inclusion bodies. Some baculoviruses appear never to be occluded and probably lack the genes required for the crystal protein and possibly the assembly process itself if this is virus-coded.

Virus particles contain one or more nucleocapsids and the packaging appears to genetically determined. The nucleocapsid contains the viral genome which is a single, double-stranded, circular DNA molecule of low superhelical density (Revet & Guelpa, 1979). Occasionally multiple concatenated forms of the genome occur. The viral DNA is about 40 μm long, approximately 160 kilobases, equivalent to about 80 nonoverlapping genes. The DNA is intimately associated with a basic DNA binding protein and this deoxyribonucleoprotein is enmeshed within a protein sheath (Tweeten et al., 1980; Kelly et al., 1983). The core is about 32 nm in width and about 350 nm long. The DNA and protein are uniformly distributed within the core. The protein sheath is about 4 nm thick and forms a discrete shell around the core. The shell or capsid comprises subunits, 4.5 nm in diameter, arranged in an open-stacked ring structure (Burley et al., 1982). At the ends of the nucleocapsids discrete structures are found which may be involved in the release of DNA from the nucleocapsid (Bud & Kelly, 1980b).

Two forms of virus particles exist with nuclear polyhedrosis and granulosis viruses. One form is packaged in polyhedra or granules within the cell nucleus whereas others are assembled at the plasma membrane while egressing from the cell. The polyhedron-derived virus (PDV) is structurally, biochemically and antigenically different to the other form – the cell-released virus (CRV) (Volkman et al., 1976; Volkman, 1983; Adams et al., 1977; Roberts, 1983; Wang & Kelly, 1985a, b). Structurally, peplomers are claimed to exist on CRV but not on PDV, but the evidence for this is poor (Summers & Volkman, 1976; Adams et al., 1977). Serologically the two forms of the virus are unrelated although the nucleocapsids contained within are structurally and antigenically identical. The major difference appears to be the possession of a major glycoprotein present in CRV and lacking in PDV which causes the differences in antigenicity, tissue tropism, and infectivity of the two virus forms. For cells in culture, CRV is 100 000 times more infectious than PDV (Volkman et al., 1976; Wang & Kelly, 1985a).

The occluded PDV form of the virus is presumably designed to interact with the proteinaceous crystal constituting the polyhedron or granule. The crystals are a face-centred cubic lattice of 4 nm, within which virus particles are orientated apparently at random with the protein lattice. The protein is a single glycoprotein of molecular weight 34 000 and this probably occurs as octomer repeats within the polyhedron. Polyhedra contain as many as 100 virus particles. Granules, on the other hand, rarely contain more than one. Fig. 1 shows an electron micrograph of a sectioned polyhedron. A

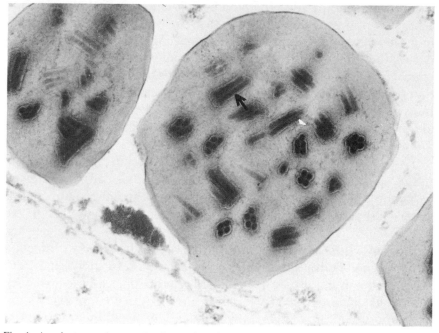

Fig. 1. An electron micrograph of a section through a polyhedron of *Spodoptera frugiperda* nuclear polyhedrosis virus showing virus particles (arrowed) containing multiple nucleocapsids occluded at random orientation within the cubic lattice of the polyhedron. Magnification 50 000 ×.

detailed review of the structure and physical characteristics of baculoviruses can be found elsewhere (Kelly, 1984).

The occluded form of the virus enables virus particles to survive from one generation of insects to another and also facilitates lateral spread of the virus in epidemics (Evans & Harrap, 1982). Polyhedra also provide protection against environmental factors. Non-occluded baculoviruses such as that from the rhinoceros beetle, which are not associated with crystals, survive amongst overlapping generations of insects.

THE DISEASE

Overt signs of baculovirus infection are seen mainly in the larval forms of insects although adults may sometimes be infected. The disease was known in the sixteenth century as a jaundice. Insects infected with a nuclear polyhedrosis virus will stop eating within 24 hours of infection and frequently they show signs of disorientation within that time period. Infected larvae later become sluggish and a reduction in growth occurs. As the disease progresses the integument may change colour and/or assume a glossy sheen. When this occurs the integument becomes fragile and the body contents will exude from ruptures in the integument. The exudate is predominantly virus polyhedra and frequently appears white in colour. In most lepidopterous larvae infected with a nuclear polyhedrosis virus, about five days elapse before the gross pathology is observed. Some insect larvae then hang head down to die.

Tissues affected include the fat body, hypodermis, tracheal matrix, muscle and nerve fibres, ganglia and pericardial cells. In infected hymenoptera the virus is limited to the gut and the pathogenesis differs in that virus polyhedra are sloughed out of the gut in damaged cells.

Progeny virus can be detected in larvae 24 hours after infection although polyhedra are found from 48 hours. An individual larva may contain as many as 10E9 polyhedra at death (Kelly *et al.*, 1978). Infectivity of virus for larvae diminishes as later larval instars are encountered. Evans (1981) showed that about 10 polyhedra comprise a lethal dose for first instar larvae whereas the sixth and final instar requires the massive dose of 10E5 polyhedra. Larvae infected in late instars which pupate, may emerge as crippled and paralysed adults chronically and systemically infected with virus. Mazzone (1985) has recently outlined the pathology of baculovirus infection in detail.

THE INFECTION CYCLE

With nuclear polyhedrosis and granulosis viruses, infection is initiated in the gut upon ingestion of polyhedra from contaminated leaf surfaces by the larvae. The gut of most leaf-eating insects is alkaline and at this pH the polyhedra are unstable and dissolve to release virus particles within the gut. This is facilitated in naturally occurring polyhedra by an alkaline protease activity present in the polyhedra. This enzyme may not be essential for dissolution since a simlar activity has not been found in polyhedra grown in insect cells in culture although the polyhedra are equally infectious (Zummer & Faulkner, 1979). The virus released from polyhedra is naturally PDV and is presumably designed to survive and be taken up at the alkaline pH of the insect gut.

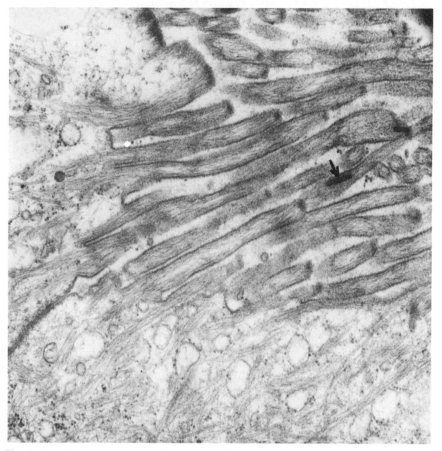

Fig. 2. An electron micrograph of *Aglais urticae* nuclear polyhedrosis virus nucleocapsids (arrowed) in gut microvilli. Magnification 60 000 ×.

Virus particles released from polyhedra are next found in association with microvilli of columnar cells in the insect midgut (Fig. 2). This is the only site where there is a free plasma membrane. The fore and hind gut, together with the larval exterior, are enclosed within a layer of cuticle impervious to virus invasion. Virus gains entry to the cell by fusion of the virus envelope with the microvillus plasma membrane (Summers, 1969; Harrap, 1970; Kawanishi *et al.*, 1972; Granados, 1978). The nucleocapsid is observed intact in the microvillus, in the cytoplasm peripheral to the nucleus, and aligned at right angles to nuclear pores (Summers, 1971). This is interpreted as uncoating at the nuclear membrane. Granados (1980) claims from a limited survey that granulosis viruses uncoat within the nucleus.

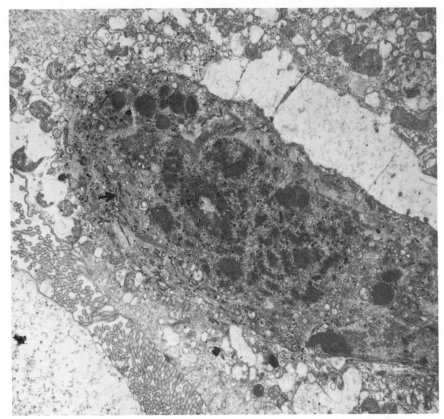

Fig. 3. An electron micrograph of *Aglais urticae* nuclear polyhedrosis virus replicating in the columnar midgut cells showing a virogenic stroma with associated nucleocapsids, virus particles (containing single nucleocapsids) (arrowed), and small polyhedra lacking virus particles. Magnification 10 000 ×.

Once within the midgut epithelium the virus replicates in the nucleus of these cells without forming substantial amounts of the polyhedra, although the polyhedron protein is synthesized. Enveloped virus accumulates at the basement membrane and appears within the hae- molymph – whereupon the haemocytes become the next focus of infection. Virus can be seen budding across the membrane of haemo- cytes (Hess & Falcon, 1981). The CRV interacts with other cells within the body at the neutral pH of haemolymph and does not interact with polyhedra intracellularly. Acquisition of the plasma membrane by nucleocapsids has been observed in columnar cells, haemocytes, and fat body cells (Robertson *et al.*, 1974; Hess & Fal- con, 1977). The cell-released virus appears to be solely concerned with cell-to-cell infection within the organism whereas transport of virus from insect to insect occurs primarily via polyhedra. Cell-released virus is, however, infectious orally (Volkman & Summers, 1977; Siobhan Clark & Kelly, unpublished) although, with the exception of non-occluded baculoviruses, there is no evidence that it plays an important role in lateral or vertical transmission. Figs 4 and 5 show fat body cells at late and very late stages of infection with an abundance of polyhedra within the cell nucleus.

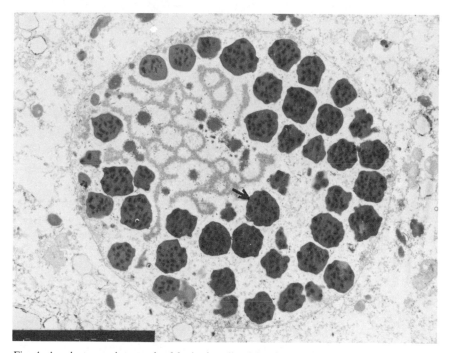

Fig. 4. An electron micrograph of fat body cells of *Spodoptera frugiperda* cells infected with a nuclear polyhedrosis virus showing a virogenic stroma surrounded by virus particles and numerous polyhedra (arrowed). Magnification 7000 ×.

CELL INFECTION

The replication of baculoviruses in cells in culture has been studied in depth on just one system – the *Trichoplusia ni* or closely related *Autographa californica* nuclear polyhedrosis virus in *Spodoptera frugiperda* or *T. ni* cells. Other viruses such as some of the Spodoptera viruses, *Bombyx mori* virus and the *Oryctes* non-occluded baculovirus have been studied to a limited extent.

CRV is highly infectious for cells in culture. Uptake of virus into cells is by cell fusion, and uncoating and transport of the viral genome

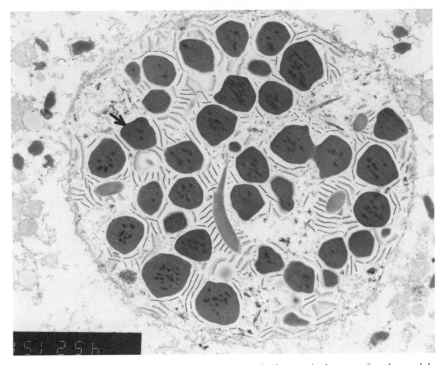

Fig. 5. An electron micrograph of degenerate nucleus in the terminal stages of nuclear polyhedrosis virus replication in the fat body of a *Spodoptera frugiperda* larva showing numerous
polyhedra (arrowed). Magnification 9000 ×.

to the cell nucleus is rapid (Wang & Kelly, 1985a). Volkman &
Goldsmith (1985) have interpreted their data to indicate that only
adsorptive endocytosis occurs with CRV. Their evidence rests solely
on inhibition by chloroquine. The virus nucleocapsids are not ordi-
narily infectious indicating that the envelope is essential for infection
(Kelly & Wang, 1981).

A glycopeptide, commonly referred to as the 64 K polypeptide
(64 Kp), appears to be central in the interaction of the CRV with
cells in culture (Volkman *et al.*, 1986; Volkman & Goldsmith 1985).
This peptide is not found in PDV (Kelly & Lescott, 1983).

Nuclear polyhedrosis virus DNA is infectious (Bud & Kelly, 1980a;
Burand *et al.*, 1980; Carstens *et al.*, 1980; Kelly & Wang, 1981).
The virus DNA must be circular to be infectious, although the presence of nicks in the covalently closed form does not diminish infectivity. The infectivity is enhanced by DNA binding proteins including
the major DNA-binding nucleocapsid protein (Kelly & Wang, 1981;
Kelly *et al.*, 1983). The host range of the virus DNA in cells in

culture reflects the true *in vivo* range, suggesting that cell transcription and replication factors are important in defining host range.

Replication of the virus occurs in the cell nucleus. A cascade induction of four, possibly five, phases of protein synthesis occurs (Wood, 1980; Kelly & Lescott, 1981). Two early phases precede virus DNA synthesis and the remainder accompany DNA synthesis. The two forms of the virus are formed at different subcellular locations. Virus destined to become occluded within polyhedra are formed within an apparently intact nucleus. Consequently it has been suggested that *de novo* membrane synthesis occurs within the nucleus. The mechanism of transport of membrane material to the virus assembly site in the nucleus is unknown. PDV becomes occluded within polyhedra and so the particles are withdrawn from the intracellular pool and await liberation when the polyhedra are ingested by a marauding insect. Cell-released virus (CRV), on the other hand, acquires its membrane by budding at the plasma membrane.

The morphogenesis of baculovirus assembly is complex. Initially, within a few hours, the cell chromatin marginates and a 'virogenic stroma' forms. This probably results from early protein synthesis since it is formed in cells in which viral DNA synthesis is blocked (Kelly, 1981). Initially, nucleocapsids appear at the periphery, and later within the virogenic stroma. Eventually some of these become enveloped within the nucleus. Virus is observed budding at the plasma membrane before occlusion with polyhedra and this presumably facilitates intercellular spread of virus.

The replication of baculoviruses has been a subject of reviews recently (Kelly 1982; Vaughn & Dougherty, 1985).

MOLECULAR ASPECTS OF INFECTION

No defined antibody response is known in invertebrates although there are a variety of non-specific antimicrobial substances present in the insect haemolymph. It is probably incorrect therefore to consider the interaction of virus with antibodies in the true context of pathogenesis. Nevertheless the interaction of insect viruses with antibodies generated in vertebrates provides a useful understanding of molecular aspects of pathogenesis. The two forms of virus (PDV and CRV) can be distinquished by a variety of serological techniques including neutralization. Both polyclonal and monoclonal antibodies have been used in this context. Antisera raised to nucleocapsids fail to neutralize virus infectivity whereas antisera to whole virus which has been preabsorbed with nucleocapsids does neutralize, confirming that the envelope is essential for virus infectivity. Monoclonal antibodies have confirmed that in the case of CRV the major envelope glycoprotein, 64 Kp, plays a crucial role in infectivity (Volkman & Goldsmith, 1985).

PDV infects gut cells at an alkaline pH. It can infect cells in culture inefficiently at neutral pH and this inefficiency may be due to the fact that this form of the virus accumulates in cell vacuoles and does not efficiently undergo further fusion with vacuole membranes to enable the nucleocapsids to enter the cell cytoplasm. This requirement for adaptation for alkaline pH fusion may serve to prevent effective spread of virus within other tissues of the larva and this localization may contribute to the disease process. The alkaline dependence has not been demonstrated *in vitro* (Wang & Kelly, 1985*a*). *In vitro* there appears to be little difference in the mode of entry of PDV and CRV – both forms requiring divalent cations in similar molar proportions. Uncoating also occurs in the cytoplasm not the nucleus and it may be that the electron microscope observations of Granados (1980) represent the exceptional, rather than the routine, mode of entry and uncoating.

HOST RANGE AND VIRULENCE

Viral insecticides ideally should have a defined host range and virulence so that their efficacy can accurately be predicted. Preliminary studies with infectious baculovirus DNA indicates that the host range of DNA matches that of live virus (Kelly & Wang, 1981; Burand *et al.*, 1980) implying that transcription and translation of the virus genome is important in defining host range rather than entry of virus into cells. However, now that recombinant baculoviruses are being produced by molecular techniques (Maeda *et al.*, 1985; Carbonell *et al.*, 1985; Smith *et al.*, 1983) it should be possible to interchange envelope proteins between different viruses and evaluate the effect on both host range and virulence.

One approach currently being considered to enhance nuclear polyhedrosis virus pathogenicity is to incorporate toxin genes into the virus genome. The genes for the *d* toxin of *Bacillus thuringiensis* have been cloned (Carlton & Gonzalez, 1985) and are available for insertion into the baculovirus genome. The toxin is insecticidal, mainly for lepidoptera (the main baculovirus reservoir) and diptera. When viruses containing the *B. thuringiensis* toxin gene become available it will be interesting to determine if the virus spread is curtailed, since it is possible that the insect is killed by the toxin before sufficient virus is produced to infect secondary contacts. A second generation of virulence modified insecticides will probably contain genes for various invertebrate bioregulators affecting behaviour, particularly foraging.

Although the thrust of baculovirus molecular biology centres on gene expression and replication, there is a requirement for antidotes to baculovirus disease in commercial invertebrate production (particularly crustacea and silkmoths). Baculoviruses induce enzymes concerned with DNA synthesis including DNA polymerase and possibly deoxypyrimidine kinase (Kelly, 1981; Miller *et al.*, 1981; Wang & Kelly, 1983). Other DNA viruses such as Herpes viruses can be controlled by antiviral drugs such as Acyclovir and bromovinyldeoxyuridine which are effective in directly or indirectly inhibiting virus-coded DNA polymerase activity. Preliminary evidence suggests that BVDU suppresses baculovirus disease in insect larvae (Kelly, 1984) as well as in cells in culture (Wang *et al.*, 1983).

CONCLUSIONS

The pathogenesis of baculoviruses is a neglected aspect of current baculovirology yet a true understanding of its molecular base is essential so that competent virus insecticides may be produced in quantity and with the quality desired. The current fashion amongst molecular virologists to develop baculoviruses as expression vectors for a variety of antigens and bioactive compounds should, however, provide a sound base for future investigations of the diseases caused by baculoviruses and insect viruses generally.

ACKNOWLEDGEMENTS

I thank Miss Margaret Arnold and Dr K. A. Harrap of the Natural Environment Research Council for the electron micrographs, and N. F. Moore for his comments on this manuscript.

REFERENCES

ADAMS, J. R., GOODWIN, R. H. & WILCOX, T. A. (1977). Electron microscopic investigations on invasion and replication of insect baculoviruses *in vivo* and *in vitro*. *Biologie Cellulaire*, **28**, 261–8.

BUD, H. M. & KELLY, D. C. (1980a). Nuclear polyhedrosis virus DNA is infectious. *Microbiologica*, **3**, 103–8.

BUD, H. M. & KELLY, D. C. (1980b). An electron microscope study of partially lysed baculovirus nucleocapsids: the intranucleocapsid packaging of viral DNA. *Journal of Ultrastructure Research*, **73**, 361–80.

BURAND, J. P., SUMMERS, M. D. & SMITH, G. E. (1980). Transfection with baculovirus DNA. *Virology*, **101**, 286–90.

BURLEY, S. K., MILLER, A., HARRAP, K. A. & KELLY, D. C. (1982). Structure of the baculovirus nucleocapsid. *Virology*, **120**, 433–40.

CARBONELL, L. F., KLOWDEN, M. J. & MILLER, L. K. (1985). Baculvirus mediated expression of bacterial genes in dipteran and mammalian cells. *Journal of Virology*, **56**, 153–60.

CARLTON, B. C. & GONSALEZ, J. M. (1985). The genetics and molecular biology of *Bacillus thurigiensis*. In *The Molecular Biology of the Bacilli*, vol. 2, ed. D. A. Dubnau, pp. 211–50. New York: Academic Press.

CARSTENS, E. B., TJIA, S. T. & DOEFLER, W. (1980). Infectious DNA from *Autographa californica* nuclear polyhedrosis virus. *Virology*, **101**, 311–12.

EPPSTEIN, D. A. & THOMA, J. A. (1975). Alkaline protease activity associated with the matrix protein of a virus infecting the cabbage looper. *Biochemical and Biophysical Research Communications*, **62**, 478–84.

EVANS, H. F. (1981). Quantitative assessment of the relationships between dosage and response of the nuclear polyhedrosis virus of *Mamestra brassicae*. *Journal of Invertebrate Pathology*, **37**, 101–9.

EVANS, H. F. & HARRAP, K. A. (1982). Persistence of insect viruses. In *Virus Persistence*, eds. B. W. J. Mahy, A. Minson & G. K. Darby, pp. 57–96. Cambridge: Cambridge University Press.

GRANADOS, R. R. (1978). Early events in baculovirus infection of *Heliothis zea*. *Virology*, **108**, 297–308.

GRANADOS, R. R. (1980). Infectivity and mode of action of baculoviruses. *Biotechnology and Bioengineering*, **22**, 1377–405.

GRANADOS, R. R. & LAWLER, K. A. (1981). *In vivo* pathway of infection of *Autographa californica* baculovirus invasion and infection. *Virology*, **108**, 297–308.

HARRAP, K. A. (1970). Cell infection by a nuclear polyhedrosis virus. *Virology*, **42**, 311–18.

HESS, R. T. & FALCON L. A. (1977). Observations on the interaction of baculoviruses with the plasma membrane. *Journal of General Virology*, **36**, 525–30.

HESS, R. T. & FALCON, L. A. (1981). Electron microscope observations of *Autographa californica* (Noctuidae) nuclear polyhedrosis virus replication in the midgut of the saltmarsh caterpillar *Estigme acrea* (Arctiidae). *Journal of Invertebrate Pathology*, **37**, 86–90.

KAWANISHI, C. Y., SUMMERS, M. D., STOLTZ, D. B., & ARNOTT, H. J. (1972). Entry of an insect virus *in vivo* by fusion of viral envelope and microvillus membrane. *Journal of Invertebrate Pathology*, **20**, 104–8.

KELLY, D. C. (1981a). Baculovirus replication: electron microscopy of the sequence of infection of *Trichoplusia ni* nuclear polyhedrosis virus in *Spodoptera frugiperda* cells. *Journal of General Virology*, **52**, 209–19.

KELLY, D. C. (1981b). Baculovirus replication: stimulation of thymidine kinase and DNA polymerase activities in *Spodoptera frugiperda* cells infected with *Trichoplusia ni* nuclear polyhedrosis virus. *Journal of General Virology*, **52**, 313–19.

KELLY, D. C. (1982). Baculovirus replication. *Journal of General Virology*, **63**, 1–13.

KELLY, D. C. (1984). Baculovirus replication: effects of inhibitors of macromolecular synthesis. *Journal of Antimicrobial Chemotherapy*, **14**, 43–55.

KELLY, D. C. (1985). The structure and physical characteristics of baculoviruses. In *Viral Insecticides for Biological Control*, ed. K. Maramorosch & K. E. Sherman, pp. 469–88. New York: Academic Press.

KELLY, D. C. & LESCOTT, T. (1981). Baculovirus replication: protein synthesis in *Spodoptera frugiperda* cells infected with *Trichoplusia ni* nuclear polyhedrosis virus. *Microbiologica*, **4**, 35–57.

KELLY, D. C. & LESCOTT, T. (1983). Baculovirus replication: glycosylation of polypeptides synthesised in *Trichoplusia ni* nuclear polyhedrosis virus infected cells and the effect of Tunicamycin. *Journal of General Virology*, **64**, 1915–26.

KELLY, D. C., EDWARDS, M.-L., EVANS, H. F. & ROBERTSON, J. S. (1978). The use of the enzyme linked immunosorbent assay to detect a nuclear polyhedrosis virus in *Heliothis armigera* larvae. *Journal of General Virology*, **40**, 465–9.

KELLY, D. C., BROWN, D. A., AYRES, M. D., ALLEN, C. J., & WALKER, I. O. (1983). Properties of the major nucleocapsid protein of *Heliothis zea* singly enveloped nuclear polyhedrosis virus. *Journal of General Virology*, **64**, 399–408.

KELLY, D. C. & WANG, X. (1981). The infectivity of nuclear polyhedrosis virus DNA. *Annales de Virologie*, **132E**, 247–59.

KISLEV, N., HARPAZ, I. & ZELCER, A. (1969). Electron microscope studies on hemocytes of the Egyptian bollworm infected with a nuclear polyhedrosis virus. *Journal of Invertebrate Pathology*, **14**, 25–257.

MAEDA, S., KAWAI, T., OBINATA, M., FUUIWARA, M., HORUICHI, T., SAEKI, Y., SATO, Y., & FURUSAWA, M. (1985). Production of human interferon in silkworms using a baculovirus vector. *Nature*, **315**, 592–5.

MAZZONE, H. M. (1985). Pathology associated with baculovirus infection. In *Viral insecticides for biological control*, ed K. A. Marmorosch & K. E. Sherman, pp. 81–120. New York: Academic Press.

MILLER, L. K., JEWELL, J. E., & BROWN, D. (1981). Baculovirus induction of a DNA polymerase. *Journal of Virology*, **40**, 305–8.

REVET, B. M. J. & GUELPA, B. (1979). The genome of a baculovirus infecting *Tipula paludosa* (Meig.) (Diptera): a high molecular weight closed circular DNA of zero superhelical density. *Virology*, **96**, 633–9.

ROBERTS, P. L. (1983). Neutralisation studies on *Autographa californica* nuclear polyhedrosis virus. *Archives of Virology*, **47**, 147–50.

RITTER, K. S., TANADA, Y., HESS, R. T. & OMI, E. M. (1982). Eclipse period of baculovirus infection in larvae of the Armyworm *Pseudaletia unipuncta*. *Journal of Invertebrate Pathology*, **39**, 203–9.

ROBERTSON, J. S., HARRAP, K. A. & LONGWORTH, J. F. (1974). Baculovirus morphogenesis: the acquisition of the virus envelope. *Journal of Invertebrate Pathology*, **23**, 248–51.

SMITH, G. E., SUMMERS, M. D. & FRASER, M. J. (1983). Production of human beta interferon in insect cells infected with a baculovirus expression vector. *Molecular and Cellular Biology*, **3**, 2156–65.

SUMMERS, M. D. (1969). Apparent *in vivo* pathway of granulosis virus invasion and infection. *Journal of Virology*, **4**, 188–90.

SUMMERS, M. D. (1971). Electron microscope observations on granulosis virus entry, uncoating, and replication processes during the infection of midgut cells of *Trichoplusia ni*. *Journal of Utrastructure Research*, **35**, 606–25.

SUMMERS, M. D. & VOLKMAN, L. E. (1976). Comparison of biophysical and morphological properties of occluded and extracellular non-occluded baculovirus from *in vivo* and *in vitro* host systems. *Journal of Virology*, **17**, 962–72.

TWEETEN, K. A., BULLA, L. A. & CONSIGLI, R. A. (1980). Characterisation of an extremely basic protein derived from granulosis virus nucleocapsids. *Journal of Virology*, **33**, 866–76.

VAUGHN, J. L. & DOUGHERTY, E. W. (1985). The replication of baculoviruses. In Viral Insecticides for Biological Control, eds. K. Maramorosch & K. E. Sherwood, pp. 569–633. New York: Academic Press.

VOLKMAN, L. E. (1983). Occluded and budded *Autographa californica* nuclear polyhedrosis virus: immunological relatedness of structural proteins. *Journal of Virology*, **46**, 221–9.

VOLKMAN, L. E. & GOLDSMITH, P. A. (1985). Mechanism of neutralisation of budded *Autographa californica* nuclear polyhedrosis virus by a monoclonal antibody: inhibition of entry by adsorptive endocytosis. *Virology*, **143**, 185–95.

VOLKMAN, L. E., GOLDSMITH, P. A., & HESS, R. T. (1986). Alternate pathway of entry of *Autographa californica* nuclear polyhedrosis virus: fusion at the plasma membrane. *Virology*, **148**, 288–97.

VOLKMAN, L. E. & SUMMERS, M. D. (1977). *Autographa californica* nuclear polyhedrosis virus: comparative infectivity of the occluded, alkali-liberated, and non-occluded forms. *Journal of Invertebrate Pathology*, **30**, 102–3.

VOLKMAN, L. E., SUMMERS, M. D., & HSIEH, C. H. (1976). Occluded and non-occluded nuclear polyhedrosis virus grown in *Trichoplusia ni*: compariative neutralisation, infectivity and *in vitro* growth studies. *Journal of Virology*, **19**, 820–32.

WANG, X. & KELLY, D. C. (1983). Baculovirus replication: purification and identification of the *Trichoplusia ni* nuclear polyhedrosis virus-induced DNA polymerase. *Journal of General Virology*, **64**, 2229–36.

WANG, X. & KELLY, D. C. (1985a) Baculovirus replication: uptake of *Trichoplusia ni* nuclear polyhedrosis virus particles by insect cells. *Journal of General Virology*, **66**, 541–50.

WANG, X. & KELLY, D. C. (1985b). Neutralisation of *Trichoplusia ni* nuclear polyhedrosis virus with antisera to two forms of the virus. *Microbiologica*, **8**. 141–49.

WANG, X., LESCOTT, T., DE CLERCQ, E., & KELLY, D. C. (1983). Baculovirus replication: inhibition of *Trichoplusia ni* multiple nuclear polyhedrosis virus by E-5-(2-bromovinyl)-2'-deoxyuridine. *Journal of General Virology*, **64**, 1221–7.

WOOD, H. A. (1980). *Autographa californica* nuclear polyhedrosis virus induced proteins in tissue culture. *Virology*, **102**, 21–7.

ZUMMER, M. & FAULKNER, P. (1979). Absence of protease in baculovirus polyhedra grown *in vitro*. *Journal of Invertebrate Pathology*, **33**, 383–4.

SYMPTOMS OF PLANT DISEASE MEDIATED BY PLANT VIRUS GENE PRODUCTS

LOUS VAN VLOTEN-DOTING

Department of Biochemistry, State University of Leiden, Leiden, The Netherlands

INTRODUCTION

Plant disease symptoms are known to be influenced by both host and pathogen gene products. The symptoms induced by viroids or by plant viruses, alone, or in combination with other viruses or with satellites, may vary from rapid death to hardly any disease signs. Between these two extremes, a wide variety of disease may be manifested. Since it is economically important to breed plants resistant, or tolerant, to pathogens, the genetics of the plant genes involved in these two processes have been studied extensively (Fraser, 1985). Plants reacting with a localized hypersensitive response to a pathogen are often as useful in agriculture as plants showing a real resistance to the pathogen. A well-known example of a plant gene responsible for such a hypersensitive reaction is the N gene of tobacco. Although the N gene has been relatively well studied, its gene product and the mechanism of its action are still unknown. Even less is known about plant genes involved in other disease symptoms.

Knowledge of virus genes involved in plant disease symptoms is also very limited. In the last few years molecular plant virology (Davies, 1985 and refs. therein), together with the cytopathological effects induced by plant viruses (Francki *et al.*, 1985 and refs. therein) have received considerable attention. Unfortunately, little attention has been given to the molecular basis of disease. A discussion about virus disease symptoms mediated by virus gene products therefore, could be extremely short, since we know so little about the virus genes and/or their products involved and even less about the molecular interaction between host- and virus-encoded products. On the other hand, a discussion summing up all papers describing changes in disease symptoms due to changes in virus genetic information could be very long. This article will avoid these extremes and describe a number of, somewhat speculative ideas about how disease symptoms are mediated by virus gene products. No attempt will be made to produce a comprehensive list of references, but each point will be illustrated by one or a few examples. For convenience, most examples are chosen from the author's own work with alfalfa mosaic virus (AlMV).

VIRUS PRODUCTS INVOLVED IN MOVEMENT THROUGH THE HOST

Visible disease symptoms will only appear in a plant after at least a few hundred cells have responded to the virus infection. This means that, even for the formation of a local lesion, both virus replication and transport of newly produced virus from cell to cell is required. Studies with a thermosensitive (ts) mutant of TMV (Nishiguchi et al., 1978) suggest that, for this virus, a 30 000 dalton protein is involved in cell-to-cell movement (Leonard & Zaitlin, 1982). For several multipartite viruses, analysis of incomplete infections in protoplasts has indicated that the larger RNA from the viruses with a bipartite genome and the combination of the two larger RNAs of viruses with a tripartite genome contain all information required for virus RNA replication (Lister, 1968; Goldbach et al., 1980; Robinson et al., 1980; Kibertsis et al., 1981; Nassuth et al., 1981). In plants inoculated with only the larger genome part of tobacco rattle virus (TRV), viral RNA replication and symptoms can be detected (Lister, 1968), indicating that this RNA contains the information for its own replication and transport. Apparently the replication and expression of RNA 1 of TRV is sufficient to induce the characteristic host response. In contrast to the situation with TRV it has been shown that plants inoculated with the larger RNA from the como- or the nepoviruses, or plants inoculated with a mixture of the two larger RNAs from the bromo-, the cucumo- or the ilarviruses* contained no detectable viral RNA and showed no symptoms. This suggests that some factor required for transport of the newly synthesized RNA from the initially infected cells to the neighbouring cells is located on the part of the genome which is missing in such infections. In all cases cited above this part of the genome is known to encode the coat protein(s) and at least one other protein. Since several virus isolates lacking (functional) coat proteins are able to spread from cell to cell it is unlikely that this transport function is provided by the coat protein itself. Therefore, it is generally assumed that, for the viruses with a tripartite genome, the 35 000

* Including AlMV, as proposed by van Vloten-Doting et al., 1981.

dalton protein encoded by the smallest genome part is involved in cell-to-cell transport.

Even when there is cell-to-cell movement, newly synthesized virus or RNA is often confined to a local lesion. For a large number of viruses, there are both isolates which can infect plants systemically and isolates which can only induce local lesions. Local lesions may be chlorotic or necrotic, and little is known about the factors which cause the virus to remain localized. We do not know whether confinement is due to triggering of a host response blocking virus transport, or whether it is due to the loss of a virus-coded function actively involved in virus transport beyond the first local lesion. The latter explanation seems to be more plausible since it is easy to obtain a local lesion isolate from a systemic isolate, while the reverse is a rare event (Van Vloten-Doting, 1985). For AlMV it has been shown by pseudorecombinant analysis (Dingjan-Versteegh, 1972; Roosien et al., 1983) that the virus information involved in systemic spread in leguminosa is located on RNA 2. This map position proves that the factor involved in systemic invasion differs from the factor involved in cell-to-cell transport, because the latter is encoded by RNA 3.

RELATIONSHIP BETWEEN VIRUS SYMPTOMS AND VIRUS CONCENTRATION

For a given virus strain–host species combination, a relationship exists between the concentration of the virus and the number or the intensity of the symptoms induced. It has been known for a long time (Holmes, 1929) that when the host reacts with the formation of a local lesion at the point of initial infection, the number of local lesions can be used to estimate the virus concentration of the inoculum. When using this test one should be aware of the fact that the relationship between virus concentration and local lesions is only linear over a short concentration range, and that the slope of the dilution curve increases with increasing genome parts (Matthews, 1981).

When the host becomes infected systemically, the severity of the disease symptoms will often reflect the virus content of the host. In some cases the initial reaction consists of a necrotic local response, and whether or not this response is followed by a systemic response depends on the number of primary infection loci. Figure 1 illustrates the effects of low versus normal inoculum of AlMV on tobacco

Fig. 1. Tobacco plants infected with a yellow strain of alfalfa mosaic virus (YSMV). The left-hand plant was inoculated at a low virus concentration. A limited number of primary infection centres appeared but this was not followed by a systemic infection. The right-hand plant was inoculated at a virus concentration approximately five times higher. Two weeks post-inoculation, virus symptoms are not only present on the primary infected leaves but also on the newly formed upper leaves.

plants. The mechanism(s) which prevent(s) low doses of virus giving rise to a systemic infection in such cases is unknown. This type of resistance may be related to the systemic acquired resistance observed after a localized necrotic response (Ross, 1961). A discussion about systemic acquired resistance is beyond the scope of this paper and the reader is referred to Fraser (1985).

Comparison of the disease symptoms induced in a particular host by different strains of a single virus species reveal an enormous variation in their type and intensity (Matthews, 1981; Bos, 1978). From strain to strain there is *no* correlation between the intensity of symptoms and the amount of virus present in the host. On the contrary, strains inducing no or only very mild symptoms can often be produced in high yields, e.g. for AlMV in tobacco, virus content may reach $2 \, mg \, g^{-1}$ of fresh leaf for mild strains, while virus concentrations in plants infected with AlMV strains inducing more virulent symptoms are often less than $0.02 \, mg \, g^{-1}$ fresh leaf. The lack of correlation between virus production and disease symptoms in these cases indicates that disease symptoms are not only due to competition between host and pathogen for metabolites, but must also result from interference in host functions by virus products.

MOLECULAR BASES OF MOSAIC, RINGSPOT AND STREAK PATTERNS

Unrelated viruses with very different types of genome structure and organization may induce similar or even identical symptoms. A particular pattern of symptoms is therefore not necessarily due to one particular molecular mechanism. On the contrary, it is highly probable that a certain type of symptom (especially necrosis) may be caused by any of a number of completely different molecular interactions.

Mosaic, ringspot, or streak symptoms, etc. can be produced by a large number of viruses among which are representatives of nearly all plant virus groups (Matthews, 1981; Bos, 1978). These types of symptoms are characterized by areas with clearly visible disease interspersed with areas which appear normal. This distribution of symptoms suggests that, in the leaf, susceptible areas alternate with resistant areas.

Detailed research has only been performed on the light green/dark green mosaic induced by tobacco mosaic virus (TMV) (reviewed by Matthews, 1981). It was indeed found that the light green areas were loaded with virus, while the dark green islands contained no or only minute amounts of virus. Somehow these islands had escaped the infection. The dark green islands are resistant to superinfection with the same, or with closely related, virus strains. This type of resistance, which is a form of cross-protection, is limited to related virus strains (Matthews, 1981) and differs fundamentally from the systemic acquired resistance, which is not virus-specific. The mechanism (or mechanisms) responsible for cross-protection are unknown. Recently Powell Abel et al. (1986) and Loesch-Fries and coworkers (1986) have shown that transgenic plants expressing the coat protein of TMV and AlMV, respectively, are protected against infection with several other TMV or AlMV strains. These two examples suggest that coat protein is involved in cross-protection, but this is probably not the case for all viruses. TRV RNA 1 (which does not encode the coat protein) can induce cross-protection, moreover it is well known that viroids (not encoding any protein) can also induce cross-protection (Niblett et al., 1978). The pattern of light and dark green patches in TMV-induced mosaic is not genetically determined, but depends on the stage of development of the leaves at the time of infection. Plants regenerated from these dark green islands are susceptible to infection with TMV, thus the resistance of cells in these areas to TMV infection is apparently the result of signals coming from the light green areas. The cellular population making up the susceptible light green areas is not genetically different from that is the resistant dark green areas.

Similarly, in the case of ringspot symptoms, the pattern of diseased to healthy tissue is not due to a genetically determined difference in susceptibility of host cells. The site of entry of the infecting virus determines the location of the first spot, and the growing conditions determine the distances between the (necrotic) rings.

In monocotyledonous plants, virus infection often results in stripes or streaks. The development of stripe disease in these plants follows a similar general pattern to that found in mosaic diseases in dicotyledonous plants (Matthews, 1981).

All these disease patterns suggest that virus replication in the initially infected cell leads to the induction of an antiviral state in neighbouring cells. The newly produced virus has to move beyond this circle or patch of resistant cells before it can initiate a second round of replication. This second round of virus replication induces in its turn resistance in the surrounding cells, etc. The resistant areas may be very small, e.g. the narrow rings of 'healthy' material surrounding infection centres, or very large, e.g. parts of leaves in a mosaic disease. The reproducible variation in size of these resistant areas is an intriguing phenomenon for which there is presently no explanation.

Since variations in the type of symptoms are determined by variations in virus information (mutants), virus gene products must play a role in the level of resistance induced.

VIRUS PRODUCTS AFFECTING HORMONE BALANCE

The title of this section is not to imply that hormone production in healthy and virus infected material has been intensively studied. Unfortunately, there is little information on this topic. However, plant virus disease symptoms are sometimes strikingly similar to growth abnormalities induced by hormones.

The best-known example of this is shoestring disease, which is characterized by an almost complete suppression of mesophyll development. Such plants bear a striking resemblance to plants treated with 2,4-dichlorophenoxyacetic acid (2,4-D). Another example is leaf-curl disease, characterized by growth retardation of the veins. Again virus infection and hormone can induce similar phenotypes (Matthews, 1981).

Since plant viruses contain only a very limited amount of genetic information (varying from four to ten proteins, Davies & Hull, 1982; Van Vloten-Doting & Neeleman, 1982; Davies, 1985), it is unlikely that viruses upset the hormone balance by producing enzymes involved in the hormone metabolism as is done by *Agrobacterium tumefaciens* upon induction of gall formation. The effect of virus on the hormone balance, if any, is probably much more indirect. In this context, it is noteworthy to bear in mind that different mutants of a virus may induce different symptoms. Apparently small variations in virus gene products may lead to a completely different response of the plant (Fig. 2).

VIRUS PRODUCTS AFFECTING PLANT DEVELOPMENT

There are numerous reports concerning the effect of virus infection on plant development (Bos, 1978; Matthews, 1981, and refs. therein). Often development is retarded as in dwarfing, stunting and witch broom diseases, etc. and this is obvious upon cursory inspection of the plant.

However, there are (a very few) cases where virus infection may lead to a healthy-looking plant but one which shows a morphology different from that of uninfected plants. A very fascinating example is the appearance of toothed leaves on tobacco plants infected with tobacco streak virus (TSV) (Fulton, 1972) (Fig. 3). Apparently, a gene product of TSV has (via several intermediates?) affected the leaf development programme such that toothed leaves are formed instead of smooth-edged leaves.

MAPPING OF VIRUS GENES INVOLVED IN SYMPTOMS FORMATION

For viruses with a monopartite genome, mapping of the genes involved in symptom formation is difficult and only for TMV is information available. Kado & Knight (1966) used a combination of protein stripping and nucleic acid mutagenesis to map the position of the local lesion gene. Recently, this approach has been refined by

Fig. 2. Tobacco plants inoculated with a mild strain of alfalfa mosaic (AMV 425) and three mutants thereof. A. Plant infected with the wild type, showing faint chlorotic spots (not visible on the photograph) on primary and secondary leaves. B., C. and D. Plants infected with three different mutants showing B. necrotic spots on the primary infected leaves as well as on the upper leaves, C. necrotic spots on the primary leaves only, D. systemic necrosis and reduction of vein elongation.

Mundry & coworkers. They found that when the RNA had been stripped for 83% or more from the 5' terminus, the local lesion gene became more susceptible to HNO_2, suggesting that mutagenesis of the 30 K protein is involved in the production of the local lesion variants of TMV (Nitschko *et al.*, submitted for publication).

For viruses with a multipartite genome, analysis of pseudorecombinants has resulted in mapping of the information involved in symptom formation, lesion size and lesion colour on several host plants. These data have been reviewed extensively by Bruening (1977) and

Fig. 3. Toothed leaves induced in tobacco by infection with tobacco streak virus (TSV).

by Van Vloten-Doting & Jaspars (1977). In a large number of pseudorecombinants it has been found that a certain symptom type is determined by just one RNA species, indicating that this viral RNA encodes a gene product responsible for its induction. In other cases it has been found that information present on two or even three viral RNAs (e.g. cucumber mosaic virus (CMV), Rao & Francki, 1982) contributed to the induction of a particular type of symptoms, indicating that in these cases the induction of symptoms is multifactorial. The symptoms are due to the summation of two or more separate interactions between the host and the virus genes products. Unfortunately, the molecular basis of symptom formation is unknown and these data are therefore difficult to interpret.

VIRUS DISEASE SYMPTOMS MODIFIED BY SATELLITES

Viruses belonging to very different groups may carry satellites. The first satellite described in plant viruses was satellite virus (STNV) of tobacco necrosis virus (TNV). Although STNV encodes its own capsid protein, it is fully dependent on TNV for its replication. The presence of STNV in TNV infections reduces both the amount of TNV as well as the size of the necrotic lesion (Jones & Reichman, 1973).

Nepoviruses contain one of two different types of satellite RNA: a small one of 1×10^5 dalton or a larger one of about 5×10^5 dalton. The small RNA which has been found in tobacco ringspot virus (TRSV) does not display any messenger activity in *in vitro* cell-free translation systems. For its replication it is completely dependent on TRSV. Replication of satellite RNA interferes with replication of the helper virus (Schneider, 1971). The larger RNA 5×10^5 dalton has been found in several isolates of tomato blackring. This RNA is translated both *in vitro* and *in vivo* (protoplasts) into a protein of about 48 000 d; the function of this protein, if any, is unknown. The satellite has little or no quantitative influence on the type of symptoms produced, its only effect seems to be to decrease the number of local lesions produced by the helper virus (Fritsch *et al.*, 1978).

The most complicated situation is found with the satellite RNAs present in several cucumoviruses. These satellites have been called SATRNA (*satellite RNA*) (Mossop & Francki, 1978) or CARNA 5 (which stands for *c*ucumber mosaic virus *a*ssociated RNA 5) (Kaper *et al.*, 1976). There are a number of strains of CARNA 5 known, all of which are fully dependent upon mosaic virus for their replication. The biological effect of CARNA 5 depends both on the strain of CARNA 5 and on the host infected. In tobacco most CARNA 5 isolates are replicated to a high concentration, and the yield of CMV as well as the disease symptoms are significantly depressed. The attenuation of CMV by satellite is observed in a number of hosts and it is so marked that preinoculation of plants with CMV containing satellite is used as cross protection against infection with CMV itself. In contrast to these 'beneficial' CARNA 5s, some isolates have been described which induce, in combination with CMV, top necrosis in infected plants (Fig. 4).

DISEASE SYMPTOMS CAUSED BY INTERACTION OF A PATHOGENIC RNA WITH THE HOST

The previous sections did not address the question of whether the molecules directly responsible for disease symptoms of a particular pathogen are nucleic acids or proteins. Generally it is assumed that virus-coded proteins are involved and in several cases this is

Fig. 4. Effect of CARNA 5 on disease symptoms induced by cucumber mosaic virus (CMV) on tomato (left) and pepper (right). Upper row of plants was inoculated with CMV clones. Lower row with CMV containing CARNA 5. Photograph reproduced with permission from Kaper and Tousignant (1984).

very likely; for instance, the proteins involved in cell-to-cell movement and in systemic spreading of the virus as discussed above.
However, the possibility that pathogenic RNAs interfere directly
with host functions should not be overlooked. That RNA itself –
in the absence of protein – can cause all manner of symptoms, is
clearly illustrated by the effect of viroids, which do not encode any
proteins (Fig. 5).

PROSPECTS

Recently, two new methods for the analysis of the effects of virus
gene products on disease symptoms have become available: site-
directed mutagenesis of viral information and construction of transgenic plants expressing only one or a few viral genes. Both methods
are based on recombinant DNA technology.

Using site-directed mutagenesis, defined changes may be introduced at predetermined locations in the genome of DNA viruses
or in cloned cDNAs of the genome of RNA viruses. Because plant
viral cDNAs, in contrast to viroid cDNAs, are not infectious themselves, the cDNA must be transcribed *in vitro* into RNA (Ahlquist
et al., 1984). This approach to studying the determinants of disease
symptoms has two limitations:

1. The potential number of base sequence changes is enormous and
   it will be difficult to find non-lethal variations (Owens *et al.*, 1986)
   showing a clear-cut effect on symptom development. We are try-
   ing to circumvent this problem by using base changes found in
   'classically' isolated mutants to locate interesting areas.
2. Since replication of RNA viruses is error-prone (Holland *et al.*,
   1982; Van Vloten-Doting *et al.*, 1985) and RNA genomes show
   more recombination than previously thought (Bujarski *et al.*,
   1986), it may be necessary to prove that the effect observed is
   due to the mutation introduced rather than to a spontaneous
   mutation. This may mean that the complete base sequence of
   mutants will have to be checked.

In the second approach, the viral DNA, or the cloned DNA copy
equipped with start and stop signals, is inserted into a plasmid (in
most cases derived from the Ti plasmid of *Agrobacterium tumefaciens*) and transferred to plant cells. Upon regeneration the transformed cells will give rise to transgenic plants. Depending on the
genomic part inserted, these plants may show symptoms. In the cases

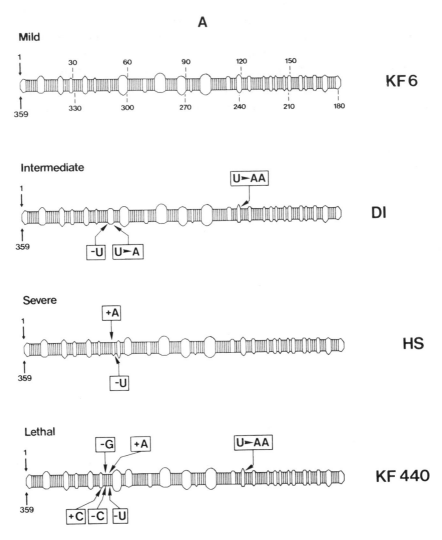

where viral gene(s) can induce symptoms in the absence of virus replication these genes can be analysed in more detail by site-directed mutagenesis. This approach faces the same problem about the large number of choices in base substitutions discussed above, but, since the information is kept at the DNA level, it avoids the problem created by the high intrinsic variability of RNA. A disadvantage of this approach is that, in transgenic plants, in principle all cells may express the viral information, while disease symptoms observed after virus infection are the result of virus replication in some cells

Control   KF6   DI   HS KF440

Fig. 5. Nucleotide sequence A. and pathogenic effect B. of several isolates of potato spindle tuber viroid (PSTV). Photograph reproduced with permission from Schnölzer *et al.* (1985).

and host reaction occurring in infected and non-infected cells.

In this context it is very interesting that Rogers *et al.* (1985*a*) reported that transgenic plants expressing one of the two parts of the genome of the white-fly-transmitted geminivirus, tomato golden mosaic virus, appeared phenotypically normal. However, when plants containing one part of the genome were crossed with plants containing the complementing part of the viral genome, one-quarter of the progeny showed phenotypic expression of symptoms of viral infection (Rogers *et al.*, 1985*b*).

The type of experiments described above are presently being performed in several laboratories, including our own, and progress may be expected in the near future. However, in order to understand the molecular mechanism leading to the different type of symptoms, these studies need to be complemented by a detailed analysis of the interactions between virus- and host-encoded products.

REFERENCES

Ahlquist, P., French, R., Janda, M. & Loesch-Fries, L. S. (1984). Multicomponent RNA plant virus infection derived from cloned viral cDNA. *Proceedings of the National Academy of Sciences, USA*, **81**, 7066–70.

Bos, L. (1978). *Symptoms of virus disease in plants*. 3rd edn. Wageningen: Pudoc.

Bruening, G. (1977). Plant covirus systems: Two component systems. In *Comprehensive Virology*, vol. II, ed. H. Fraenkel-Conrat & R. Wagner, pp. 55–141. New York: Plenum Press.

Bujarski, J. J. & Kaesberg, P. (1986). Genetic recombination between RNA components of a multipartite plant virus. *Nature*, **321**, 520–31.

Davies, J. W. (1985). *Molecular Plant Virology*, vol. 1 & 2, Boca Raton, Florida: CRC Press.

Davies, J. W. & Hull, R. (1982). Genome expression of positive strand RNA viruses. *Journal of General Virology*, **61**, 1–19.

Dingjan-Versteegh, A., Van Vloten-Doting, L. & Jaspars, E. M. J. (1972). Alfalfa mosaic virus hybrids constructed by exchange of nucleoprotein components. *Virology*, **49**, 716–22.

Francki, R. I. B., Milne, R. G. & Hatta, T. (1985). *Atlas of Plant Viruses*, vol. 1 & 2, Boca Raton, Florida: CRC Press.

Fraser, R. S. S. (1985). *Mechanism of Resistance to Plant Diseases*. Martinus Nijhoff/Dr W. Dordrecht–Boston–Lancaster: Junk Publishers.

Fritsch, C., Mayo, M. A. & Murant, A. F. (1978). Translation of the satellite RNA of tomato blackring virus *in vitro* and in tobacco protoplasts. *Journal of General Virology*, **40**, 587–93.

Fulton, R. W. (1972). Inheritance and recombination of strain-specific characters in tobacco streak virus. *Virology*, **50**, 810–20.

Gardner, R. C., Chonoles, K. R. & Owens, R. A. (1986). Potato spindle tuber viroid infections mediated by the Ti plasmid of *Agrobacterium tumefaciens*. *Plant Molecular Biology*, **6**, 221–8.

Goldbach, R., Rezelman, G. & Van Kammen, A. (1980). Independent replication and expression of B-component RNA of cowpea mosaic virus. *Nature*, **286**, 297–300.

Holland, J., Spindler, K., Horodyski, F., Grabau, E., Nichol, S. & Van der Pol, S. (1982). Rapid evolution of RNA genomes. *Science*, **215**, 1577–85.

Holmes, F. O. (1929). Local lesions in tobacco mosaic. *Botanical Gazette*, (Chicago), **87**, 39–55.

Jones, I. M. & Reichmann, M. E. (1973). The proteins synthesized in tobacco leaves infected with tobacco necrosis virus and satellite tobacco necrosis virus. *Virology*, **52**, 49–56.

Kado, C. I. & Knight, C. A. (1966). Location of a local lesion gene in tobacco mosaic virus RNA. *Proceedings of the National Academy of Sciences, USA*, **55**, 1276–83.

Kaper, J. M., Tousignant, M. E. & Lot, H. (1976). A low molecular weight replicating RNA associates with a divided genome plant virus: Defective or satellite RNA? *Biochemical and Biophysical Research Communications*, **72**, 1237–43.

Kaper, J. M. & Tousignant, M. E. (1984). Viral satellites: parasitic nucleic acids capable of modulating disease expression. *Endeavour New Series*, vol. 8, 194–200.

Kiberstis, P., Loesch-Fries, L. S. & Hall, T. C. (1981). Viral protein synthesis in barley protoplasts infected with native and fractionated brome mosaic virus RNA. *Virology*, **112**, 804–8.

Leonard, D. A. & Zaitlin, M. (1982). A temperature-sensitive strain of tobacco mosaic virus defective in cell to cell movement generates an altered viral-coded protein. *Virology*, **117**, 416–26.

Lister, R. M. (1968). Functional relationship between virus-specific products of infection by viruses of the tobacco rattle type. *Journal of General Virology*, **2**, 43–58.

LOESCH-FRIES, L. S., HALK, E., MERLO, D., JARVIS, N., NELSON, S., KRAHN, K. & BURHOP, L. (1986). Expression of alfalfa mosaic virus coat protein gene and anti-sense cDNA in transformed tobacco tissue. *UCLA Symposia Proceedings on Molecular and Cellular Biology*, in press.

MATTHEWS, R. E. F. (1981). *Plant Virology*. 2nd edn., New York, London, Toronto, Sydney, San Fransisco: Academic Press.

MOSSOP, D. W. & FRANCKI, R. I. B. (1978). Survival of a satellite RNA *in vivo* and its dependence on cucumber mosaic virus for replication. *Virology*, **86**, 562–6.

NASSUTH, A., ALBLAS, F. & BOL, J. F. (1981). Localization of genetic information involved in the replication of alfalfa mosaic virus. *Journal of General Virology*, **53**, 207–14.

NIBLETT, C. L., DICKSON, E., FERNOW, K. H., HORST, R. K. & ZAITLIN, M. (1978). Cross protection among four viroids. *Virology*, **91**, 198–203.

NISHIGUCHI, M., MOTOYOSHI, F. & OSHIMA, N. (1978). Behaviour of a temperature-sensitive strain of tobacco mosaic virus in tomato leaves and protoplasts. *Journal of General Virology*, **39**, 53–61.

OWENS, R. A., HAMMOND, R. W., GARDNER, R. C., KIEFER, M. C., THOMPSON, S. M. & CRESS, D. E. (1986). Site specific mutagenesis of potato spindle tubes viroid cDNA. Alterations within premelting region 2 that abolish infectivity. *Plant Molecular Biology*, **6**, 179–92.

POWELL ABEL, P., NELSON, R. S., DE, B., HOFFMANN, N., ROGERS, S. G., FRALEY, R. T. & BEACHY, R. N. (1986). Delay of disease development in transgenic plants that express the tobacco mosaic virus coat protein. *Science*, **232**, 738–43.

RAO, A. L. N. & FRANCKI, R. I. B. (1982). Distribution of determinants for symptom production and host range on the three RNA components of cucumber mosaic virus. *Journal of General Virology*, **61**, 197–205.

ROBINSON, D. J., BARKER, H., HARRISON B. D. & MAYO, M. A. (1980). Replication of RNA 1 of tomato black ring virus independently of RNA 2. *Journal of General Virology*, **51**, 317–26.

ROGERS, S., HORSCH, R., HOFFMANN, N., BRAND, L., CHANG, I., SUNTER, G. & BISARO, D. (1985a). Stable integration of geminivirus genomes into Petunia plants. (Abstract), *First International Congress on Plant Molecular Biology*, Savannah, Ga.

ROGERS, D. G., BISARO, D. M., SUNTER, G., HORSCH, R. B., FRALEY, R. T., BRAND, L. A. & ELMER, J. S. (1985b). Introduction and expression of integrated copies of geminivirus DNAs in transformed plants. (Abstract), *EMBO Workshop on Plant DNA Infectious Agents*, Rorschacherberg, Switzerland.

ROOSIEN, J., SARACHU, A. N., ALBLAS, F. & VAN VLOTEN-DOTING, L. (1983). An alfalfa mosaic virus RNA 2 mutant, which does not induce a hypersensitive reaction in cowpea plants, is multiplied to a high concentration in cowpea proto-plasts. *Plant Molecular Biology*, **2**, 85–8.

ROSS, A. F. (1961). Systemic acquired resistance induced by localized virus infection in plants. *Virology*, **14**, 340–58.

SCHNEIDER, I. R. (1971). Characteristics of a satellite-like virus of tobacco ringspot virus. *Virology*, **45**, 108–22.

SCHNÖLZER, M., HAAS, B., RAMM, K., HOFMANN, H. & SÄNGER, H. L. (1985). Correlation between structure and pathogenicity of potato spindle tuber viroid (PSTV). *EMBO Journal*, **4**, 2181–90.

VAN VLOTEN-DOTING, L. (1985). *The Plant Viruses*, vol. 1, ed. R. I. B. Francki, Chapter 5, *Virus Genetics*. Plenum Publishing Corporation.

VAN VLOTEN-DOTING, L. & NEELEMAN, L. (1982). Translation of plant virus RNAs. In *Encyclopedia of Plant Physiology* New Series, vol. 14B, ed. B. Parthier & D. Boulter, pp. 337–67.

Van Vloten-Doting, L. & Jaspars, E. M. J. (1977). Plant covirus systems: three-component systems. In *Comprehensive Virology*, vol. 11, ed. H. Fraenkel-Conrat & R. R. Wagner, pp. 1–53, New York: Plenum Press.

Van Vloten-Doting, L., Bol, J. F. & Cornelissen, B. J. C. (1985). Plant-virus-based vectors for gene transfer will be of limited use because of the high error frequency during viral RNA synthesis. *Plant Molecular Biology*, **4**, 323–26.

PLANT VIRUS TRANSMISSION BY VECTORS: MECHANISMS AND CONSEQUENCES

B. D. HARRISON

Scottish Crop Research Institute, Invergowrie, Dundee DD2 5DA, UK

To survive in nature, a virus needs three attributes: the abilities to replicate in host cells, to pass from cell to cell of multicellular hosts and to spread from one host individual to another. In their passage from cell to cell and from plant to plant, plant viruses face the special problems posed by the rigid cellulose-containing cell walls, an outer cuticle which is impermeable to macromolecules and the immobility of their hosts. The two last of these problems are solved, for most plant viruses, by the involvement of mobile vectors that can breach plant cell walls in ways that enable them to acquire and deliver virus particles. These vectors are taxonomically diverse. Many are insects (such as aphids, leafhoppers, planthoppers, beetles and whiteflies). Others include eriophyid mites, dorylaimid nematodes, and chytrid and plasmodiophoromycete fungi. However, the ability to transmit a given virus typically is confined to closely allied species. A minority of plant viruses ($<10\%$) seem not to be spread by vectors in the usual sense. Some of these, such as tobacco mosaic tobamovirus, possess stable nucleoprotein particles that accumulate in large numbers ($>10^{13}$/g leaf tissue) and are transmitted through small wounds produced by mechanical contact between healthy tissue and virus-infected tissue or virus-contaminated objects. Others are transmitted through the pollen or ovules of infected plants into the seed and also, in a few instances, from virus-carrying pollen to the pollinated plant. However, the relative infrequency of these alternative methods of transmission serves to emphasise the overriding importance in the ecology of plant viruses of vector-mediated virus spread.

TYPES OF INTERACTION OF VIRUS AND VECTOR

Plant viruses have evolved an impressive variety of mechanisms of transmission by their vectors (Table 1). One type, known as propagative transmission, involves infection of the insect vector itself: viruses

Table 1. *Examples of types of interaction of plant viruses with vectors*

Type of interaction	Virus group or virus	Vector type(s)	Duration of virus persistence in vector
Propagative	Phytoreovirus	Leafhopper	Weeks
	Fijivirus	Planthopper	Weeks
	Plant rhabdovirus	Leafhopper, aphid	Weeks
	Rice stripe virus	Planthopper	Weeks
	Maize rayado fino virus	Leafhopper	Weeks
Circulative, non-propagative	Luteovirus	Aphid	Many days
	Pea enation mosaic virus	Aphid	Many days
	Geminivirus	Leafhopper, whitefly	Many days
Non-circulative	Nepovirus	Nematode	Weeks
	Cucumovirus	Aphid	Few hours
	Tobacco necrosis virus	Fungus	Few hours
Non-circulative, involving helper component	Potyvirus	Aphid	Few hours
	Caulimovirus	Aphid	Several hours
	'Plant picornavirus'	Aphid	Few days

transmitted in this way are therefore insect viruses as well as plant viruses. Typically, ingested virus particles infect cells of the insect's gut wall, and progeny virus is shed into the haemocoele and so carried by the haemolymph to a range of other tissues, including the salivary glands, which in turn become infected. Virus is then liberated into the saliva and injected into plant cells when the vector feeds. This sequence of events takes a minimum of a few days, but often longer. The vector typically remains able to transmit the virus for several weeks (or even for life); several of the viruses, such as rice dwarf phytoreovirus, are passed through the eggs of the vectors to many of their progeny. Three of the groups of plant viruses that replicate in their vectors (phytoreoviruses, fijiviruses and plant rhabdoviruses) closely resemble viruses of vertebrates on the one hand, and viruses which are confined to invertebrates on the other. However, some insect-infecting plant viruses, such as rice stripe virus, have no known equivalent among vertebrate viruses.

The second type of transmission involves the circulation of virus particles in the body of the vector, apparently without virus replication. Ingested virus particles are thought to pass through the gut wall into the haemocoele, and from there into salivary gland cells and saliva. This takes a minimum of about 10 h, but often longer. The vectors can continue to transmit virus for many days or for life, indicating that the virus particles are extremely stable in the vector's body. However, transmission from the infective vectors to their progeny has rarely been reported and, if it occurs at all, must be rare.

Variant forms of non-circulative transmission, the third type, are shown by viruses that have nematodes, fungi or aphids as their vectors. In several of these systems, the virus particles are known or thought to attach to external or exoskeletal surfaces of the vector, later to be released and delivered to plant cells. In some instances, such as with nepoviruses, the virus particles are remarkably stable in their vectors and are released slowly, so that the vector nematodes may remain infective for long periods. In other instances the virus particles are less stable, or are released rapidly or, as with fungal zoospores, the vector is short-lived, so that almost all transmission occurs within a few hours after the vector has acquired the virus.

In some kinds of non-circulative transmission by insects, the interaction of virus and vector is thought to involve a virus-induced 'helper component' which, in the potyviruses and caulimoviruses, is a non-structural virus-coded protein. Viruses transmitted in this manner persist in their insect vectors for a few hours to a few days but, like those transmitted by attachment–detachment mechanisms not known to involve a helper component, are not retained through the moult that separates one life stage of the vector from another. In contrast, viruses that circulate in the body of vector insects can be retained through a moult, whether or not they replicate in the vector. Examples of these contrasting kinds of transmission mechanism will now be described in more detail.

PROPAGATIVE TRANSMISSION

Potato yellow dwarf rhabdovirus

Two examples, both involving leafhopper vectors, will be used to illustrate propagative transmission. Potato yellow dwarf rhabdovirus (PYDV) has a ssRNA genome of about 14 kb that is packaged in enveloped particles which contain five viral proteins. By analogy with other rhabdoviruses, the RNA in the virus particles is negative-sense and is associated with viral polymerase. Two antigenically distinguishable strains of PYDV are described: sanguinolenta yellow dwarf virus (SYD), transmitted by *Aceratagallia sanguinolenta* and constricta yellow dwarf virus (CYD), transmitted by *Agallia constricta*.

The ability of PYDV to infect insects as well as plants, and the apparent lack of this ability in most other plant viruses, leads one to search for adaptations that permit the virus to replicate in each kind of host. One clue comes from experiments in which the glycoprotein (G protein) that forms the spikes on the surface of PYDV particles was removed. This treatment greatly decreased infectivity for monolayers of cells of the insect vector but had little effect on infectivity for mechanically inoculated plants (Hsu, Nuss & Adam, 1983). Moreover, the G protein seems closely involved in determination of the vector specificity of strains SYD and CYD. Thus SYD particles infect vector cell monolayers optimally at pH 5.9 whereas CYD particles infect optimally at pH 5.3, a difference that probably reflects a difference in properties of the G proteins of the two strains (isoelectric points about pH 4.8 and 4.3, respectively; Gaedigk, Adam & Mundry, 1986). Moreover, antiserum to the G protein of strain SYD neutralises the infectivity of the homologous virus for vector cells much more strongly than that of strain CYD (Gaedigk *et al.*, 1986). Evidence was also obtained that infection of vector cell monolayers is inhibited reversibly by lysosomotropic substances such as chloroquine and ammonium chloride. It was therefore proposed that PYDV particles enter cells by adsorptive endocytosis (Marsh & Helenius, 1980), with the G protein playing a key role in the initial adsorption to vector cells. This would be followed by fusion of the viral envelope with lysosomal membranes; and the release into the cytoplasm of viral RNA and the protein intimately associated with it (Adam & Gaedigk, 1986). Presumably, the function performed by G protein is not needed for infection of plants because PYDV particles enter leaf cells through the wounds made by vector leafhoppers or by mechanical inoculation, and subsequent cell-to-cell spread is via plasmodesmata.

There is little other evidence of differences in PYDV replication in plants and insects, and in cells of both the virus particles are budded into the perinuclear space (Chiu *et al.*, 1970).

Wound tumor phytoreovirus

The isometic 70 nm diameter particles of wound tumor phytoreovirus (WTV) contain 12 capped monocistronic segments of dsRNA with a total size of about 25 kbp; and 7 of the virus-coded proteins are components of the virus particles.

Two possible kinds of adaptation for replication in vector leaf-hoppers are known. The first was discovered as a result of maintaining virus isolates for long periods in vegetatively propagated sweet clover plants without transmission by leafhoppers (*Agallia constricta*). These isolates had become poorly transmissible (subvectorial) or non-transmissible (exvectorial) by *A. constricta*, and either had deletions in one or more of four of their genome segments, or produced abnormally small amounts of these segments (Reddy & Black, 1974, 1977). In several instances, genome segments were partially replaced by shorter segments that retained the original 5' and 3' ends (Nuss & Summers, 1984) but some individual isolates completely lacked segment 2 or segment 5 (Reddy & Black, 1977), which respectively contain genes encoding the proteins of 130K and 76K (Nuss & Peterson, 1980, 1981) that form the outer surface of the virus particles (Reddy & MacLeod, 1976). These two proteins seem needed for WTV replication in vector insects or vector cell monolayers but not for virus replication in plants. However, they may not be required for infection to be initiated because protease treatment of virus particles appeared to remove them without affecting infectivity for vector cells (Reddy & MacLeod, 1976), and it was suggested that they may have some other function in virus replication (D. L. Nuss, quoted by Adam, 1984). Further work is needed to resolve this question.

A requirement for a propagative virus is that it should not damage its vector and indeed there is only slight evidence that *A. constricta* is adversely affected by WTV (Hirumi, Granados & Maramorosch, 1967). One way in which damage might be prevented is by the regulation of WTV replication, as observed when a persistent infection is produced in vector cells. When the cells were infected by inoculation, cellular protein synthesis was not perceptibly affected and no cytopathic changes occurred despite virus polypeptide synthesis reaching up to 15% of total protein synthesis. Moreover, when the infected cells were subcultured, the synthesis of viral polypeptides declined to about 5% of the peak rate and remained at this level even though more than 90% of cells were still infected after 100 passages (Peterson & Nuss, 1985). This decrease in viral polypeptide synthesis was caused by a decrease in the translational activity of viral transcripts but not in the amount of virus-specific mRNA (Peterson & Nuss, 1986). Whether this change in activity reflects an alteration in the 5' end of the viral transcripts, as happens with human reovirus (Skup & Millward, 1980), remains to be seen but it is tempting to speculate that the effect may also occur in the later stages of infection of vector insects, enabling them to continue to transmit WTV for the remainder of an apparently normal life span (Hirumi *et al.*, 1967).

In summary, the data obtained with both WTV and PYDV indicate that virus infection and replication in vector insects require virus-coded proteins that are not essential for virus replication in plants. In addition, a start has been made with WTV in uncovering mechanisms that may be involved in minimising the adverse effects of propagative viruses on their insect vectors.

CIRCULATIVE TRANSMISSION

Luteoviruses

Luteoviruses have isometric particles of *c*. 25 nm diameter that contain the ssRNA genome and circulate through the body of vector aphids. For example, potato leafroll luteovirus (PLRV) can be detected in the haemolymph of virus-carrying aphids (*Myzus persicae*) and, when injected into the haemocoele of virus-free *M. persicae*, the virus can be transmitted by the recipient aphids (Heinze, 1955; Harrison, 1958). However, the PLRV content of aphids does not increase after they have left the virus source (Tamada & Harrison, 1981) indicating that little if any virus replication takes place in the vector. Evidence on the route or transfer of virus from gut to haemolymph and from haemolymph to saliva relies heavily on electron microscopy of sections of virus-carrying aphids. With isolates of barley yellow dwarf luteovirus (BYDV), Gildow (1985) considered the hindgut to be the main site of passage of virus particles into the haemolymph. Passage of virus particles from haemolymph to saliva occurs via cells of the accessory salivary glands (Gildow & Rochow, 1980; Gildow, 1982).

The specificity of transmission of luteoviruses by aphid species seems to be determined by the success of this second step. Indeed, luteovirus particles can accumulate in the haemolymph of aphid species that are poor vectors, and even when their haemocoele is injected with virus these species rarely transmit (Harrison, 1958; Rochow, 1969). In good and poor vectors alike, the virus particles become associated with the basal lamina of cells of the accessory salivary glands but they are only found within the salivary gland cells of efficient vector species (Gildow & Rochow, 1980).

The virus particle surface is thought to possess the signal that leads to transport of virus through salivary gland cells into the saliva. This was first demonstrated with two vector-specific forms of BYDV, called rho palosiphum padi virus (RPV) and macrosiphum avenae virus (MAV), which are serologically only distantly related (Aapola & Rochow, 1971; Rochow et al., 1971). The aphid *Rhopalosiphum padi* normally transmits RPV efficiently and MAV inefficiently, whereas another aphid, *Sitobion* (*Macrosiphum*) *avenae*, transmits MAV efficiently and RPV inefficiently. However, both RPV and MAV were readily transmitted by *R. padi* from mixed infections (Rochow, 1965), and also by *R. padi* injected with virus from doubly infected plants. Moreover, only RPV was transmitted by *R. padi* injected with mixtures of the two viruses from singly infected plants, and treatment of the virus from doubly infected plants with RPV antiserum before injecting it into virus-free aphids prevented all transmission, whereas MAV antiserum had no effect. These findings led Rochow (1970) to conclude that MAV RNA can be packaged in RPV coat protein to produce particles that are transmissible by *R. padi*, and therefore that luteovirus coat protein has a key function in transmission. The ultrastructural evidence already quoted indicates that this function concerns the passage of virus particles through salivary gland cells.

These ideas have received strong support from other kinds of experiments with luteoviruses. For example, lettuce speckles mottle and carrot mottle, two aphid non-transmissible viruses, both become aphid transmissible when they occur in mixed infection with a luteovirus, beet western yellows (BWYV) and carrot red leaf (CRLV) respectively, and their RNA becomes packaged in particles that contain the luteovirus 25K coat protein (Falk, Duffus & Morris, 1979; Waterhouse & Murant, 1983). In addition, the specific vector of carrot mottle virus can be changed from *Cavariella aegopodii* to *Myzus persicae* by associating it with PLRV or BWYV instead of with CRLV (Waterhouse & Murant, 1983). Further support is provided by tests on an inefficiently aphid-transmitted strain of PLRV. This strain is acquired and retained normally by *M. persicae* but was not transmitted, even when purified virus was injected into the haemocoele of virus-free aphids. At first, no antigenic difference could be found between particles of the efficiently and inefficiently transmitted strains in gel-diffusion precipitin tests with polyclonal antiserum to the readily transmitted strain (Tamada, Harrison & Roberts, 1984). However, in later tests with ten monoclonal antibodies to the efficiently transmitted strain, two of these distinguished the inefficiently transmitted strain from all but one of more than 30 British field isolates of PLRV, the rest of which were indistinguishable from the efficiently transmitted strain using any of the monoclonal antibodies. The single anomalous isolate failed to react with the same two monoclonal antibodies that did not react with particles of the inefficiently transmitted strain, and it too proved to be poorly transmissible by *M. persicae*. The impressive antigenic uniformity of most of the PLRV isolates, and their distinction from the two poorly transmitted isolates, therefore suggest that the epitope(s) missing in these two isolates may constitute, or form part of, the recognition signal for transport of virus particles through salivary gland cells (Massalski, Thomas & Harrison, 1986 and unpublished results).

Pea enation mosaic virus

This virus, like luteoviruses, has small isometric particles which circulate within the body of vector aphids (*Acyrthosiphum pisum*) and are retained for long periods, apparently without multiplying (Fargette, Jenniskens & Peters, 1982). Evidence that viral particle protein has a key function in this system too was obtained by comparing the composition of particles of aphid transmissible and aphid non-transmissible isolates. Particles of the aphid transmissible isolates proved to contain a protein of 56K in addition to the predominant 22K protein found in particles of all isolates (Hull, 1977; Adam, Sander & Shepherd, 1979). The 22K protein cistron forms part of the larger (RNA-1) of the two genomic ssRNA species. However, in aphid transmissible isolates, RNA-1 has a molecular weight that is larger by about 1.2×10^5 than that of RNA-1 of aphid non-transmissible isolates, suggesting that the non-transmissible isolates have lost part of their genome, very possibly a portion of the 56K protein cistron (Adam et al., 1979). A further difference between the two kinds of isolate is that, whereas particles of a transmissible isolate were found embedded in the basal lamina of accessory salivary gland cells of *A. pisum*, the particles of a non-transmissible isolate were not observed at this site (Harris et al., 1975). As with luteoviruses, the particle protein of pea enation mosaic virus seems to have a key function in transport of virus particles through salivary gland cells of the aphid vector.

Geminiviruses

Geminiviruses have circular ssDNA genomes, and fall into two subgroups. Members of one subgroup have leafhopper vectors and a one-part genome whereas viruses in the other subgroup have whitefly vectors and a bipartite genome. In both subgroups the vectors can retain the viruses for many days and, where tests have been done, the viruses were found to circulate but not to multiply in vector insects (Harrison, 1985). All the whitefly-transmitted geminiviruses tested have serologically related particles and the same vector, *Bemisia tabaci*, whereas the leafhopper-transmitted geminiviruses are antigenically diverse and are transmitted by different leafhopper species, here again suggesting a role for virus particle protein in transmission by vectors (Roberts, Robinson & Harrison, 1984).

Only one protein species (about 30K) is known to occur in geminivirus particles, and comparison of the nucleotide sequences of several geminiviruses shows that the 30K cistron is conserved more strongly than other cistrons in the whitefly-transmitted geminiviruses but less strongly in those transmitted by leafhoppers (Hamilton, *et al.*, 1984; Harrison, 1985). Indeed, it is remarkable that the particle proteins of whitefly-transmitted geminiviruses that apparently have no hosts in common may share almost 90% of their amino acid residues. Thus circumstantial evidence supports the notion that virus particle proteins have an important function in this third example of circulative transmission by vectors.

Recent work with monoclonal antibodies to African cassava mosaic geminivirus, which has whitefly vectors, has identified at least one epitope in the virus particle protein that resembles an epitope in all six other whitefly-borne geminiviruses tested and therefore could be involved in transmission by *B. tabaci* (Thomas, Massalski & Harrison, 1986 and unpublished results).

NON-CIRCULATIVE TRANSMISSION

Nepoviruses

Nepoviruses have ssRNA genomes in two parts which are packaged in separate isometric particles of *c.* 25 nm diameter that contain the same 55K protein. Natural transmission is by soil-inhabiting longi-dorid nematodes that feed on roots and can retain the viruses for weeks or months unless the animals moult. The viruses are also transmitted through a proportion of the seeds produced by infected plants. As with the examples of circulative transmission described above, so too with nepoviruses the virus particle surface seems to have an important function in vector transmission. Some of the vir-uses, such as raspberry ringspot (RRV) and tomato black ring nepo-viruses, occur as serologically distinctive strains that are adapted for transmission by different nematode species (Harrison, 1964; Taylor & Murant, 1969). Furthermore, pseudo-recombinant isolates produced by reassorting the two genome parts of such strains have the vector and serological specificities of the parental strain donating the smaller genome segment, RNA-2 (Harrison et al., 1974a; Harrison & Murant, 1977), which contains the particle protein gene (Harrison, Murant & Mayo, 1972).

Ultrastructural studies on virus-carrying vector nematodes show that the virus particles attach to specific surfaces in the feeding apparatus. For example, RRV particles attach to the wall of the spear lumen and guiding sheath of Longidorus elongatus and those of arabis mosaic nepovirus (AMV) are found on the cuticular lining of the odontophore and oesophagus of Xiphinema diversicaudatum (Taylor & Robertson, 1969, 1970). Moreover, in different nematode species, the attachment of virus particles at such sites is closely associated with ability of the species to transmit the virus in question (Harrison, Robertson & Taylor, 1974b). In addition, differences between popu-lations of X. diversicaudatum in ability to transmit AMV are mir-rored by differences in numbers of virus particles that attach to the odontophore and oesophagus linings (Brown & Trudgill, 1983).

The specificity of attachment of virus particles at these sites in vector nematodes suggests that structural features of the virus parti-cle surface are recognised by complementary features at the attach-ment site in the nematode. The genetic control of transmission efficiency and, apparently, of the occurrence of these complementary structural features in the nematode has been examined by crossing individuals from the populations of Xiphinema diversicaudatum that differ in efficiency as vectors of AMV and in the frequency with which virus particles bind at the specific retention site in the nema-tode. The inheritance of transmission efficiency (Brown, 1986) seems compatible with control by a single partially dominant gene, although various other possibilities are not excluded.

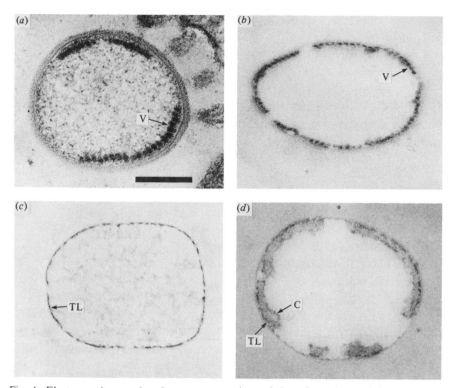

Fig. 1. Electron micrographs of transverse sections of the odontophore of the nematode *Xiphinema diversicaudatum*: (*a*) and (*b*) stained with uranyl acetate and lead citrate, (*c*) and (*d*) stained for carbohydrate using periodic acid-thiosemicarbazide-silver proteinate. (*a*) Fixed with glutaraldehyde and post-fixed with osmium tetroxide; (*b*)–(*d*) fixed with formaldehyde. (*a*) and (*b*) Particles of arabis mosaic virus (V) lining the odontophore lumen of virus-carrying nematodes; (*c*) lumen of a virus-free nematode showing the thin discontinuously stained lining (TL); (*d*) odontophore of a virus-carrying nematode showing cloud-like material (C) associated with regions where virus particles (barely visible) are attached to the stained areas (TL) of the lumen lining. Bar represents 200 nm (from Robertson & Henry, 1986).

Histochemical tests have provided evidence of two sorts of carbo-hydrate-containing material at the site of retention of AMV particles in the oesophagus (Fig. 1; Robertson & Henry, 1986). One sort consists of a thin layer of stained material that is patchily distributed on the lining of the food canal of all *X. diversicaudatum* examined. The other type, which was found only in AMV-carrying individuals, has a cloud-like appearance and surrounds AMV particles that are associated with the patches of stained material (Fig. 1). The first sort of material is probably of nematode origin and could be concerned in binding AMV particles to the oesophageal lining. Whether the cloud-like material is of plant or nematode origin, and whether it is involved in particle attachment and/or in preventing particle detachment is unclear, but it too may well have an important function in transmission.

Little is known about the mechanism of detachment of the virus particles from their retention site as a preliminary to injection into punctured root cells and the initiation of infection. However, the fact that salivary secretions flow forward over the virus-retention sites during nematode feeding indicates that materials in these secretions have the opportunity to play a key role in particle detachment.

'Plant picornaviruses'

The 30 nm diameter isometric particles of two previously described viruses, anthriscus yellow (AYV) and parsnip yellow fleck (PYFV), seem to have a composition resembling that of particles of vertebrate- and invertebrate-infecting picornaviruses (Hemida & Murant, 1986). For convenience, these plant viruses will therefore be referred to here as 'plant picornaviruses'. Of the two viruses, AYV is transmitted by aphids (*Caveriella* spp.) in the semi-persistent manner whereas PYFV is not transmitted unless the virus source plants also contain AYV or the aphids have previously fed on a source of AYV, when PYFV too is transmitted semi-persistently (Murant & Goold, 1968; Elnagar & Murant, 1976). AYV-infected plants therefore contain something that is needed for aphid transmission of PYFV.

The nature and function of this helper factor are unclear but electron microscopy of AYV-carrying aphids and AYV-infected plants has provided some possible clues. Ultrastructural examination of AYV-carrying aphids has shown that clusters of virus particles accumulate at the junction of the sucking pump and oesophagus close to the tentorial bar. These particles are attached to a pad of lightly stained 'M-material' that is also found in virus-free aphids, and closer inspection shows that each particle is surrounded by an ill-defined layer of more darkly stained material (Murant, Roberts & Elnagar, 1976). It seems likely that this darkly staining material is derived from the cytoplasmic inclusion bodies in which the virus particles are embedded in plant cells (Murant & Roberts, 1977). A plausible but speculative interpretation of these observations (Murant, 1978) is that the AYV helper factor is a virus-coded inclusion protein that is involved in the reversible attachment of AYV and PYFV particles to mucus-like material that binds to the ventral intima of the anterior foregut. Further work is needed to test this hypothesis.

Caulimoviruses

Evidence of another kind is beginning to suggest that there are similarities between the mechanisms of aphid transmission of caulimoviruses and AYV. It is well established that leaves infected with cauliflower mosaic virus (CaMV) contain a helper component which is needed by aphids to enable them to transmit virus particles that are acquired subsequently from purified preparations (Lung & Pirone, 1974). Other caulimoviruses have helper components that can aid the transmission of CaMV (Markham & Hull, 1985), but some isolates and laboratory mutants of CaMV lack an active helper component (Lung & Pirone, 1973; Armour et al., 1983; Woolston et al., 1983). Examination of their dsDNA genomes shows that all such variants of CaMV have mutations in open reading frame II, which normally is thought to code for an 18K protein, and that aphid transmissibility can be restored by substituting for the mutated sequence a region that contains open reading frame II from a transmissible isolate (Woolston et al., 1983). Interestingly, this 18K protein co-purifies with, and apparently affects the firmness with which virus particles are held in, the cytoplasmic inclusion bodies produced in CaMV-infected plant cells (Woolston et al., 1983; Givord et al., 1984). Whether the 18K protein interacts directly with CaMV particles is unclear but any such interaction would have obvious implications for the mechanism of aphid transmission. The site where caulimovirus particles are held in aphids is also not known but the fact that virus is not retained when the insects moult points to a location in the feeding apparatus or foregut.

Potyviruses

Considerable progress has been made in recent years in understanding the mechanism of transmission of potyviruses by aphids. These viruses can be acquired efficiently and inoculated within a few minutes but most aphids lose their ability to transmit within an hour after leaving the virus source. Potyvirus particles are flexuous filaments c. 750 nm long containing the monopartite genome of ssRNA. Govier & Kassanis (1974) found that aphids fed on purified preparations of such particles did not transmit virus unless they had previously fed on infected cells, or on extracts from such cells that contained a helper component needed for transmission. The helper component of one potyvirus typically can assist the aphid transmission of some other but not all members of the group (Kassanis &

Govier, 1971; Pirone, 1981; Sako & Ogata, 1981). Purified preparations of helper component contain a virus-specific polypeptide of
$c.\,55K$ and have been used to prepare antisera that inhibit helper
component action (Thornbury & Pirone, 1983; Thornbury *et al.*,
1985). Immunoabsorption of the helper activity in preparations from
infected leaves resulted in removal of a polypeptide of $c.\,55K$ that
was detectable by electrophoresis in SDS/polyacrylamide gels. However, analyses by gel filtration chromatography or HPLC-gel permea-
tion chromatography indicate that native helper component has a
molecular weight of 100–150K, suggesting that it is a dimer of the
55K molecule (Govier, Kassanis & Pirone, 1977; Thornbury *et al.*,
1985). Antisera to helper component also react with a 75K polypeptide produced by *in vitro* translation of potyvirus RNA (Hellmann
et al., 1983) but not with the 5' terminal third of this polypeptide
(Hellmann, Shaw & Rhoads, 1985).

These findings, together with the evidence that helper factors of
different potyviruses have a degree of helper and antigenic specificity
(Thornbury & Pirone, 1983; Thornbury *et al.*, 1985), indicate (*a*)
that helper component is virus-coded and (*b*) that it is derived from
the 75K product which is translated from the 5' end of the viral
RNA (Hellmann *et al.*, 1985). Helper component seems to occur
in substantial amounts in cells infected with some potyviruses and
there is evidence that it is the main constituent of the amorphous
cytoplasmic inclusions produced by infection with papaya ringspot
potyvirus (De Mejia *et al.*, 1985).

The importance of helper component in aphid transmission is also
illustrated by experiments with the C strain of potato virus Y. This
strain is not normally aphid transmissible but can be transmitted
by aphids that have previously acquired the helper component of
a normally aphid transmissible strain (Kassanis & Govier, 1971).
Evidently the C strain lacks a functional helper component. In other
instances, aphid non-transmissible strains of potyviruses produce
functional helper components but seem to have altered 30K particle
proteins that are unsuitable for aphid transmission. Nucleotide
sequencing of the RNA of such a non-transmissible strain and of
a transmissible strain of tobacco etch potyvirus has located the coat
protein coding region at the 3' end of the viral RNA and shown
that the coat proteins of the two strains differ in only 6 out of 263
amino acids. Three of these six changes are near the N-terminus,
a hydrophilic region (Allison *et al.*, 1985*a*, *b*). Treatment of virus
particles with trypsin removes the N-terminus (Hiebert, Thornbury

& Pirone, 1984), suggesting that it lies at the particle surface. In addition, tests with ten monoclonal antibodies to particles of the aphid transmissible strain showed that all but one reacted with intact particles of the non-transmissible strain (Dougherty, Willis & Johnston, 1985). It therefore seems possible that an epitope near the N-terminus of the particle protein which is missing in the non-transmissible strain has a key function in transmission.

Experiments in which aphids were allowed to acquire virus particles and helper component by feeding through a membrane on radioactive solution have shown that about one-third of the individual aphids that later transmitted tobacco vein mottling potyvirus to test plants had taken up volumes estimated to contain <100 virus particles (Pirone & Thornbury, 1985). This apparent high efficiency of transmission may in part explain previous difficulties in locating potyvirus particles in the feeding apparatus of virus-carrying aphids. However, Taylor & Robertson (1974) found particles resembling those of tobacco etch potyvirus associated with the distal 20 μm of the maxillae of *Myzus persicae*, although the relation of these particles to the transmission process has remained unclear. Recent tests, in which aphids were fed through a membrane on helper component and radio-iodinated potyvirus particles, have shown that these particles became bound to regions in the maxillary stylets and foregut whereas no particles were bound when the aphids were fed on virus only (Berger & Pirone, 1986). This strongly suggests that the biological action of helper component involves the binding of virus particles to surfaces in the aphid's maxillae and/or foregut. Further work is needed to ascertain whether helper component acts by binding to the N-terminal portion of the virus particle protein, as might be suspected from the comparisons of the particle proteins of aphid transmissible and non-transmissible virus strains, and whether it also has other effects such as protecting virus particles from inactivation in the vector.

THREE MAIN TRANSMISSION MECHANISMS

Although earlier work has emphasised the wide range of mechanisms of vector transmission that are used by different viruses, comparison of the examples described in this review suggests that the different systems within each of the three main categories have more in common than realised hitherto. For instance, in propagative transmission there is now evidence that proteins which form the particle surface in a phytoreovirus and a rhabdovirus are involved primarily in infection of vector cells and seem less necessary, and possibly dispensable, for infection of plant cells. Similarly, in both systems, there are mechanisms that prevent the viruses damaging their vectors seriously. In WTV this may be achieved by the decrease in translational activity of viral transcripts observed later in infection. No comparable data are available for PYDV but in this system defective interfering particles, which apparently can occur in infected plants (Adam, Gaedigk & Mundry, 1983), may perhaps play a part in restricting virus replication in vector insects.

In all three examples quoted of circulative transmission – in luteoviruses, geminiviruses and pea enation mosaic virus – the virus coat protein also plays a crucial role, especially in the passage of virus particles through salivary gland cells. Pea enation mosaic virus differs from the other two examples in being known to possess an extra particle protein that is needed for aphid transmission but not for packaging of the genomic nucleic acid. In contrast to pea enation mosaic virus, both luteoviruses and geminiviruses tend to occur in greatest amounts in phloem tissue, and several viruses in these groups are virtually confined to phloem cells although they replicate when inoculated to isolated mesophyll protoplasts (Kubo & Takanami, 1979; Townsend, Watts & Stanley, 1986). Perhaps the very attributes of the virus particle surface that enable the virus particles to pass through salivary gland cells of virus vectors militate against their transport into mesophyll cells of intact plants.

The examples given of non-circulative transmission at first sight seem a heterogeneous collection, with the usual persistence of virus in the vector ranging from less than 1 h (potyviruses) to more than 1 month (nepoviruses). However, at both these extremes, virus coat protein is critically involved in the process and specificity of transmission. Moreover in both instances, and in the 'plant picornaviruses', there is strong evidence that transmission involves binding of virus particles at sites in the feeding apparatus or in the anterior part of the alimentary canal. Particles of nepoviruses appear to be bound to carbohydrate-containing material of non-viral origin, and in the 'plant picornaviruses' the comparable material is also non-viral and may perhaps contain carbohydrate. In three of the four examples of non-circulative transmission cited, a helper factor has also been implicated and there is evidence that it may be a constituent of cytoplasmic inclusion bodies; in two of these examples it is known to be a virus-coded protein. The helper component appears to cause potyvirus particles to bind to surfaces in the aphid mouthparts, and a similar function is plausible in 'plant picornaviruses' and caulimoviruses. In contrast, no helper component for nepovirus transmission has been found. However, in another group of nematode-transmitted viruses, the tobraviruses, the great antigenic variation of the coat protein in different strains of the same virus (Robinson & Harrison, 1985; Harrison & Robinson, 1986), and the availability of a viral gene without any other known function (Boccara, Hamilton & Baulcombe, 1986), make the existence of a tobravirus helper factor an interesting possibility.

The state of knowledge of some of the salient features of individual transmission mechanisms is summarised in Table 2. The feature common to all the mechanisms examined in detail is the critical importance of the protein at the surface of virus particles.

TRANSMISSION MECHANISMS AND VIRUS EPIDEMIOLOGY

In general, vector specificity is most strongly developed for viruses that are retained for long periods by their vectors, whether the transmission mechanism is propagative, circulative or, as in nepoviruses, non-circulative. As a result, the ecology and epidemiology of the majority of plant viruses that are transmitted in the persistent manner is intimately tied up with the population dynamics and behaviour of one or a few closely related vector species. Long persistence in the vector has the advantage that it enables highly mobile vectors, such as some leafhoppers, to spread virus to fresh hosts from distant sources, and poorly mobile vectors, such as nematodes, to retain virus until they encounter an infectible plant root. Moreover, individual infective vectors can inoculate a series of plants. Propagative transmission has the disadvantage of an extended latent period between virus acquisition by the vector and inoculation of the virus to healthy plants. This limits the number of cycles of transmission and amplification of disease incidence that can occur in a limited period, and may be the reason why viruses with propagative transmission seem not to reach epidemic proportions in crops in countries with cool temperate climates and cold winters. They are more important in tropical and sub-tropical areas, although when propagative transmission is complemented, as with rice dwarf phytoreovirus, by transovarial transmission to the progeny of infective vectors, they

Table 2. *Identity and roles of virus coat proteins and helper components involved in transmission of plant viruses by invertebrate vectors*

Type of transmission	Virus	Virus coat protein(s) involved	Role of virus coat protein(s)	Occurrence of helper component	Role of helper component
Propagative	Potato yellow dwarf rhabdovirus	G protein	Attachment to vector cells	—	—
	Wound tumor phytoreovirus	130K and 76K proteins	Infection of vector cells	—	—
Circulative	Luteoviruses	25K protein	Uptake by salivary gland cells	—	—
	Pea enation mosaic	56K protein	Uptake by salivary gland cells	—	—
Non-circulative	Geminiviruses	? 30K protein	?	—	—
	Nepoviruses	55K protein	Temporary attachment to surface of feeding apparatus	—	—
	'Plant picornaviruses'	?	?	+	Probably aids attachment of virus particles to M-material in fore-gut
	Caulimoviruses	?	?	+ (18K polypeptide)	?
	Potyviruses	30K protein	? Temporary attachment to maxillae with aid of helper component	+ (55K virus-coded polypeptide)	Aids attachment of virus particles to maxillae

may also be prevalent in regions with warm temperate climates.

Circulative, non-propagative viruses cause many of the world's most damaging epidemic diseases in crops in both tropical and temperate regions. As they lack the ability to be transmitted transovarially, in temperate regions they are most important where overwintering infected plants are common and so act as sources of inoculum in the following spring.

Non-circulative viruses that persist in vector nematodes for long periods, such as nepoviruses, have the disadvantage that their longidorid vectors can move only short distances. However, most nepoviruses have the advantages of a wide host range and in being transmitted through a proportion of the seeds of many plant species. They consequently survive for long periods at the same site in dormant weed seed, and are dispersed to other sites in seed (Murant & Lister, 1967). They are therefore difficult to eliminate once established at a site.

Vector specificity seems less well developed where transmission is dependent on a helper component, and several or many aphid species may be able to transmit individual potyviruses or caulimoviruses. Most viruses transmitted in this way have the counterbalancing disadvantages of being retained by vector aphids for only relatively short times and in having more restricted host ranges. They therefore tend usually to spread only short distances to another plant of the same species, such as from plant to plant in a crop. However, their relative lack of vector specificity among aphids, and their ability (especially potyviruses) to be acquired and inoculated in brief probes, enables aphids of non-colonising species, where numerous, to make a substantial contribution to virus spread. For example, much of the spread of potato Y potyvirus in Northern Ireland is thought to be caused by the aphid *Brachycaudus helichrysi*, which migrates in large numbers in spring from hedgerow blackthorn trees and makes brief probes on nearby potato plants without colonising them (Edwards, 1963; Bell, 1983).

The ease with which virus spread can be prevented by chemical control of vectors depends on many factors, including the number of vector species involved, their population dynamics and activity patterns, the virus transmission mechanism, the abundance and location of virus sources, and the susceptibility to infection of the healthy plants. However, other things being equal, it is easier to control virus spread within than into a crop, and easier to prevent the spread of viruses whose vectors need prolonged acquisition and inoculation access periods for virus to be transmitted than of non-persistently transmitted viruses, which can be acquired and inoculated in less than a minute – too short a time for pesticides to exert much effect.

CONSERVATION OF VIRUS PARTICLE PROTEIN STRUCTURE IN VECTOR-TRANSMITTED PLANT VIRUSES

The central role of virus particle proteins in the transmission mechanisms described above indicates that in many, probably most, plant viruses they have a dual function: genome protection and the potentiation of vector transmission. The adaptation of virus particle proteins for transmission is probably most highly developed in viruses with the greatest vector specificity. Any mutant protein has to maintain both this adaptation, and its genome-protecting function, if the mutant virus is not to have less biological fitness than the parental form. The result is that field isolates of such viruses tend to show relatively little antigenic variation, and therefore that serological tests are valuable for virus identification, and antigenic specificity is useful as an aid to the delineation of plant virus taxa. However, in a few instances, as in the geminiviruses transmitted by *Bemisia tabaci* (Roberts *et al.*, 1984), the adaptation of the particle protein for transmission may be so extreme that a close antigenic relationship is found between viruses that have no hosts in common and probably do not share a gene pool. Such conservation of virus particle protein structure in plant viruses contrasts with the situation in many vertebrate viruses that spread without vectors. Here selection is for variants that do not react with antibodies elicited by previously prevalent virus strains, and antigenic variants are commonplace.

Antigenic conservation seems less evident among plant viruses that need a helper component for transmission. For example, antigenic variants may abound among virus isolates considered to be strains of the same potyvirus. Here the feature of the helper component that confers its biological activity is perhaps at least as strongly conserved as virus particle protein structure. Indeed the helper component of one potyvirus typically can assist the transmission of several others, some of which may have particles that are serologically only distantly related, if at all. With viruses of such kinds, a standard antiserum may be less reliable for detecting infection by variant strains, and antigenic affinities provide a less clear cut guide to virus taxa.

CONCLUDING COMMENTS

A theme emerging from this discussion of transmission of plant viruses by invertebrate vectors is that the different systems grouped within each of the three main categories of mechanism have much in common. Moreover, in all three categories, virus particle proteins have important functions in transmission. The resulting dual function of virus particle proteins – in transmission and in genome protection – appears to limit the range of variation that is consistent with biological fitness, with consequential implications for serological tests as an aid to virus detection, and for antigenic relatedness as a criterion in virus classification.

Mechanisms of vector transmission also have an important influence on patterns of virus spread, and they interact with other biological properties of the viruses to produce a variety of strategies for virus survival, and with differing potentials for causing damaging epidemics in crops. Gaining a deeper understanding of the intricate biological adaptations involved in transmission by vectors is therefore not only scientifically challenging but also of great potential practical importance as a possible approach to devising novel methods of preventing some of the enormous crop losses that are caused by vector-borne plant viruses.

ACKNOWLEDGMENTS

I am most grateful to G. Adam, W. G. Dougherty, D. L. Nuss, and T. P. Pirone, and to my colleagues D. J. F. Brown, P. R. Massalski, A. F. Murant, W. M. Robertson and J. E. Thomas for giving me access to their data before publication; and to W. M. Robertson for permission to reproduce the illustrations used in Fig. 1.

REFERENCES

AAPOLA, A. I. E. & ROCHOW, W. F. (1971). Relationships among three isolates of barley yellow dwarf virus. *Virology*, **46**, 127–41.

ADAM, G. (1984). Plant virus studies in insect vector cell cultures. In *Vectors in Virus Biology*, ed. M.A. Mayo & K. A. Harrap, pp. 37–62. London: Academic Press.

ADAM, G. & GAEDIGK, K. (1986). Inhibition of potato yellow dwarf virus infection in vector cell monolayers by lysosomotropic substances. *Journal of General Virology*, **67**, 2775–80.

ADAM, G., GAEDIGK, K. & MUNDRY, K. W. (1983). Alterations of a plant rhabdovirus during successive mechanical transfers. *Zeitschrift für Pflanzenkrankheiten und Pflanzenschutz*, **90**, 28–35.

ADAM, G., SANDER, E. & SHEPHERD, R. J. (1979). Structural differences between pea enation mosaic virus strains affecting transmissibility by *Acyrthosiphon pisum* (Harris). *Virology*, **92**, 1–14.

ALLISON, R. F., DOUGHERTY, W. G., PARKS, T. D., WILLIS, L., JOHNSTON, R. E., KELLY, M. & ARMSTRONG, F. B. (1985a). Biochemical analysis of the capsid protein gene and capsid protein of tobacco etch virus: N-terminal amino acids are located on the virion's surface. *Virology*, **147**, 309–16.

ALLISON, R. F., SORENSON, J. C., KELLY, M. E., ARMSTRONG, F. B. & DOUGHERTY, W. G. (1985b). Sequence determination of the capsid protein gene and flanking regions of tobacco etch virus: evidence for synthesis and processing of a polyprotein in potyvirus genome expression. *Proceedings of the National Academy of Sciences, USA*, **82**, 3969–72.

ARMOUR, S. L., MELCHER, U., PIRONE, T. P., LYTTLE, D. J. & ESSENBERG, R. C. (1983). Helper component for aphid transmission encoded by region II of cauliflower mosaic virus DNA. *Virology*, **129**, 25–30.

BELL, A. C. (1983). The life-history of the leaf-curling plum aphid *Brachycaudus helichrysi* in Northern Ireland and its ability to transmit potato virus $Y^{C(AB)}$. *Annals of Applied Biology*, **102**, 1–6.

BERGER, P. H. & PIRONE, T. P. (1986). The effect of helper component on the uptake and localization of potyviruses in *Myzus persicae*. *Virology*, **153**, 256–61

BOCCARA, M., HAMILTON, W. D. O. & BAULCOMBE, D. C. (1986). The organisation and interviral homologies of genes at the 3′ end of tobacco rattle virus RNA1. *EMBO Journal*, **5**, 223–9.

BROWN, D. J. F. (1986). Transmission of virus by the progeny of crosses between *Xiphinema diversicaudatum* (Nematoda: Dorylaimoidea) from Italy and Scotland. *Revue de Nématologie*, **9**, 71–4.

BROWN, D. J. F. & TRUDGILL, D. L. (1983). Differential transmissibility of arabis mosaic and strains of strawberry latent ringspot viruses by three populations of *Xiphinema diversicaudatum* (Nematoda: Dorylaimida) from Scotland, Italy and France. *Revue de Nématologie*, **6**, 229–38.

CHIU, R. J., LIU, H. Y., MACLEOD, R. & BLACK, L. M. (1970). Potato yellow dwarf virus in leafhopper cell culture. *Virology*, **40**, 387–96.

DE MEJIA, M. V. G., HIEBERT, E., PURCIFULL, D. E., THORNBURY, D. W. & PIRONE, T. P. (1985). Identification of potyviral amorphous inclusion protein as a nonstructural virus-specific protein related to helper component. *Virology*, **142**, 34–43.

DOUGHERTY, W. G., WILLIS, L. & JOHNSTON, R. E. (1985). Topographic analysis of tobacco etch virus capsid protein epitopes. *Virology*, **144**, 66–72.

EDWARDS, A. R. (1963). A non-colonizing aphid vector of potato virus diseases. *Nature*, **200**, 1233–4.

ELNAGAR, S. & MURANT, A. F. (1976). The role of the helper virus, anthriscus yellows, in the transmission of parsnip yellow fleck virus by the aphid *Cavariella aegopodii*. *Annals of Applied Biology*, **84**, 169–81.

FALK, B. W., DUFFUS, J. E. & MORRIS, T. J. (1979). Transmission, host range, and serological properties of the viruses that cause lettuce speckles disease. *Phytopathology*, **69**, 612–7.

FARGETTE, D., JENNISKENS, M.–J. & PETERS, D. (1982). Acquisition and transmission of pea enation mosaic virus by the individual pea aphid. *Phytopathology*, **72**, 1386–90.

GAEDIGK, K., ADAM, G. & MUNDRY, K. W. (1986).The spike-protein of potato yellow dwarf virus and its functional role in the infection of insect vector cells. *Journal of General Virology*, **67**, 2763–73.

GILDOW, F. E. (1982). Coated-vesicle transport of luteoviruses through salivary glands of *Myzus persicae*. *Phytopathology*, **72**, 1289–96.

GILDOW, F. E. (1985). Transcellular transport of barley yellow dwarf virus into the haemocoel of the aphid vector, *Rhopalosiphum padi*. *Phytopathology*, **75**, 292–7.

GILDOW, F. E. & ROCHOW, W. F. (1980). Role of accessory salivary glands in aphid transmission of barley yellow dwarf virus. *Virology*, **104**, 97–108.

GIVORD, L., XIONG, C., GIBAND, M., KOENIG, I., HOHN, T., LEBEURIER, G. & HIRTH, L. (1984). A second cauliflower mosaic virus gene product influences the structure of the viral inclusion body. *EMBO Journal*, **3**, 1423–7.

GOVIER, D. A. & KASSANIS, B. (1974). A virus-induced component of plant sap needed when aphids acquire potato virus Y from purified preparations. *Virology*, **61**, 420–6.

GOVIER, D. A., KASSANIS, B. & PIRONE, T. P. (1977). Partial purification and characterization of the potato virus Y helper component. *Virology*, **78**, 306–14.

HAMILTON, W. D. O., STEIN, V. E., COUTTS, R. A. & BUCK, K. W. (1984). Complete nucleotide sequence of the infectious cloned DNA components of tomato golden mosaic virus: potential coding regions and regulatory sequences. *EMBO Journal*, **3**, 2197–205.

HARRIS, K. F., BATH, J. E., THOTTAPPILLY, G. & HOOPER, G. R. (1975). Fate of pea enation mosaic virus in PEMV-injected pea aphids. *Virology*, **65**, 148–62.

HARRISON, B. D. (1958). Studies on the behavior of potato leaf roll and other viruses in the body of their aphid vector *Myzus persicae* (Sulz.). *Virology*, **6**, 265–77.

HARRISON, B. D. (1964). Specific nematode vectors for serologically distinctive forms of raspberry ringspot and tomato black ring viruses. *Virology*, **22**, 544–50.

HARRISON, B. D. (1985). Advances in geminivirus research. *Annual Review of Phytopathology*, **23**, 55–82.

HARRISON, B. D. & MURANT, A. F. (1977). Nematode transmissibility of pseudo-recombinant isolates of tomato black ring virus. *Annals of Applied Biology*, **86**, 209–12.

HARRISON, B. D., MURANT, A. F. & MAYO, M. A. (1972). Two properties of raspberry ringspot virus determined by its smaller RNA. *Journal of General Virology*, **17**, 137–41.

HARRISON, B. D., MURANT, A. F., MAYO, M. A. & ROBERTS, I. M. (1974a). Distribution of determinants for symptom production, host range and nematode transmissibility between the two RNA components of raspberry ringspot virus. *Journal of General Virology*, **22**, 233–47.

HARRISON, B. D., ROBERTSON, W. M. & TAYLOR, C. E. (1974b). Specificity of retention and transmission of viruses by nematodes. *Journal of Nematology*, **6**, 155–64.

HARRISON, B. D. & ROBINSON, D. J. (1986). Tobraviruses. In *The Plant Viruses*, Vol. 2: *The Rod-Shaped Plant Viruses*, ed. M. H. V. van Regenmortel & H. Fraenkel-Conrat, pp. 339–69. New York: Plenum Publishing Company.

HEINZE, K. (1955). Versuche zur Übertragung des Blattrollvirus der Kartoffel in den Überträger (*Myzodes persicae* Sulz.) mit Injektionsverfahren. *Phytopathologische Zeitschrift*, **25**, 103–8.

HELLMANN, G. M., SHAW, J. G. & RHOADS, R. E. (1985). On the origin of the helper component of tobacco vein mottling virus: translational initiation near the 5′-terminus of the viral RNA and termination by UAG codons. *Virology*, **143**, 23–34.

HELLMANN, G. M., THORNBURY, D. W., HIEBERT, E., SHAW, J. G., PIRONE, T. P. & RHOADS, R. E. (1983). Cell-free translation of tobacco vein mottling virus RNA. II. Immunoprecipitation of products by antisera to cylindrical inclusion, nuclear inclusion, and helper component proteins. *Virology*, **124**, 434–44.

HEMIDA, S. K. & MURANT, A. F. (1986). Aphid-transmitted viruses resembling picornaviruses. *Report of the Scottish Crop Research Institute for 1985*, 148–9.

HIEBERT, E., THORNBURY, D. W. & PIRONE, T. P. (1984). Immunoprecipitation analysis of potyviral *in vitro* translation products using antisera to helper component of tobacco vein mottling virus and potato virus Y. *Virology*, 135, 1–9.

HIRUMI, H., GRANADOS, R. R. & MARAMOROSCH, K. (1967). Electron microscopy of a plant-pathogenic virus in the nervous system of its insect vector. *Journal of Virology*, 1, 430–44.

HSU, H. T., NUSS, D. L. & ADAM, G. (1983). Utilization of insect tissue culture in the study of the molecular biology of plant viruses. In *Current Topics in Vector Research*, Vol. 1, ed. K. F. Harris, pp. 189–214. New York: Praeger Publishers.

HULL, R. (1977). Particle differences related to aphid-transmissibility of a plant virus. *Journal of General Virology*, 34, 183–7.

KASSANIS, B. & GOVIER, D. A. (1971). The role of the helper virus in aphid transmission of potato aucuba mosaic virus and potato virus C. *Journal of General Virology*, 13, 221–8.

KUBO, S. & TAKANAMI, Y. (1979). Infection of tobacco mesophyll protoplasts with tobacco necrotic dwarf virus, a phloem-limited virus. *Journal of General Virology*, 42, 387–98.

LUNG, M. C. Y. & PIRONE, T. P. (1973). Studies on the reason for differential transmissibility of cauliflower mosaic virus isolates by aphids. *Phytopathology*, 63, 910–14.

LUNG, M. C. Y. & PIRONE, T. P. (1974). Acquisition factor required for aphid transmission of purified cauliflower mosaic virus. *Virology*, 60, 260–4.

MARKHAM, P. G. & HULL, R. (1985). Cauliflower mosaic virus aphid transmission facilitated by transmission factors from other caulimoviruses. *Journal of General Virology*, 66, 921–3.

MARSH, M. & HELENIUS, A. (1980). Adsorptive endocytosis of semliki forest virus. *Journal of Molecular Biology*, 142, 439–54.

MASSALSKI, P. R., THOMAS, J. E. & HARRISON, B. D. (1986). Production and properties of monoclonal antibodies to potato leafroll luteovirus. *Report of the Scottish Crop Research Institute for 1985*, 165–6.

MURANT, A. F. (1978). Recent studies on association of two plant virus complexes with aphid vectors. In *Plant Disease Epidemiology*, ed. P. R. Scott & A. Bainbridge, pp. 243–9. Oxford: Blackwell Scientific Publications.

MURANT, A. F. & GOOLD, R. A. (1968). Purification, properties and transmission of parsnip yellow fleck, a semi-persistent, aphid-borne virus. *Annals of Applied Biology*, 62, 123–7.

MURANT, A. F. & LISTER, R. M. (1967). Seed-transmission in the ecology of nematode-borne viruses. *Annals of Applied Biology*, 59, 63–76.

MURANT, A. F. & ROBERTS, I. M. (1977). Virus-like particles in phloem tissue of chervil (*Anthriscus cerefolium*) infected with anthriscus yellows virus. *Annals of Applied Biology*, 85, 403–6.

MURANT, A. F., ROBERTS, I. M. & ELNAGAR, S. (1976). Association of virus-like particles with the foregut of the aphid, *Cavariella aegopodii*, transmitting the semi-persistent viruses anthriscus yellows and parsnip yellow fleck. *Journal of General Virology*, 31, 47–57.

NUSS, D. L. & PETERSON, A. J. (1980). Expression of wound tumor virus gene products *in vivo* and *in vitro*. *Journal of Virology*, 34, 532–41.

NUSS, D. L. & PETERSON, A. J. (1981). Resolution and genome assignment of mRNA transcripts synthesized *in vitro* by wound tumor virus. *Virology*, 114, 399–404.

NUSS, D. L. & SUMMERS, D. (1984). Variant dsRNAs associated with transmission-defective isolates of wound tumor virus represent terminally conserved remnants of genome segments. *Virology*, **133**, 276–88.

PETERSON, A. J. & NUSS, D. L. (1985). Wound tumor virus polypeptide synthesis in productive noncytopathic infection of cultured insect vector cells. *Journal of Virology*, **56**, 620–4.

PETERSON, A. J. & NUSS, D. L. (1986). Regulation of expression of the wound tumor virus genome expression in persistently infected vector cells is related to change in translational activity of viral transcripts. *Journal of Virology*, **59**, 195–202.

PIRONE, T. P. (1981). Efficiency and selectivity of the helper-component-mediated aphid transmission of purified potyviruses. *Phytopathology*, **71**, 922–4.

PIRONE, T. P. & THORNBURY, D. W. (1986). Number of potyvirus particles required for transmission by aphids. *Phytopathology*, **75**, 1324.

REDDY, D. V. R. & BLACK, L. M. (1974). Deletion mutations of the genome segments of wound tumor virus. *Virology*, **61**, 458–73.

REDDY, D. V. R. & BLACK, L. M. (1977). Isolation and replication of mutant populations of wound tumor virions lacking certain genome segments. *Virology*, **80**, 336–46.

REDDY, D. V. R. & MACLEOD, R. (1976). Polypeptide components of wound tumor virus. *Virology*, **70**, 274–82.

ROBERTS, I. M., ROBINSON, D. J. & HARRISON, B. D. (1984). Serological relationships and genome homologies among geminiviruses. *Journal of General Virology*, **65**, 1723–30.

ROBERTSON, W. M. & HENRY, C. E. (1986). An association of carbohydrates with particles of arabis mosaic virus retained within *Xiphinema diversicaudatum*. *Annals of Applied Biology*, **109**, 299–305.

ROBINSON, D. J. & HARRISON, B. D. (1985). Unequal variation in the two genome parts of tobraviruses and evidence for the existence of three separate viruses. *Journal of General Virology*, **66**, 171–6.

ROCHOW, W. F. (1965). Apparent loss of vector specificity following double infection by two strains of barley yellow dwarf virus. *Phytopathology*, **55**, 62–8.

ROCHOW, W. F. (1969). Biological properties of four isolates of barley yellow dwarf virus. *Phytopathology*, **59**, 1580–9.

ROCHOW, W. F. (1970). Barley yellow dwarf virus: phenotypic mixing and vector specificity. *Science*, **167**, 875–8.

ROCHOW, W. F., AAPOLA, A. I. E., BRAKKE, M. K. & CARMICHAEL, L. E. (1971). Purification and antigenicity of three isolates of barley yellow dwarf virus. *Virology*, **46**, 117–26.

SAKO, N. & OGATA, K. (1981). Different helper factors associated with aphid transmission of some potyviruses. *Virology*, **112**, 762–5.

SKUP, D. & MILLWARD, S. (1980). Reovirus-induced modification of cap-dependent translation in infected L cells. *Proceedings of the National Academy of Sciences, USA*, **77**, 152–6.

TAMADA, T. & HARRISON, B. D. (1981). Quantitative studies on the uptake and retention of potato leafroll virus by aphids in laboratory and field conditions. *Annals of Applied Biology*, **98**, 261–76.

TAMADA, T., HARRISON, B. D. & ROBERTS, I. M. (1984). Variation among British isolates of potato leafroll virus. *Annals of Applied Biology*, **104**, 107–16.

TAYLOR, C. E. & MURANT, A. F. (1969). Transmission of strains of raspberry ringspot and tomato black ring viruses by *Longidorus elongatus* (de Man). *Annals of Applied Biology*, **64**, 43–8.

TAYLOR, C. E. & ROBERTSON, W. M. (1969). The location of raspberry ringspot and tomato black ring viruses in the nematode vector, *Longidorus elongatus* (de Man). *Annals of Applied Biology*, **64**, 233–7.

TAYLOR, C. E. & ROBERTSON, W. M. (1970). Sites of virus retention in the alimentary tract of the nematode vectors, *Xiphinema diversicaudatum* (Micol.) and *X. index* (Thorne and Allen). *Annals of Applied Biology*, **66**, 375–80.

TAYLOR, C. E. & ROBERTSON, W. M. (1974). Electron microscopy evidence for the association of tobacco severe etch virus with the maxillae in *Myzus persicae* (Sulz.). *Phytopathologische Zeitschrift*, **80**, 257–66.

THOMAS, J. E., MASSALSKI, P. R. & HARRISON, B. D. (1986). Relationships among geminiviruses assessed with polyclonal and monoclonal antibodies. *Report of the Scottish Crop Research Institute for 1985*, 151–3.

THORNBURY, D. W. & PIRONE, T. P.(1983). Helper components of two potyviruses are serologically distinct. *Virology*, **125**, 487–90.

THORNBURY, D. W., HELLMANN, G. M., RHOADS, R. E. & PIRONE, T. P. (1985). Purification and characterization of potyvirus helper component. *Virology*, **144**, 260–7.

TOWNSEND, R., WATTS, J. & STANLEY, J. (1986). Synthesis of viral DNA forms in *Nicotiana plumbaginifolia* protoplasts inoculated with cassava latent virus (CLV); evidence for the independent replication of one component of the CLV genome. *Nucleic Acids Research*, **14**, 1253–65.

WATERHOUSE, P. M. & MURANT, A. F. (1983). Further evidence on the nature of the dependence of carrot mottle virus on carrot red leaf virus for transmission by aphids. *Annals of Applied Biology*, **103**, 455–64.

WOOLSTON, C. J., COVEY, S. N., PENSWICK, J. R. & DAVIES, J. W. (1983). Aphid transmission and a polypeptide are specified by a defined region of the cauliflower mosaic virus genome. *Gene*, **23**, 15–23.